AIR POWER

IN THREE WARS

GENERAL WILLIAM W. MOMYER

USAF, RET.

EDITORS:

MANAGING EDITOR - LT COL A.J.C. LAVALLE, MS

TEXTUAL EDITOR - MAJOR JAMES C. GASTON, PHD

ILLUSTRATED BY:

LT COL A.J.C. LAVALLE

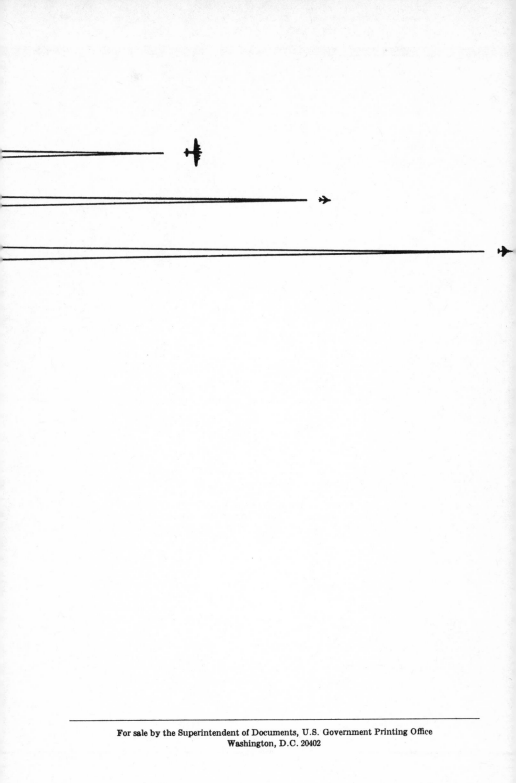

TO . . . .
all those brave airmen who fought their battles in the skies for control of the air in World War II, Korea, and Vietnam.

ACKNOWLEDGEMENTS

I am indebted to many colleagues who have encouraged me during the writing of this book. On numerous occasions I have sought their views on their perspective of airpower from their tours of duty in command and staff in peace and war. It should be understood that the views herein expressed are solely those of the author and are not those of the Department of the Air Force.

Without the use of the extensive materials from the USAF Archives and the close assistance of the Office of History, this story of airpower in three wars would have been seriously incomplete. The CORONA HARVEST reports were particularly valuable in documenting many of the operational aspects of the Vietnam war. Additionally, the services of the USAF photo depository and the English and History departments at the Air Force Academy were welcomed participants in the production of this book. My special thanks to Major James C. Gaston, professor of English at the Academy, for his extensive assistance in the textual editing of the manuscript. Of course, there were others—my thanks to them also.

Finally, I wish to express special appreciation to my assistant and managing editor, Lt Col A. J. C. Lavalle, for his untiring efforts in every phase of the preparation of this book, and to Mrs. Donna Caldwell for her patience in typing the many drafts it took to reach the final product.

WILLIAM W. MOMYER
General, USAF (Retired)
1 January 1978

FOREWORD

I began working on this book some months ago when a number of colleagues asked me to record my thoughts about the employment of airpower, especially tactical airpower, after 35 years in the profession. I hadn't any illusions of being blessed with special wisdom, but, as they said, no one else shared exactly my perspective on tactical airpower, and other professional airmen might find it useful to know how I saw things, particularly during the Vietnam years, whether they happened to approve of my perceptions or not.

Very soon I realized that my perspective was in fact several perspectives, and none of them could be maintained in perfect isolation from the others. I had watched strategy, tactics, and technology evolve, and all three of these evolutions fascinated me in recollection. I had seen tactical airpower from the viewpoints of the greenest fighter pilot (in 1939), the senior air commander in our longest war, and almost every position in between: dozens of perspectives there, and all of them seemed valid and important to me. So my problem became one of choosing from among my many perspectives the few that seemed likely to offer the most to other airmen.

Although I take most delight in recalling my experiences as a young fighter pilot, I had to admit that there's probably nothing unique about that perspective. Hundreds of others shared about the same experiences and could describe them as well or better than I. Thus I turned away (fellow fighter pilots will understand how difficult this was) from the temptation to spin stories about those days.

On the other hand, if there's little justification for my discussing many of the things I do recall from World War II, there's little point, either, in attempting to analyze what I didn't know (or knew only by reading about it later) about airpower in World War II. My experience was in North Africa and Italy; I didn't participate in, for instance, the combined bomber offensive against Germany or the B-29 offensive against Japan. I have some strong opinions about the mistakes and successes of those campaigns, opinions which I'll share with other airmen in private, but I don't want those judgments lying around in a book like this one where future airmen might see them and suppose they were based on authoritative, firsthand observation.

I examined and discarded many other approaches using this same filtering process—avoid discussing what I don't know from my own experience and the experiences of my companions, and consider telling what I do know only if future airmen might profit from seeing how those events looked from a perspective that was uniquely, or almost uniquely, mine. This filtering process kept me away from perspectives that would include such large topics as our employment of the atomic bomb in World War II (no firsthand knowledge of the decision process), and such personal topics as the ways in which President Johnson seemed to have aged between December 1967 when I talked with him at length about the bombing campaign and the defense of Khe Sahn as we flew from Korat to Cam Ranh Bay and October 1968 when I spoke with him for the last time at the White House (not likely to be of professional interest to future airmen).

What the filtering left me with were the perspectives you find in this book, the major preoccupations of my years as a senior commander: strategy, command and control, counter air operations, interdiction, and close air support. Most of my unique opportunities to perceive airpower occurred during my tenure as Commander of 7th Air Force in Vietnam from July 1966 until August 1968, and you'll see here mostly what I saw then. But some of my perceptions from earlier and later years must be recorded, too, to place my observations from the Vietnam years in context. My perspective on command and control when I ran 7th Air Force was certainly affected by my earlier observations in 1942–1944 when I was a fighter group commander in North Africa and those in 1944–1946 when I was Chief of the Army Air Forces Board for Combined Operations. While I was Assistant Chief of Staff at Tactical Air Command headquarters between 1946–1949, I undoubtedly picked up many of the ideas reflected in my approach to close air support in Vietnam. Also, as a member of the faculty of the Air War College from 1950–1953, I was ideally situated to observe the command and control relationships and the complexities of the interdiction, close air support, and counter air missions during the Korean War.

After a series of tours in which I commanded the 8th Fighter-Bomber Wing and the 314th Air Division in Korea, and the 312th Fighter-Bomber Wing and the 832nd Air Division in the U.S., I served as Director of Plans, Headquarters Tactical Air Command, from 1958 until 1961. There I saw firsthand the effects on our tactical air forces of both the Eisenhower administration's emphasis on nuclear weapons and the Kennedy administration's enthusiasm for the weapons and techniques of sub-limited war. During my tour in the Air Staff from 1961–1964 I was directly involved in the discussion of counterinsurgency and the forces that were needed for the developing war in Vietnam. My assignment in

Vietnam was preceded by a two-year tour as Commander of Air Training Command. As Commander of Tactical Air Command from the time I returned from Vietnam in 1968 until I retired in 1973, I remained intimately involved in the planning for all of our tactical air operations in Vietnam.

What I offer in this book, as fairly and as clearly as I can, is an account of the way airpower looked to me from the perspectives I think will matter most to airmen. I don't record these views in the hope that airmen, even my friends, will approve them. In fact I hope that all of our airmen who examine them will do so critically. We mustn't rely entirely upon yesterday's ideas to fight tomorrow's wars, after all, but I hope our airmen won't pay the price in combat again for what some of us have already purchased.

TABLE OF CONTENTS

LIST OF ILLUSTRATIONS

Page

Chapter I

STRATEGY

My vantage point in World War II, as Commander of the 33rd Fighter Group in North Africa, Sicily, and Italy, gave me a good view of more German and Italian Fighters than I really cared to see, but not many opportunities to witness the making of Allied air strategy. However, every pilot knew that our strategy embraced two fundamental features: attacks against the enemy heartland (with which I had little to do, either in Europe or the Pacific) and participation with surface forces to destroy the opposing forces or cause them to surrender. The first priority of our air strategy was to gain control of the air. Then we concentrated our efforts on isolating the battlefield and providing close air support. This air strategy provided flexibility to the Allied armies in their ground campaigns and guaranteed a minimum of interference from the German Air Force. By the time I returned to the U.S. in 1944 to become Chief of Combined Operations on the Army Air Forces Board, our airpower had virtually destroyed the *Luftwaffe* in the Mediterranean through air-to-air engagements and attacks on airfields and logistical bases; and we had repeatedly cut the enemy's air, sea, and land lines of communication, enabling our armies to capture North Africa and Sicily and to invade southern Italy.

At about the time I was leaving Europe, our B–29s in the Pacific were beginning their attacks against Japan from bases in China. In November 1944, B–29s from China and the Marianas raided Tokyo, and in March 1945, Major General Curtis E. Lemay began the decisive campaign of night, low-level incendiary attacks. The air war in the Pacific culminated with the dropping of atomic bombs on 6 and 9 August, events which profoundly affected U.S. air strategy.

NUCLEAR WEAPONS AND TACTICAL AIR FORCES

With nuclear weapons a reality in the late forties and early fifties, many strategists urged that we evaluate all military forces in light of their ability to contribute to a general nuclear war.[1] But other planners disagreed. A reduction in the size of U.S. armed forces, and our increasingly heavy emphasis on nuclear weapons, prompted a debate which brought out basic differences among the service chiefs and within the Air Force itself, I was uniquely situated to view this debate. Having been assigned as Assistant Chief of Staff of Tactical Air Command in 1946, I was at Hq TAC when the Air Force separated from the Army in 1947, and I remained with TAC until going to the Air War College in 1949.

The Army maintained that substantial conventional forces would be needed to fight limited wars. To evaluate all forces on the basis of their contribution to a general nuclear war with the Soviet Union would be imprudent, they said. Several air strategists replied that with nuclear weapons, it no longer made sense to maintain large conventional forces since such forces couldn't survive in a nuclear war. Furthermore, airpower's capacity to eliminate the command centers of an enemy made extensive surface campaigns unnecessary. Airmen conceded that some conventional forces would be needed for limited wars, but said that these forces need only be large enough to force the enemy into tactics that would produce a target for our nuclear weapons. They doubted, too, that a limited war could remain limited indefinitely. Either the employment or the threat of nuclear weapons would halt the conflict, or the conflict would rapidly expand to a general war.

But even within the Air Force during the late 1940s and early 1950s, there was fundamental difference of views on limited war. Many tactical airmen, including Lieutenant General Elwood R. Quesada and Major General Otto P. Weyland, believed that non-nuclear war was the most probable type of future conflict. These airmen argued that limited wars of the future would be fought without nuclear weapons because national leaders would realize that once nuclear weapons were introduced, it would become impossible to prevent the escalation of any conflict into general nuclear war: If the initial employment of small nuclear weapons didn't produce the desired effects, commanders would surely strike additional targets with more and larger weapons. With the explosion of a nuclear device by the Soviet Union in 1949, it was clear that nuclear weapons were no longer a U.S. monopoly, and tactical airmen argued that we had to prepare for limited wars in which both sides would voluntarily refrain from using nuclear weapons. We had to maintain sizeable tactical forces capable of fighting with conventional weapons.

At a time when the Air Force was shrinking and funds were short, though, it wasn't easy to find money for conventional tactical weapon systems. Understandably, most of the Air Force budget was earmarked for that part of the force which would have to deter or win a general nuclear war with the Soviet Union.[2] Strategic forces received most of the Air Force dollars, and only those tactical forces that had a nuclear capability could demand and get substantial funding. Other elements of the tactical force had to forego modernization.

2

Despite our national emphasis on strategic nuclear forces, tactical airmen continued to press for the restoration of a non-nuclear capability such as we had possessed during World War II. They stressed that the type of command and control system needed in a theater nuclear war was the same as that needed for non-nuclear war. If the tactical air force were to conduct a theater nuclear campaign, it would require a modernized command and control system and procedures for close coordination with ground forces, irrespective of the intensity and duration of the conflict. To carry out a theater nuclear strategy, precise control of airpower would be essential to prevent fallout and casualties to our own air and ground forces.

It seemed to these airmen that the essential elements of a tactical air force would be the same whether the force were designed for a nuclear or non-nuclear situation. They believed, further, that although additonal aircrew training would be necessary for some aspects of nuclear operations, basic tactical skills would remain the same. Tactical training would simply omit certain aspects of non-nuclear weapons delivery and emphasize a few basic techniques such as dive bombing and low altitude bombing which were common to tactical nuclear and non-nuclear weapons delivery. Thus it would be feasible to maintain non-nuclear proficiency without degrading an aircrew's ability to deliver tactical nuclear weapons.

In the years preceding the Korean War, tactical air forces were being cut back in accordance with the overall national policy following World War II. Even with these reduced forces and the emphasis on nuclear operations, however, there remained a high residuum of experience in non-nuclear operations from World War II. Despite a shortage of equipment, the high level of experience permitted expansion and modernization of the tactical air forces when they were needed in Korea.

KOREAN WAR—A DILEMMA

When the North Koreans invaded South Korea on 25 June 1950,[3] U.S. defense planners carefully evaluated our strategy for conducting limited nuclear war: Was the strategy feasible in Korea? Would it be acceptable to our allies? On both counts the strategy was deficient. There were few attractive targets for tactical nuclear weapons because of the lack of concentration of North Korean forces and the many alternative routes of advance afforded the enemy by the Korean terrain. Further, the Allied forces were retreating in such disarray that it was unrealistic to suppose that we could promptly turn them around for a counterattack in which nuclear weapons could provide the basic firepower.

By the time the Allied forces had withdrawn into the Pusan perimeter, the employment of nuclear weapons was not a realistic option because of the poor targets and the attitude of our allies toward these weapons. Air strategy, then, was based on non-nuclear weapons, and it comprehended the same missions that tactical air forces had performed in World War II.

3

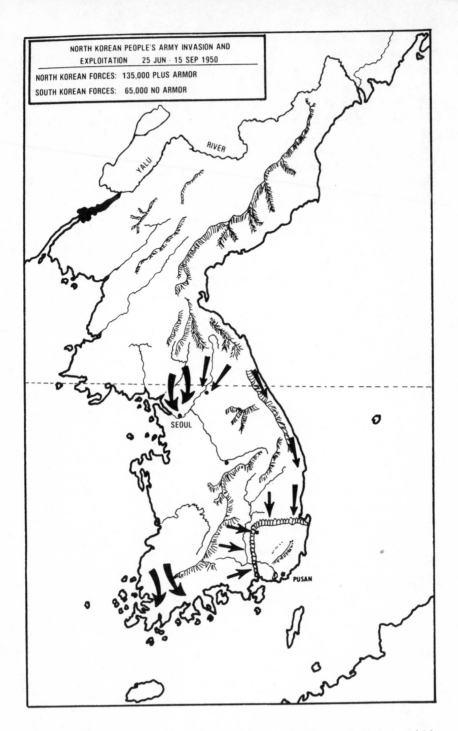

NORTH KOREAN PEOPLE'S ARMY INVASION AND
EXPLOITATION   25 JUN - 15 SEP 1950

NORTH KOREAN FORCES: 135,000 PLUS ARMOR
SOUTH KOREAN FORCES: 65,000 NO ARMOR

YALU

RIVER

SEOUL

PUSAN

With the North Korean Air Force neither a significant threat nor within
range of the retreating Allied forces, air strategy focused initially on

chopping off the supply lines to the North Korean ground forces, making it impossible for these divisions to mount a sustained offensive against the Pusan perimeter. Also a part of this strategy, of course, was a direct attack against assaulting ground forces. American airmen maintained complete control of the air for the Inchon invasion and the subsequent advance into North Korea. Air strategy was an essential part of the joint strategy.

When the Chinese communists invaded Korea in October 1950,[4] however, the Allies had to make major revisions in their strategy. As the enemy forces moved across the border, it appeared that airpower would have to be employed much more broadly to reduce the numerical superiority of the Chinese. MacArthur proposed that the bridges and lines of communication used by Chinese entering North Korea be subjected to sustained air attack. He felt it imperative to deny these forces the sanctuary they then enjoyed.

Among airmen the question of how Chinese and Soviet airpower could be contained along the Yalu was debated with vigor. Some airmen, including Major General Emmett O'Donnell, Jr., believed it would be necessary to strike the airfields and engage the fighters deep in the rear areas if control of the air were to be established. (All agreed that such control was absolutely essential to our retreating ground forces, who were so badly outnumbered that many Americans were questioning whether the Allies could hold any position in Korea.)[5] O'Donnell and others insisted that the enemy must not be permitted a sanctuary from which to attack the Allied air forces and our forward bases.

After considerable deliberation, the Joint Chiefs recommended that Far East Command's air offensive not be extended beyond the Yalu into Manchuria unless the enemy launched massive air attacks against our forces, in which event American airmen would destroy the airfields from which the attacks originated.[6] For airmen in Korea, the recognition of an enemy sanctuary across the Yalu posed a terrific problem: How were we to contain a numerically superior enemy fighter force when all of our forward bases and lines of communication were open to attack?

YALU—CONTAINMENT OF MIGS

Clearly we had to shift from an air strategy oriented primarily toward close support of our ground forces to a new strategy featuring (1) offensive fighter patrols along the Yalu, (2) attacks against forward staging bases from which MIGs might strike 5th Air Force airfields and the 8th Army, and (3) intensive attacks against the main supply lines of the advancing Chinese army. These air operations became the primary means of preventing the enemy's air and ground forces from pushing the Allied army out of Korea. The 8th Army's objective was to hold, rather than to defeat or destroy, the opposing ground forces. This objective evolved from the pragmatic observation that a much larger ground force would be needed to defeat the enemy. Such a ground campaign would be too long and too costly.

Maintaining continuous pressure on the enemy's rear area, his lines of communication, and his engaged troops, airpower helped persuade the enemy to cut his losses. The North Koreans were finally persuaded that they should seek an end to the war at the conference table rather than on the battlefield, and negotiations ended the conflict on 27 July 1953 after three years of fighting. [7]

IMPACT OF KOREAN WAR

With the end of the Korean War, defense planners reevaluated our strategy for employing airpower. Perhaps the paramount question of the time was whether we should prepare to fight limited as well as general wars. After the agony and expense of Korea, an understandably popular positon was that we would never fight, nor should we prepare to fight, another war like Korea. Adding to the popularity of this position was the fact that it could be used to justify a reduction in defense forces and expenditures. If a limited war should break out, proponents said, nuclear weapons could end it quickly. But the way to prevent such wars would be to maintain military and political pressure against potential instigators. If the outside support for a limited conflict were neutralized, the conflict itself would soon die for lack of weapons and other resources. Most airmen consented to the idea that nuclear weapons should be the basis of our defense strategies, but the Army and Navy maintained that limited conflict was most likely and that limited wars would, at least initially, be fought with conventional weapons.

Once again nuclear forces were accepted as the dominant element of our national defense, and all forces were evaluated in light of their usefulness in the event of nuclear conflict. Resources allocated to non-nuclear forces were sufficient only to fight a brief, very limited war. Throughout the mid-fifties, all of the services accepted the nuclear war premise in their yearly budget arguments. The Army, however, continued to press for sizeable forces capable of fighting a limited non-nuclear conflict. Army spokesmen feared that the dominant concern about nuclear war was overshadowing the need for ground forces capable of fighting in situations other than nuclear battlefields in Europe. Nevertheless, the survivability of forces on a nuclear battlefield continued to be a major concern of most strategists during the period.

THE FRENCH IN INDOCHINA

In 1953 on the eve of Dien Bien Phu, U.S. defense planners differed widely in their opinions about the appropriate role of airpower in low scale conflict. Several Army planners felt that airpower could operate only as a supporting force. The main role of airpower was, in their view, the delivery of supplies, equipment, and personnel, and the support of civic action measures. Whatever firepower was required to deal with guerrilla actions wouldn't demand the sophisticated weapons of our airpower arsenal. Based on these views of airpower, and the experience of ground warfare in Korea, the prevailing view in the U.S. military establishment was that U.S. forces should not become engaged in Vietnam and Laos; rather, we should continue to support the French in their expansion of Vietnamese forces to counterbalance the Viet Minh threat.

However, some elements of the U.S. military were not convinced the French were making sufficient progress in building self-sufficiency into the Vietnamese armed forces. They felt that the French were placing too much emphasis on training for a conventional conflict rather than a counterinsurgency war. These views were based largely on Britain's experience in Malaya where there were no large, conventional ground actions. Almost all engagements were small, brief counter-guerrilla actions. The success of the British in containing and eventually eliminating the insurgents in this conflict convinced many in the U.S. military that this was the strategy the French should pursue in containing the Viet Minh. Airpower had played a limited role in the Malayan insurgency, and this fact was used as evidence that airpower would not be critical for the success of the French in Indochina.

We should have learned from the French defeat at Dien Bien Phu on 7 May 1954, however, that the French were fighting a much different foe than the British had faced in Malaya. For the British, it was relatively simple to shut off most of the external support to the Malayan insurgents. [8] On the other hand, the Viet Minh were a much larger force and were equipped with the weapons of an organized army rather than those of a guerrilla band. The tactics and strategy required to defeat the Viet Minh, therefore, were closer to those of conventional warfare than to those of counterinsurgency operations. Militarily, the loss of Dien Bien Phu was not disastrous. Many battles have been lost in campaigns that were eventually successful, and the French strength in Southeast Asia was still formidable after the loss of Dien Bien Phu. The battle was, however, a crushing blow to French morale. In France, news of the loss further inflamed those who already deeply resented the fact that their country had been in a constant state of war since World War II. Thus, Dien Bien Phu set the stage for the total disengagement of the French, and for the involvement of the United States.

SEATO

Having failed to achieve collective participation in an effort to save Dien Bien Phu, the United States sought a regional arrangement that would provide a basis for future collective action. The Southeast Asia Treaty Organization (SEATO) came into existence in February 1955,

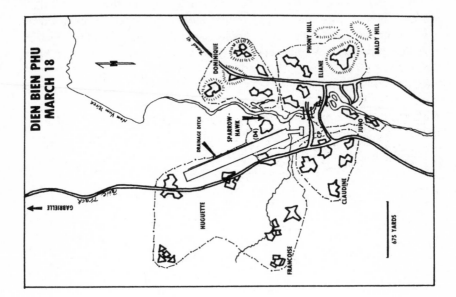

DIEN BIEN PHU
MARCH 18

GABRIELLE

Nam Yum River

DOMINIQUE

PHONY HILL

ELIANE

BALDY HILL

DRAINAGE DITCH

SPARROW-
HAWK

(D4)

C.P.

JUNO

HUGUETTE

CLAUDINE

FRANCOISE

675 YARDS

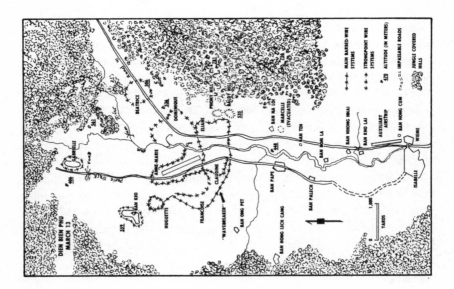

DIEN BIEN PHU
MARCH 13

GABRIELLE

BEATRICE

DOMINIQUE

PHONY HILL
FOOT HILL

ELIANE

ANNE-MARIE

CLAUDINE

FRANCOISE

HUGUETTE

"WAVE BREAKER"

BAN KEO

BAN ONG PET

BAN NA LOI

MARCELLE
(EVACUATED)

BAN HONG LICH CANG

BAN TEN

BAN DON LA

BAN PAPE

BAN PALIGH

BAN NHONG NHAI

BAN KHO LAI

AUXILIARY
AIRSTRIP

BAN HONG CUM

ISABELLE

WIEME

MAIN BARBED-WIRE
SYSTEMS

STRONGPOINT WIRE
SYSTEMS

570 ALTITUDE (IN METERS)

IMPASSABLE ROADS

JUNGLE COVERED
HILLS

1,000

YARDS

8

following the signing of the Geneva Accord in July 1954.* Although this treaty did not obligate the U.S. to commit military forces, it left the way open if the Congress supported such an action. In one respect, the treaty was essentially a warning to the North Vietnamese that if they attempted military action in South Vietnam, Laos, or Cambodia, the members of the treaty organization might respond with military force. U.S. policy at this point, therefore, was to provide assistance to the French while threatening to use greater force if the North Vietnamese continued their effort to undermine the governments of South Vietnam and Laos by covert and overt military actions. Our strategy was based on the hope that the insurgency would be contained by the South Vietnamese and Laotian forces and that these countries of Southeast Asia would eventually achieve a peaceful political accommodation among themselves.

Contrary to our hopes, it soon became clear that the Soviets, Chinese, and North Vietnamese were not about to discontinue their support of the Pathet Lao and the Viet Cong. Military supplies and weapons from the People's Republic of China and the Soviet Union made their way through North Vietnam into Laos and South Vietnam with increased frequency. By the beginning of 1961, the situation in Laos and South Vietnam had become so critical that deployment of U.S. forces was under serious consideration.[9] Our level of logistical and training support was no longer adequate to halt the enemy's advances in Laos and South Vietnam, and the threat of retaliation contained in the SEATO treaty was having little influence on the North Vietnamese-backed insurgents.

WARS OF LIBERATION

In 1961, Khrushchev's speech proclaiming "wars of liberation" as the wars of the future and President Kennedy's confrontation at Vienna with the Soviet leader led to a vigorous re-examination of U.S. military strategy.[10] President Kennedy then directed the expeditious development of U.S. forces with special skills in the conduct of counterinsurgency or sub-limited wars. The intention of his planners was that indigenous forces would suppress guerrilla activities while our specialists did the training and assisted in "nation building."

This reorientation of our defense priorities toward smaller conflicts prompted considerable debate about how best to cope with these wars. In the Army, most believed that it was necessary to create Special Forces (Green Berets) specifically trained and organized for counterinsurgency activities. In the Air Force, many believed that existing tactical forces could adjust to counterinsurgency warfare without major changes, while others argued that counterinsurgency was the combat of the future and that the Air Force should build a special force for such conflicts.

---

*The United States did not sign the 1954 Geneva Accord, but did sign the Geneva Agreement. See Senate Foreign Relations Committee Print, Background Information Relating to Southeast Asia and Vietnam (7th Revised Edition), December 1974.

To most Department of Defense (DOD) analysts, the argument of those who favored specialized forces seemed to be supported by the circumstances in Southeast Asia. These circumstances apparently didn't demand sophisticated equipment or massive firepower, but rather called for weapons consistent with the abilities of the Laotian and South Vietnamese forces who would operate them. Further, our employment of sophisticated aircraft would raise the level of violence and could promote a larger war. But appearances in 1961 were deceptive. There was little to remind us that the North Vietnamese who conquered Dien Bien Phu were a highly trained, conventional army employing relatively sophisticated weapons and tactics. In fact, while we considered the merits of various approaches to counterinsurgency warfare, the fighting in parts of Southeast Asia had already passed through that stage of conflict. Soon we would confront an enemy who was trained and ready to employ sophisticated weapons and to fight in large, highly organized units.

In the U.S. in 1961, however, the training of special forces for counterinsurgency operations was proceeding with the utmost speed. On 4 May 1961, Secretary of State Rusk said that U.S. forces would not then be sent to Vietnam in a combat role, but he made no promises about the future.[11] In the meantime, the situation in South Vietnam was deteriorating rapidly. By the end of May, the Joint Chiefs were telling the Secretary of Defense that if South Vietnam were to remain free, it would be necessary to deploy U.S. combat forces. So far, the South Vietnamese had not demonstrated an ability to stop the Viet Cong in the countryside. The larger cities were relatively free of the Viet Cong, but the countryside, to a large extent, was under enemy control. Thus the plan to assist without actively participating was proving unsuccessful.

"JUNGLE JIM" FIRST COMBAT DEPLOYMENT

With the deployment of U.S. Army Special Forces to train South Vietnamese late in 1961, the direct involvement of U.S. forces in combat was virtually assured.[12] Even though our forces were sent primarily to teach, they were sure to become involved in fighting at their isolated camps deep in territory dominated by the Viet Cong.

While the Army was sending the first of its Special Forces to Vietnam, we in the Air Force were activating our first special unit for guerrilla warfare since World War II. Many senior airmen still questioned the wisdom of investing in such units, but Secretary McNamara stated that the Vietnamese conflict should be a "laboratory for the development of organizations and procedures for the conduct of sub-limited war,"[13] and we responded with an all-out effort to put together a unit of World War II aircraft capable of fighting sub-limited wars. Our efforts were spurred, too, by the visit of Walt W. Rostow and General Maxwell D. Taylor to South Vietnam in October 1961. They recommended more aid for South Vietnam and supported the decision made earlier in the month to deploy U.S. troops for logistical support and training. By November, we were ready to deploy a combat unit of "Air Commandos" equipped with T–28s, B–26s, and other vintage aircraft. JUNGLE JIM was the nickname given this unit, and the detachment that deployed to South Vietnam was dubbed FARM GATE.

FARM GATE's purpose was to train South Vietnamese pilots, but our crews soon found themselves flying combat missions in response to emergency requests. The South Vietnamese Air Force (VNAF) simply could not provide all of the help that was urgently needed by the South Vietnamese Army. FARM GATE, therefore, was engaged in combat operations before the close of 1962, and U.S. forces passed from the gray area of training into a limited combat role in a "sub-limited war."

JOINT TASK FORCE 116—RESPONSE TO CRISIS IN LAOS

Most U.S. defense planners considered Laos to be an area of strategic importance because of its location between China and Thailand. In the event of war between the U.S. and China, Laos could delay Chinese forces seeking to overrun Thailand. Realistically, we could not expect Laos to be our ally, but Laotian neutrality, at least, was essential to the security of Thailand. However, by the spring of 1962, the North Vietnamese supported Pathet Lao appeared about to capture most of the important areas of the country. Threats of U.S. intervention, implicit in the SEATO Treaty, had no significant impact on the conflict or on the peace negotiations then taking place in Geneva.

An Air Commando poses in front of his U–10 "Psywar" aircraft somewhere in the heart of South Vietnam.

11

The Commander-in-Chief Pacific (CINCPAC) had developed various contingency plans for Southeast Asia, one of which addressed precisely the situation then existing in Laos. Increasingly alarmed by events in Laos, President Kennedy activated the plan, thereby directing the deployment of a Joint Task Force to Thailand. JTF–116 consisted mostly of air units, and the possibility that this force might be employed against the Pathet Lao was, and was intended to be, apparent to all. Whether a causal relationship or mere coincidence was at work, the Communist negotiators in Geneva found the proffered peace agreement increasingly attractive as JTF–116 moved into Thailand. An agreement to abide by the 1954 Accord was signed on 23 July 1962. The U.S. had not been a signatory to the Accord, but we agreed to accept its provisions.

Unfortunately, the North Vietnamese willingness to avoid JTF–116 far exceeded their willingness to comply with the Geneva Agreement. Our own efforts to comply now seem almost pathetically naive in comparison with the open contempt for the agreement demonstrated by the North Vietnamese. While we began withdrawing our advisors from Southeast Asia in accordance with the agreement, the North Vietnamese were withdrawing none of their forces. Furthermore, based on our interpretation of Articles 17–19 of the original accord, we deferred the modernizing of FARM GATE units in South Vietnam with jet aircraft. Basically, the articles restricted the replacement of worn-out equipment with new types of arms and materiel to a piece-by-piece basis. These articles were particularly troublesome to our Air Force planners, but much less so to the North Vietnamese who simply ignored them from the outset.

So despite the Geneva Agreement, it was evident that the situation in Southeast Asia was not improving. By mid-1962, many other senior airmen and I were of the opinion that air strikes against the North Vietnamese homeland would be necessary if the war in South Vietnam were to be ended. The only alternative, in our view, would be the deployment of numerous American ground forces. But even as early as 1962, opinion was sharply divided on the issue of airpower's ability to stop the fighting in South Vietnam. Among those who disagreed with our position was Secretary McNamara, who said that "while naval and air support are desirable, they won't win the war." [14] Paradoxically, a figure who offered considerable support to our contention about the importance of the revolutionary base in North Vietnam was General Vo Nguyen Giap, the architect of the victory at Dien Bien Phu and in 1962 Commander of the North Vietnamese Army. He was always clear about the facts that North Vietnam was the revolutionary base and that the success of communist military operations in Laos and South Vietnam depended directly upon the support and employment of North Vietnamese forces. In 1962, however, relatively few senior DOD officials thought seriously about a strategic air offensive against North Vietnam. For the time being, our efforts and our strategy were limited to South Vietnam.

Throughout 1963, the North Vietnamese continued to infiltrate personnel and equipment. Furthermore, the improved quality of their weapons indicated that more modern arms were being shipped from China and the Soviet Union. Most senior U.S. commanders with whom I talked in 1963 felt sure that the war in South Vietnam was rapidly expanding into a conventional conflict, although a number of DOD planners still believed the war was primarily an insurgency that could be brought under control within South Vietnam if we trained and equipped the South Vietnamese properly. Secretary McNamara said that "South Vietnam is a test case for the new Communist strategy,"[15] by which he seemed to mean that South Vietnam would be a test case, too, for his strategy of graduated response to provocation.

A CHANGING ROLE—THE U.S. TO FIGHT

Early in 1964, the Joint Chiefs of Staff recommended that the U.S. take over the fighting in South Vietnam. They had previously directed Admiral Harry D. Felt, CINCPAC, to update contingency plans and to propose a strategy for an air campaign against North Vietnam. Felt proposed a series of measures designed to seal off North Vietnam by mining harbors and attacking shipping and selected lines of communication.[16] Senior airmen agreed with Felt's proposals but added that it would also be necessary to cut off the infiltration of men and equipment into South Vietnam by attacking the North Vietnamese homeland. Cutting lines of communication (LOCs) would be relatively ineffective because most of them were hidden by jungle growth and because the North Vietnamese could multiply them almost indefinitely simply by pressing more porters into service. In accordance with DOD policy, however, Felt's strategy was required to be one of graduated response: Air strikes would begin on targets close to the DMZ, gradually working toward the North Vietnamese heartland with an increasing sortie rate.

Although the Joint Chiefs agreed that the U.S. would have to intervene if South Vietnam were to be saved, they were not in complete agreement on precisely how we should intervene. General Curtis E. LeMay, USAF Chief of Staff, argued for a concentrated attack against targets in the heart of North Vietnam. Indirect attacks in South Vietnam and Laos, in his judgment, were not apt to be decisive.[17] He recommended that a minimum number of troops be deployed to South Vietnam immediately to secure the main airfields and other strategic areas. Then we should conduct a swift, devastating air offensive against North Vietnam's strategic targets. All of his experience had taught him that such a campaign would end the war. If this strategy failed, he said, we should then have to consider whether we were willing to deploy a large ground force to Southeast Asia.

On the other hand, General Earle G. Wheeler, Army Chief of Staff, thought it necessary for U.S. troops in South Vietnam to take on more of the combat role. An air campaign, he believed, should be directed at the

13

lines of communication near the border of South Vietnam, but not at the heartland of North Vietnam. The main emphasis should be on the Ho Chi Minh Trail and the logistical network south of Vinh. Most senior Army officials believed that the war had to be won in South Vietnam and that the air campaign should support the in-country war chiefly through close air support. [18]

The Army view was essentially that of Secretary McNamara. He believed that the war should be fought in South Vietnam and that the main roles of airpower should be close air support and interdiction of lines of communication south of the 20th parallel and in Laos. The Secretary felt that the threat of air attacks on military and industrial targets could influence the North Vietnamese to restrain their support of the Viet Cong, but he disagreed with LeMay and other senior airmen who insisted that the only way to end the North Vietnamese pressure on South Vietnam was to destroy the war-related installations in North Vietnam.

On 1 June 1964 a top-level strategy conference convened at CINCPAC Headquarters in Honolulu. Those in attendance included General William C. Westmoreland who was about to replace General Paul D. Harkins as Chief of Military Assistance Command, Vietnam (MACV); General Maxwell D. Taylor, Chairman of the Joint Chiefs of Staff; Ambassador Henry Cabot Lodge; Secretary of State Dean Rusk; and Secretary McNamara. Before the conference, LeMay and General Wallace M. Green, Jr., Commandant of the Marine Corps, had been pressing the view that air attacks against the North Vietnamese were essential to halting the war in South Vietnam. [19] Admiral David L. McDonald, Chief of Naval Operations, agreed essentially with LeMay's and Greene's view, but he would have placed more emphasis on sealing off North Vietnam from external support, and less emphasis on attacks against industrial targets. At the conference, however, Taylor questioned whether we should attack North Vietnam at all. He agreed with the Secretary of Defense that our main efforts should be designed to bolster the forces of South Vietnam and to cut the lines of communication in Laos. If attacks were to be made against North Vietnam, Taylor believed they should be near the DMZ, using U.S. and South Vietnamese aircraft to demonstrate our joint resolve to expand the conflict if it continued in Laos and South Vietnam.

BOMB THE NORTH?—NO AGREEMENT

The U.S. strategy that emerged from the June 1964 Honolulu Conference differed in no important way from our pre-Conference strategy: We would build the South Vietnamese armed forces; provide combat support when the South Vietnamese were unable to handle the situation; and, if air attacks against North Vietnamese targets should be necessary, we would select only targets near the DMZ and would use both U.S. and South Vietnamese aircraft.

14

Admiral U.S. Grant Sharp, who replaced Admiral Felt as CINCPAC shortly after the Honolulu Conference, believed that this strategy would not force the North Vietnamese to stop the fighting in South Vietnam and Laos. Thus he added his voice to those of LeMay and Greene, urging that airpower and naval power be applied directly against North Vietnam.* His position differed from theirs only in his preference for a more gradual application of power.

THE GULF OF TONKIN

At the direction of the JCS, the Defense Intelligence Agency (DIA) had developed a list of strategic targets in North Vietnam. Of the 94 targets, 82 were fixed, and 12 were railroad routes.[20] These 94 targets were considered to have a most direct relationship to the North Vietnamese war-making capacity and will to fight.** Air Staff planners had also designed an air campaign plan based on the 94-target list.

On 2 August 1964, North Vietnamese torpedo boats attacked the U.S. destroyer *Maddox* in the Gulf of Tonkin, and on 4 August they attacked both the *Maddox* and the destroyer *C. Turner Joy*. After these attacks, LeMay argued that now was the time to execute at least part of the 94-target plan.[21] However, the Joint Chiefs decided that a more limited retaliation would be sufficient to demonstrate the serious consequences of continued aggressive acts by North Vietnam.

STRATEGY—RETALIATORY ATTACKS IN NORTH VIETNAM

On 5 August 1964, U.S. carrier aircraft retaliated against the North Vietnamese torpedo boat anchorages and oil dumps just above the 17th parallel in a strike that set the pattern for our future air strategy. Until the bombing halt of 1968, our overall air strategy was one of "tit for tat," or graduated escalation, with targets being released for attack a few at a time depending upon the activities of the North Vietnamese. Furthermore, the rationale for selecting targets was oriented toward achieving some particular effect upon the ground war in South Vietnam, not toward destroying the will of the North Vietnamese to fight.

There was little evidence that the Navy's limited retaliatory strikes of 5 August (known as PIERCE ARROW) had caused the North Vietnamese to discard their aggressive intentions. If anything, the strikes seemed to have had the opposite effect. On 7 August, the North Vietnamese responded by moving 30 MIGs from China into Phuc Yen airfield, indicating thereby that they intended to continue fighting and to challenge our air attacks. Rather than shocking their leaders and disrupting their

---

*For the view of Admiral U.S.G. Sharp (CINCPAC) on strategy and the organization for command of forces in the Vietnam War, see Strategic Direction of the Armed Forces by Adm U.S.G. Sharp, U.S. Navy (Ret), Naval War College, Newport, R.I., 1977.
**By 1967, the list had grown to 244 active and 265 contingency targets. As Commander of 7th Air Force in 1967, I believed that there were about 165 especially significant targets on the list, with over 90% of them above the 20th parallel.

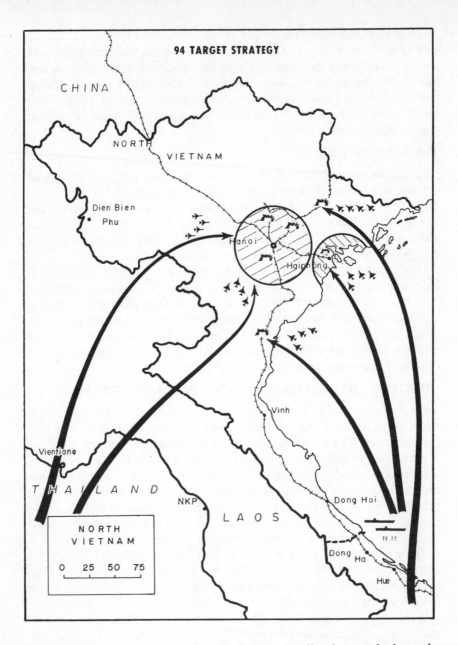

CHINA

NORTH

VIETNAM

Dien Bien
 • Phu

Hanoi

Haiphong

Vinh

Vientiane

THAILAND

NKP

LAOS

Dong Hoi

TF 77

Dong
Ha

Hue

NORTH
VIETNAM

0   25   50   75

war machinery with a concentrated, strategic air offensive, we had merely
alerted them to start work on what would become a superb air defense
system of MIGs, surface-to-air missiles (SAMs), and anti-aircraft-artillery
(AAA).

With the war moving toward higher levels of violence, the tactical air
forces in the United States were rapidly being trained and equipped for
conventional operations. Since the Korean War, relatively little attention

had been given to refining or building non-nuclear weapons or to training aircrews for delivery of non-nuclear weapons. Consequently, weapons for the initial bombing attacks in Southest Asia would have to come from the stockpiles of the Korean War. However, a hard core of combat experience still existed in the Air Force from the Korean War and in some cases from World War II, and it was only a matter of months until fighter units were thoroughly proficient in the delivery of non-nuclear weapons.

By November 1964, the situation in South Vietnam had become so alarming that Taylor (who had replaced Lodge as Ambassador shortly before the Tonkin Gulf attacks) proposed "graduated military response against the lines of communication in Laos as a means of increasing the morale of the South Vietnamese and reducing the flow of materiel to the Viet Cong forces."[22] However, there was not enough airpower in the theater to take care of the increasing demands for close air support and at the same time carry out an interdiction campaign against the LOCs in Laos. Although T–28s from both the 1st Air Commando Squadron* and VNAF were directed to attack the trail network below Tchepone in Laos, they were too few to significantly reduce the flow of materiel to the Viet Cong. Major General Joseph H. Moore, Commander of the 2nd Air Division (Second Air Division controlled USAF operations in Vietnam from 8 October 1962 until it was replaced by Seventh Air Force on 1 April 1966.) requested more forces, but at the same time he advised that the only way to stop the flow of North Vietnamese men and supplies into South Vietnam was to open up the targets in North Vietnam. He agreed with LeMay that the 94-target list should be authorized for attack.

GRADUAL ESCALATION REPLACES RETALIATION

Responding to the gradually increasing threat of our airpower, the North Vietnamese and Viet Cong stepped up attacks against airfields throughout South Vietnam late in 1964. The intense mortar and sapper attack on Bien Hoa Air Base in November, resulting in the loss of five B–57s, was an indication that our strategy of limited retaliatory strikes was not reducing the enemy's desire to fight. Further, the South Vietnamese Army was unable to cope with the increasingly aggressive thrusts of the Viet Cong and North Vietnamese forces. It was apparent that a change in strategy had to be made, and soon, if South Vietnam were to be held.

Immediately after the attack on Bien Hoa, the Joint Chiefs proposed a series of strikes against North Vietnam based on the 94-target list. These attacks would be expanded to include more of the targets if the enemy's attacks in South Vietnam continued. President Johnson declined to accept the proposal, but it served to put all of the JCS on record in favor of more aggressive strikes against targets in North Vietnam.

---

*FARM GATE was inactivated in June 1963, and the 34th Tactical Group absorbed its aircraft and personnel. The squadrons of the 34th were called Air Commando Squadrons.

Early in December 1964, the President decided to take more deliberate action to convince the North Vietnamese that the United States would not let South Vietnam be taken over by force. In doing so, he altered the policy of exchanging "tit for tat" retaliatory strikes and moved toward a strategy of steadily increasing pressure. As a part of this new strategy, American air strikes were authorized against the LOCs in Laos. Also, we were to begin striking targets just above the DMZ in North Vietnam and would move gradually northward if the North Vietnamese continued their aggressive activities.

The first of these strikes, termed FLAMING DART, were conducted in February 1965 against targets above the DMZ in response to a series of attacks by the North Vietnamese and Viet Cong during that month.[23] The enemy's February attacks on airfields, headquarters, and advisory compounds throughout South Vietnam made it apparent that they were making an all-out effort to collapse the military and political structure, convinced that the U.S. would not be able to halt their advance. Athough the FLAMING DART strikes were essentially reprisals for attacks on American installations at Pleiku and Qui Nhon (and thus appeared to resemble our reprisals after the Tonkin Gulf attacks), these February raids were followed on 2 March by the first strikes of a continuing, systematic air campaign termed ROLLING THUNDER.

ROLLING THUNDER was originally authorized to hit LOCs in Laos and North Vietnam below the 19th parallel, although it was understood that some targets in the Hanoi area would also be released from time to time. General LeMay, General John P. McConnell (who succeeded LeMay on 1 February) and other senior airmen felt strongly that the initial conception of ROLLING THUNDER was too restrictive. They agreed that the LOCs below the 19th parallel were important parts of the North Vietnamese logistical network, but the vital elements of the system were North Vietnam's ports, railroads, marshalling yards, bridges, and supply centers; there were relatively few of these in the southern part of North Vietnam. Furthermore, as supplies funneled southward, it became increasingly difficult to destroy them in large quantities because of the absence of open terrain and natural choke points. The dissemination of supplies among hundreds of jungle trails and thousands of porters guaranteed that air attacks in the south would be less efficient than attacks against the Kep Marshalling Yard, the Paul Doumer Bridge, or the ports at Haiphong. Thus senior airmen pressed for the expansion of ROLLING THUNDER into an air strategy focused upon the heart of North Vietnam.

But neither the President, the Secretary of State, nor the Secretary of Defense yet conceived of ROLLING THUNDER as a strategic air offensive. The Secretary of Defense continued to maintain that the primary role for airpower should be to support ground forces in South Vietnam, as it was here that the enemy must be denied a military victory.[24] On the eve of the April 1965 Honolulu Conference attended by the Secretary of Defense, the Ambassador to Vietnam, the Chairman of the Joint Chiefs, CINCPAC, and others, Secretary McNamara still believed that ROLLING THUNDER should be a limited application of airpower against logistics targets relatively close to the DMZ. Further, the size and frequency of these strikes, as well as the targets, should be selected in Washington.

AIR CAMPAIGN PLAN—PROPOSED

When McConnell succeeded LeMay on 1 February, he was already familiar with JCS requests for a brief, intense bombing campaign, since he had been Vice Chief of Staff for the six months preceding his assignment as Chief.[25] Thus McConnell was ready to propose, little more than one month after becoming Chief, that the JCS recommend a 28-day air campaign against the 94 targets.[26] Like LeMay and Greene, he felt that the only way to end the war was to employ airpower intensively against strategic targets in North Vietnam. McConnell agreed with McNamara that the South Vietnamese troops needed direct air support, and they would receive it. But among the alternatives of a prolonged war of attrition in South Vietnam, an invasion of North Vietnam, and a concentrated strategic air offensive, the most sensible strategy seemed clearly to be the third.

Before submitting McConnell's proposal to the Secretary of Defense, the JCS expanded it into a four-phase plan: In Phase I, attacks would be conducted below the 20th parallel for three weeks at the discretion of the field commanders. The objectives of Phase I were to reduce the flow of logistics by battering the LOCs with almost continuous attacks and to provide a clear indication to the North Vietnamese that we would increase the scope and intensity of the war if they continued their efforts to overthrow the government of South Vietnam. Phase II would be a six-week campaign to sever the northeast and northwest railroads to China. Most logistics coming from China (except for large bulk goods which traveled by ship) were carried on these lines. By cutting these rail lines, we would be hitting the logistical system at its most vulnerable points, and would be bringing the war closer to the people and the government, thereby attacking both the means and the will of the North Vietnamese to fight. Phase III would last two weeks. We would destroy the ports, mine the seaward approaches, and destroy the ammunition and supply areas in the Hanoi-Haiphong area. During Phase III, we would expect the North Vietnamese to decide that South Vietnam was no longer worth the price. By the end of Phase III, most of the targets on the 94-target list would have been struck. Phase IV, also lasting two weeks, would focus on industrial targets outside populated areas and on any earlier targets that had not been fully destroyed or had been repaired.[27]

The JCS sought at least limited approval for those portions of the plan which involved strikes north of the 20th parallel, but the President and Secretary of Defense elected only to increase the pressure on LOCs below the 20th parallel. This continued prohibition of strikes above the 20th parallel was a cause for increasing concern among McConnell and other senior airmen. As they watched the rapid improvements in North Vietnamese SAMs and AAA, they realized that it would be much more difficult and costly to penetrate these defenses in the future if targets above the 20th parallel should then be cleared for attack. In addition, by mid-1965 the North Vietnamese Air Force had acquired five or six Il–28 light jet bombers and based them at Phuc Yen airfield within range of major targets in South Vietnam. McConnell thought it only prudent to eliminate these growing defensive and offensive threats while we could still do so without losing many of our own men and planes.

By the summer of 1965, the air defense system above the 20th parallel was rapidly becoming quite formidable. Phuc Yen airfield had become a major jet base; Kep and other airfields were being expanded; and new airfields were being started. Because of this expansion, the JCS again requested authority to neutralize the airfields and SAMs, but the request was denied. Then on 24 July an F–4C was shot down, and three other aircraft were damaged. As a consequence of this enemy action, the President authorized strikes against those SAM sites that were actually firing at our aircraft. However, this authority did not extend to targets above the 20th parallel or to the main defense systems which extended outward from Hanoi and Haiphong for approximately 100 miles and were heavily concentrated with a 30-mile belt around Hanoi.

Having received approval to strike SAMs, the JCS continued to request approval for more of their proposed air campaign. The Secretary of Defense cited two principal reasons for his disapproval: First, he doubted that the campaign would make much difference to U.S. operations in South Vietnam. Second, and more importantly, he believed that the risk of a U.S.-Chinese confrontation could well be increased by a major air offensive.[28] Parenthetically, I should observe here that Secretary McNamara's misgivings were shared by many outside the military who opposed the idea of a strategic air offensive against North Vietnam. Fear that a confrontation with the Chinese would result from our expanding the war, and doubt about the ability of airpower to destroy the war-making capacity of the North Vietnamese led many to conclude that the war had to be won in South Vietnam. For those who accepted this position, it seemed apparent that airpower could contribute most by concentrating its efforts in South Vietnam and along the Ho Chi Minh Trail in Laos and the coastal rail and road systems below the 20th parallel in North Vietnam.

GROUND FORCES DEPLOYED—PRIORITY SET

The first U.S. ground combat units arrived in Vietnam in March 1965 when 3,500 Marines waded ashore to defend Danang. In April, the President authorized a substantial increase in ground forces to defend base areas and to reinforce South Vietnamese Army units where they were incapable of coping with the enemy. The Secretary of Defense decided, further, to support the deployed army forces with B–52 strikes in South Vietnam. Although most experienced airmen would have chosen to employ our strategic bombers against the enemy's major target systems and to have used them for close support only in emergencies, the use of B–52s for in-country missions was in consonance with the Secretary's view that the place to destroy the enemy was in South Vietnam. According to his strategy, the mission of in-country support took priority over all other missions in Laos or North Vietnam.

By July 1965, the buildup of U.S. ground forces was moving forward rapidly. At the same time, air units were coming to South Vietnam and Thailand, and the Navy's Task Force-77 in the Gulf of Tonkin had been expanded to include three carriers. In fact U.S. airpower in Southeast Asia had virtually acquired the strength to apply any desired level of pressure on all elements of the North Vietnamese military structure. Amid this rapid deployment of air, sea, and ground forces, the debate on strategy continued with airmen maintaining that the buildup of ground forces should be held at a level sufficient to defend airfields and the major logistical and population centers until the air campaign had been tried. If this campaign didn't persuade the North Vietnamese to end the war, then and only then should we proceed with a buildup of ground forces and accept a campaign of attrition. General Harold K. Johnson, Chief of Staff of the Army, believed it would take approximately five and one-half divisions to seal off the DMZ, and General Green thought it might eventually be necessary to move 500,000 troops into South Vietnam to stop the North Vietnamese.[29] General McConnell, supported by his commanders in the field, strongly insisted that a strategic air campaign was the only way to end the war successfully and soon. In a prolonged conflict we would risk losing the support of our allies and the public, he feared, even though we could eventually win if our country maintained the will to do so.

REQUEST TO STRIKE AIRFIELDS CONTINUES

As our air strikes hit more and more targets in North Vietnam below the 20th parallel, the North Vietnamese Air Force was growing in numbers and capability. MIG–21s at Phuc Yen were conducting limited sweeps south of Thanh Hoa, and on occasion they were using Thanh Hoa as a forward staging base for sweeps further south. In the eyes of the 2nd Air Division commander, the Commander of TF–77, and the JCS, the increasing MIG activity could soon represent a challenge to our control of the air. The JCS proposed, as they had on previous occasions, that Phuc Yen airfield, the main operating base of the North Vietnamese Air Force, be brought under attack and kept unusable. They pointed out that control of the air was essential to the security of U.S. and South Vietnamese forces. And with the massive deployments of U.S. forces, all of the ports and adjacent supply dumps in South Vietnam were congested and highly vulnerable to air attacks by IL–28s. We could not afford to ignore the growing capability of the North Vietnamese Air Force to mount attacks against our aircraft and these points of concentration.

The increasing strength of the North Vietnamese Air Force was demonstrated on 24 and 25 August when seven of our planes and one drone reconnaissance aircraft were shot down. [30] As a result of these losses, McConnell and Wheeler (who had become Chairman of the Joint Chiefs when Taylor replaced Lodge as Ambassador) vigorously sought permission to strike Phuc Yen airfield. Secretary McNamara again disapproved the request on the grounds that attacks on North Vietnamese airfields might cause the Chinese Air Force to assume the air defense mission in North Vietnam. Then an expansion of the war could easily result from the virtually inevitable confrontations between U.S. and Chinese pilots. The Secretary left the door open, however, for a re-examination of his decision depending upon the actions of the North Vietnamese and the threat of their air force to our interdiction flights.

Through the remainder of 1965, however, our strategy remained unchanged, with the ground war in South Vietnam receiving first priority for air strikes and with attacks in North Vietnam being limited to targets south of the 20th parallel. In fact most sorties into North Vietnam were flown against targets within forty to sixty miles of the DMZ.

At the January 1966 Honolulu Conference, Admiral Sharp again insisted that air and naval power had to be employed more aggressively against North Vietnam if the war were to be ended soon. Since becoming CINCPAC, Sharp had often advocated attacking targets near Hanoi and mining the Haiphong harbor. Although he believed a vigorous ground action was needed in South Vietnam to contain the North Vietnamese, he felt the full use of airpower against all suitable military targets throughout North Vietnam, Laos, and Cambodia should be the basis of our strategy. Lacking any evidence of North Vietnamese willingness to negotiate a settlement, the Joint Chiefs supported Sharp in his proposals for an expanded air war.

On 3 February 1966, Secretary McNamara stated that U.S. "objectives are not to destroy or to overthrow the Communist government of China or the Communist government of North Vietnam. They are limited to the destruction of the insurrection and aggression directed by North Vietnam against the political institutions of South Vietnam. This is a very, very limited political objective." [31] Implicit in the Secretary's statement was his belief that the war was still essentially an insurrection to be dealt with in South Vietnam. His concept differed importantly from the JCS view that this conflict was no longer an insurrection but a conventional war requiring a combined air and ground campaign that couldn't be confined to South Vietnam. CINCPAC and the JCS agreed that the only aspect of the war in which we had the initiative was our air campaign against the North Vietnamese heartland. We could control the intensity and scope of air attacks while the North Vietnamese could only attempt to blunt them when they came. On the ground in South Vietnam, the North Vietnamese had the initiative since their forces could fight when they wanted and retreat into the jungle or into sanctuaries in Laos or Cambodia when they didn't. Admiral Sharp and the JCS believed, therefore, that this war could not be treated as an insurgency.

STRATEGY SLOWLY CHANGES

These differences in concept notwithstanding, U.S. strategy early in 1966 was moving slowly in the direction advocated by LeMay and McConnell since 1964. More authority was being delegated to Sharp and his field commanders for the conduct of air operations. Whereas the number and frequency of strikes into North Vietnam had been controlled from Washington during the first half of ROLLING THUNDER, now Sharp was allowed, within certain restraints, to determine how much force would be applied and how often the targets could be struck. Sharp then delegated this authority to the Commander-in-Chief Pacific Air Force (CINCPACAF) and the Commander-in-Chief Pacific Fleet (CINC-PACFLT), subject to additional restraints included in his weekly operational intent plan.

In spite of the operational retraints, airpower was beginning to have an effect on the enemy's logistics system. The LOCs along the coast from the DMZ to the 20th parallel, although not as vulnerable as those above the 20th parallel, were better targets than the roads through the Mu Gia Pass and the jungle of the Laotian panhandle. Although still not satisfied with the changing strategy, other senior airmen and I believed that airpower was beginning to affect the enemy's logistical system as it had in Korea. We did not believe, however, that our airpower could be as effective as it had been in World War II unless we were authorized to strike the full range of interdiction targets.

The authorization to attack oil storage facilities in the closing days of June 1966 was the beginning of a new phase in the strategy of gradually increasing pressure on the North Vietnamese by attacking targets closer and closer to the vital power center of their government. But this piecemeal application of airpower was relatively ineffective because it still avoided many of the targets that were of most value to the North Vietnamese. Consequently, the message conveyed by these strikes on the oil facilities lacked the necessary ring of authority. Though harsh, the tone seemed also hesitant and uncertain to the North Vietnamese.

Throughout the remainder of 1966, additional targets above the 20th parallel were released one by one. Although the frequency of the strikes and the size of the striking forces were still very closely controlled in Washington, some change in strategy was becoming evident; and it was expected by most commanders from Sharp on down that it was only a matter of time until the most important targets would be released.

President Johnson, in his State of the Union address on 10 January 1967, endorsed the strategy of increasing pressure when he said, "Our adversary still believes . . . that he can go on fighting longer than we can. I must say to you that our pressures must be sustained . . . until he realizes that the war he started is costing more than he can ever gain. I know of no strategy more likely to attain that end than the strategy of accumulating slowly, but inexorably, every kind of material resource . . . that and patience—and I mean a great deal of patience."[32] Although the President clearly meant to increase the pressure, and although the only real pressure on the North was being applied by airpower, the ground campaign in South Vietnam remained the primary element in U.S. strategy. Thus in Southeast Asia in 1967, airpower was relegated to the role it had played in the Korean War prior to 1952 when the 8th Army was given the task of defeating the Chinese and North Korean armies and forcing negotiations. In Korea, this strategy was changed early in 1952 because of the high casualty rates. In Vietnam the strategy would eventually change, too. But this time the change would be much slower in coming, chiefly, I think, because our national policy makers in the mid-1960s believed that air strikes in the North would have little effect on the fighting in the South, and because the predictably inconclusive results of our piecemeal attacks in the North did nothing to persuade them otherwise.

At the time of the President's address, our airpower strategy had three objectives: Reduce the flow and increase the cost of infiltration; raise the morale of the South Vietnamese; and convince the North that it must pay a very high price, in the North, for its aggression in the South.[33] With these objectives in mind, the President released more targets for attack above the 20th parallel and authorized a campaign against the northeast and northwest rail lines. These intensified attacks against the rail system deep in North Vietnam were intended to make it more difficult for the North Vietnamese to support their forces in the South, but the most important elements in the system, the elements within 30 miles of Hanoi, were not released for attack during this early part of 1967.

By late spring, as the weather in North Vietnam improved and more targets were released within the 30-mile circle around Hanoi, our air campaign began to exact a heavy toll on the transportation system. As Commander of 7th Air Force, I was convinced that this was the time to release all the major targets for attack, and I was optimistic that the effects we had produced in the Korean War would be achievable in Vietnam. Admiral Sharp, too, felt strongly that this was the time to discard our strategy of gradual escalation in favor of an all-out effort to be sustained as long as necessary to end the war in South Vietnam.

While many of us in the military were encouraged by the progress being made against the North Vietnamese, Secretary McNamara and a number of his staff were concerned that we were getting deeper and deeper into the war with no end in sight. In their view, a drastic change in strategy was needed to get negotiations started. Thus at the very time that Admiral Sharp, I, and others were urging an all-out air offensive, proposals to halt the bombing of North Vietnam as a step toward initiating negotiations with the North Vietnamese were being considered by the President, Secretary of State, Secretary of Defense, and others.

Those who favored a bombing halt argued, first, that such a gesture might lead to peace negotiations and, second, that the air campaign was not apt to be conclusive in any case. This latter argument was based on interpretations of the Strategic Bombing Survey* following World War II, and on the observation that the Chinese and North Korean armies continued fighting for three years in spite of the bombing of North Korea. From this reasoning, three alternative air strategies emerged: (1) We could stop all bombing in North Vietnam, or (2) we could stop the bombing at the 20th parallel, or (3) we could continue the bombing without change.[34] Of course for those who favored a bombing halt, the only real question was whether we should stop all the bombing at once or phase back to the 20th parallel and see how negotiations progressed before reducing the bombing further.

On the eve of the Saigon Conference, 7–8 July 1967, Secretary McNamara had tentatively concluded that we should stop the bombing above the 20th parallel. In preparation for this conference, General Wheeler, as Chairman of the Joint Chiefs, had informed Admiral Sharp and his subordinate commanders that this meeting might well determine the future of the bombing campaign. If we failed to persuade the Secretary of Defense that the bombing campaign was worthwhile, the Secretary would surely recommend to the President that we halt the bombing above the 20th parallel. Wheeler and the JCS believed not only that the bombing should be continued, but that all of the 94 targets should be released and that the President should consider blockading or mining the port of Haiphong. Wheeler believed, further, that the Secretary would be more attentive to the observations of his field commanders because of their intimate involvement in the conflict than he would to the arguments of the Joint Chiefs, with whose opinions he was already well acquainted.

Admiral Sharp, Vice Admiral John Joseph Hyland (7th Fleet commander) and I all discussed the importance of the bombing campaign with the Secretary, apparently with some effect. He did not approve Sharp's proposal to release the full target list, but neither did he urge the President to stop the bombing above the 20th parallel at that time. He summarized his air strategy on 25 August 1967 in testimony to the Senate Preparedness Investigation Subcommittee: "The bombing of North Vietnam has always been considered a supplement to and not a substitute for an effective counter-insurgency land and air campaign in South Vietnam. . . . The

_____

*Strategic Bombing Survey: "Established by the Secretary of War on 3 Nov 44, pursuant to a directive from President Roosevelt, for the purpose of conducting an impartial and expert study of the effects of our aerial attack on Germany and Japan and to establish a basis for evaluating air power as an instrument of military strategy."

bombing campaign has been aimed at selected targets of military significance, primarily the routes of infiltration."[35]

Wheeler and McConnell continued to seek authority to strike all the airfields in North Vietnam, especially Phuc Yen, the main operational site of the North Vietnamese Air Force. As our losses to the MIGs and SAMs increased, McConnell, supported by Wheeler, Sharp, and me, argued that to protect Air Force and Navy pilots, we had to destroy the North Vietnamese air defenses. We needed a determined campaign of air-to-air engagements, attacks on airfields, and strikes against anti-aircraft and SAM installations. If the entire air defense system were not brought under attack, we would continue to lose pilots and planes unnecessarily, and we would continue to jettison many of our bombs because of actual or threatened MIG attacks.

The President finally responded to these arguments by releasing Phuc Yen for attack by 7th Air Force and TF–77 aircraft on 24 October 1967.[36] There were, however, limitations imposed on the number of aircraft that could make the strike and the time during which the attacks could be made. These restraints had no significant effect on the execution of the strike, but they reaffirmed that the application of airpower against the North Vietnamese homeland was not expected to be the decisive factor in halting the war. Thus, the air strategy had not changed significantly, nor did it change during the remainder of the year although more targets were released.

TET—STRATEGY CHANGED

With the Tet offensive in January of 1968, McConnell, I and other airmen again sought a reconsideration of the air strategy. From the scope and intensity of the offensive, it was evident that the North Vietnamese had no thought of negotiating an end to the war on terms that would be acceptable to the U.S. Their strategy was designed to focus on the American homefront as it had focused on public opinion in France in 1954. They hoped to create the impression in America that the U.S. and South Vietnamese military were losing and that the only sensible course for the U.S. would be to accept their terms and withdraw from the war.

Reaction in the U.S. to the Tet offensive was all that the North Vietnamese had hoped it would be. Public opinion was widely split on the issue of what the U.S. should do in Southeast Asia; and the President, convinced that he didn't have the people behind him for continued prosecution of the war, earnestly sought to bring the North Vietnamese into serious peace negotiations. One possible means of inducement was to cut back the bombing campaign as a sign of our sincere desire to negotiate in good faith. If we stopped the bombing, he hoped, the North Vietnamese would reciprocate by halting attacks on villages and cities throughout South Vietnam.

The President called for the views of the Joint Chiefs on the military effect of halting all bombing above the 20th parallel. They agreed that the bombing could be stopped for a short time to determine whether the North Vietnamese wished to negotiate seriously,[37] but they strongly suspected that the North Vietnamese would continue to fight until we applied enough force to threaten destruction of the power base in North Vietnam.

While the President considered the bombing halt, weather in Vietnam was changing from the northeast to the southwest monsoon. This meant that weather conditions over North Vietnam would be poor for another month, during April, and would then improve markedly. Thus, although I had no confidence that we would achieve a negotiated settlement at that time, I supported the proposal for a bombing halt because I realized that the weather alone would probably cause us to cancel all but a few hundred sorties and because we were not being permitted to strike the most valuable targets in any case. I felt that stopping the bombing above the 20th parallel to test the intentions of the North Vietnamese would have a minimum effect on the air campaign if the bombing halt took place in the month of April. Wheeler, McConnell, Sharp, Ryan (General John D. Ryan, CINCPACAF), and I with most of the other military leaders recommended, however, that if the North Vietnamese didn't stop the shelling of South Vietnamese cities and the assassination of village chiefs, and didn't show positive signs of de-escalating the fighting by withdrawing their regular divisions back across the DMZ, we should resume bombing with no restrictions and should mine the harbor at Haiphong. We further urged the President to set a time by which substantive discussions must begin. We believed that the U.S. should avoid long "fight and talk" negotiations such as those that developed in the Korea War, negotiations that would almost certainly be used as a propaganda forum by the enemy.

With these caveats, most of the top military commanders thought a halt to the bombing above the 20th parallel would be militarily acceptable. Thus on 31 March 1968, President Johnson said in a speech to the nation, "I am taking the first step to de-escalate the conflict . . .. Tonight I have ordered our aircraft and naval vessels to make no attacks on North Vietnam except in the area north of the demilitarized zone . . .. Our purpose in this action is to bring about a reduction in the level of violence that exists."[38]

Soon after the bombing halt, reconnaissance photos showed the restoration of all the railroad network above the 20th parallel. The marshalling yards at Thai Nguyen and Kep, and smaller ones along the northeast railroad were soon repaired, and traffic from the Chinese border to Hanoi and points south was near normal. In the CINCPAC Report on the War in Vietnam (June 1968), Admiral Sharp observed that "almost 1300 trucks were noted during Christmas and about 1800 during the slightly longer New Year stand-down. This compared with a daily average of about 170 for other days between 22 December 1967 and 4 January 1968."[39] These frantic efforts by the North Vietnamese to move as much material to South Vietnam as the system could take were indicative of their intention to settle the future of South Vietnam on the battlefield, not at the negotiating table.

By the time of the April 1968 bombing halt, defense analysts in the U.S. were already anticipating withdrawal of U.S. troops. Although the President had not announced such a move, the apparent lack of popular support for the war led many to conclude that the withdrawal of U.S. troops from Vietnam was *a fait accompli*. It would begin as soon as the South Vietnamese, with our financial help and with our airpower (to interdict supplies in Laos and support the troops in South Vietnam), were able to carry the responsibility for the fighting.

As many of us had expected, the level of violence in South Vietnam remained about the same despite the bombing halt. The North Vietnamese political assassinations and attacks against isolated villages and camps continued. And in Paris, the negotiations started very slowly amid indications that procedural matters would be employed to frustrate and delay. Nevertheless, it appeared that the trend of U.S. policy was toward less, not more, bombing as a means to advance the negotiations.

In July, the new Secretary of Defense, Clark M. Clifford, made a visit throughout Southeast Asia to determine firsthand what his field commanders thought of the combat situation. [40] In 1966, he had been a strong supporter of the war, but in mid-1968 he was dubious that the war could be won militarily. Based on his observations of our allies, he was convinced that they would not support us if the war continued. He didn't believe that the limited bombing campaign of July 1968 was effective, and he felt that resumption of attacks above the 20th parallel was not a feasible course of action. Lacking concrete evidence that we could end the war through military means by any specific date in the near future, he proposed to recommend to the President that we stop the bombing of North Vietnam and begin the withdrawal of U.S. ground forces with the South Vietnamese assuming total responsibility for the conduct of the war.

President Johnson, confronted with dissension throughout the country, asked senior officials to tell him in detail what would be the most likely results if we should stop all bombing in the North but continue our reconnaissance flights and our interdiction of the supply lines in Laos. The other field commanders and I advised against further curtailment of the bombing campaign. We felt sure that the North Vietnamese would take advantage of such a halt to move their fighters, SAMs, AAA, and logistical centers closer to the DMZ.

Not long after I returned from Vietnam in August to become Commander of Tactical Air Command, the President called each of the Joint Chiefs and me to give him our views separately. Each of us assured the President that the North Vietnamese would take advantage of the bombing halt to improve their position for a future offensive. Furthermore, we said, it would be unrealistic to suppose that airpower could control the enemy's flow of supplies into South Vietnam by striking the

LOCs in Laos if all the alternative routes in North Vietnam were immune to attack. Each of us advised that if the President were convinced that the North Vietnamese sincerely wanted substantive negotiations, he might try a brief bombing halt without unduly jeopardizing our forces in South Vietnam. But if the North Vietnamese made no prompt, visible effort to stop the fighting in South Vietnam, if they continued the infiltration of troops, if they failed to begin withdrawal of regular divisions, and if they showed no serious interest in the negotiations, then the bombing campaign should be resumed against all military targets throughout North Vietnam, and such a campaign should continue with no let-up until our demands for a cease-fire were satisfied. [41]

On 31 October 1968 the President announced his decision to stop all bombing of North Vietnam but to continue reconnaissance flights and interdiction of supplies moving through Laos. Just as the North Vietnamese had taken advantage of the 31 March bombing halt above the 20th parallel, they began immediately to take advantage of this one. Soon our reconnaissance flights showed a heavy flow of military traffic along all the major coastal routes to the DMZ. Most of the small harbors up and down the coastline were crowded with small boats being used to shuttle supplies. From these early observations, senior military officials concluded that the North Vietnamese would eventually accumulate sufficient stocks to wage a major offensive if airpower were not used to attack these supply centers and the LOCs feeding them. Even if a decision were made to strike the supply dumps, we realized that destroying them would be no simple task because camouflage would make them unusually difficult to locate.

A NEW STRATEGY

As President Nixon took office, the conflict in Vietnam presented him with the need for some hard decisions on matters of strategy.* Given the national commitment to a gradual withdrawal of American ground troops, airpower was his only means of protecting our departing troops, and providing time for the South Vietnamese to improve their fighting ability. The circumstances seemed to call more than ever for a new strategy that recognized and employed airpower's total capabilities.

As the North Vietnamese watched us prepare to withdraw from Vietnam, they hurriedly shipped materiel south for what they hoped would be their final offensive. We were able to devote a greater effort to interdiction in Laos because of the shifting of forces from targets in North Vietnam, but the North Vietnamese put more war materials into their supply lines than they had at any other period of the war. [42] As the North Vietnamese continued to expand their supply systems in Laos and the northern provinces of South Vietnam, the Joint Chiefs pressed for a

---

*For a detailed discussion on the difference of views on strategy during this period, see Congressional Record (House), vol. 118, part 13, 4–11 May 1972.

403-892 O - 83 - 4

resumption of bombing below the 20th parallel. Other senior airmen and I believed, of course, that an unrestricted campaign would have a much better chance of ending the war on favorable terms, but we understood that public opinion in the U.S. at that time would not support the President in such a decision. However, the President did adopt a more aggressive tone toward the North Vietnamese, conveying a sense that our airpower might well be employed against the North Vietnamese homeland unless the infiltration stopped and there were productive negotiations.

In preparation for the withdrawal of U.S. troops, it was critical that the supply centers in Cambodia and Laos be reduced as much as possible so that the North Vietnamese could not use these stockpiles of ammunition and other materiel to mount a sustained offensive against the departing Americans or the South Vietnamese. In 1965, General William C. Westmoreland (Commander, U.S. Military Assistance Command, Vietnam, or COMUSMACV, 1964–1968) had requested authority for B–52s to strike the supply bases in Cambodia from which the Viet Cong and North Vietnamese were staging attacks into Military Region III. Permission had been gained from Prince Sihanouk in 1967 to attack these base camps provided that the U.S. made no public announcements of the attacks. [43] In accordance with this understanding, B–52s began bombing the camps on 19 March 1969 to reduce the threat to Military Region III.

By the end of the year, the U.S. withdrawal from South Vietnam was well underway. For the protection of our remaining troops and the South Vietnamese, there was no alternative to airpower. Public opinion in this country would have strongly opposed the return of U.S. ground forces, and the South Vietnamese Army and Air Force were incapable of holding off the North Vietnamese without the backing of U.S. airpower. The North Vietnamese responded to the situation by stepping up the construction of airfields south of the 20th parallel and by moving SAM defenses further south until they covered all of the major passes and were actually a threat as far south as Dong Ha in South Vietnam. The North Vietnamese recognized that the success of our national strategy and the survival of South Vietnam depended upon U.S. airpower. To frustrate our strategy, they would have to bring U.S. aircraft under attack by MIGs, AAA, and SAMs in these southern areas in the same manner and intensity as they had earlier near Hanoi.

In November 1968, President Johnson had decided to provide fighter escorts for our reconnaissance flights going into North Vietnam. [44] Initially, the escorting fighters were authorized to attack only AAA and SAMs that were firing at the reconnaissance aircraft. The escorts were known as protective reaction flights. Their importance and their authorizations to strike defensive installations increased through 1969 and 1970. By the end of 1970, the size of the escorting forces had been expanded, and the fighters were armed with special munitions for their strikes.

Meanwhile, the North Vietnamese continued to increase their SAM coverage above the DMZ and flew frequent missions against our strike forces bombing in the Mu Gia and Ban Karai Pass area. At the same time. MIGs were attempting to shoot down B–52s that were being employed more frequently against the Laotian supply lines.

In response to the increasing enemy air activity, Admiral John S. McCain, who succeeded Admiral Sharp as CINCPAC, urged the JCS to get authority to hit the airfields below the 20th parallel. These were the fields from which the MIGs were threatening our flights below Vinh and in Laos. By keeping these forward bases out of action, we could eliminate the MIG threat to our interdiction program. (And the importance of our interdiction program was growing daily as departing U.S. ground forces shifted more and more of the combat responsibility to the South Vietnamese.) Thus Admiral Thomas H. Moorer, who had replaced General Wheeler as Chairman of the JCS, pushed vigorously for authority to hit not only the airfields but also the supply centers above Mu Gia Pass and those near Bat Lake some 30–40 miles above the DMZ. He wanted U.S. airpower to have a more visible presence in North Vietnam as positive evidence that all of North Vietnam could come under attack if the ominous buildup of North Vietnamese forces in the south didn't cease.

AIR STRATEGY ADOPTED

Throughout 1971 there was a slow but continuous expansion of air strikes in North Vietnam below the 20th parallel. Whereas our protective reaction strikes had originally been limited to active SAM and AAA sites, in 1971 special strike missions were directed against many of the same supply points that had been struck before the 1968 bombing halt. President Nixon's position was that the security of our troops in South Vietnam demanded these protective strikes.

Many of us airmen contended, however, that these small and infrequent strikes were too limited to reduce the flow of logistics along the coastal routes in North Vietnam. Although we were vigorously pressing our interdiction program in Laos, we were still under stringent limitations as to the size and duration of our missions in North Vietnam. To reduce the flow of logistics significantly, we would have to make sustained attacks on all of the LOCs. Consequently, we felt that the greatest value of these limited strikes would be a signal to the North Vietnamese that we could still intensify the conflict terrifically if we were forced to do so. But it soon became apparent that the North Vietnamese hadn't paid much attention to these strikes or to the greater threat they represented, for shortly after the first of the year, they invaded South Vietnam with more than 40,000 men.

On 30 March 1972 as the North Vietnamese rolled across the DMZ and into Military Region I with some 400 armored vehicles, anti-tank missiles, shoulder-fired infrared missiles, and 122mm and 130mm artillery, our hopes of reaching a negotiated settlement on acceptable terms grew exceedingly slim. [45] Their strategy, it appeared, was to sever the northern two provinces from South Vietnam, push a salient deeply toward Pleiku, and open the way for a future assault on Saigon. We faced the rapidly deteriorating situation in South Vietnam with virtually all of our ground forces gone and with our air force down to some 500 aircraft (as compared to more than 1000 combat aircraft that were in 7th Air Force at the time of the bombing halt in 1968.)[46] The President saw that the threat of bombing North Vietnam wouldn't dissuade the North Vietnamese and that the South Vietnamese Army (ARVN) couldn't contain the offensive without a heavy commitment of U.S. airpower.

The North Vietnamese had extended their air defense system into most of the northern two provinces. They also covered Khe Sanh, Pleiku, and the Cambodian border in Military Region III. These defenses, consisting of heavy AAA concentrations, SA–7s and SA–2s, made it impossible to operate the low performing aircraft of the Vietnamese Air Force in the northern areas without high losses.[47] The high performing fighters and SAM-killing F–105s and F–4s of 7th Air Force were needed to suppress the defenses and deliver close air support strikes to halt the advance. After the collapse of the ARVN 3rd Division, in fact, it was only the extensive employment of U.S. airpower that was able to bring the offensive to a halt.

Our airpower's first priority was to support the South Vietnamese ground forces so that the ground battle could be stabilized. This was done by staging fighters from Thailand into airfields in South Vietnam, running a sortie, and then recovering back in Thailand. By using this staging program and air refueling, and calling upon the combined efforts of 7th Air Force, TF–77, and the available Marine aircraft, we stopped the offensive after the loss of the northern half of Military Region I. In all the other areas, the offensive was ground to a halt by the extensive employment of B–52s and the magnificent performance of the tactical airlift force in delivering arms, ammunition, and food under intensive fire from AAA and SAMs.

While 7th Air Force was flying many extra sorties per day to contain the offensive, the JCS recommended and the President approved both the return of flying units previously deployed to the U.S. and a buildup of B–52s and carriers. These measures were not only intended to blunt the North Vietnamese Easter offensive but also to prepare for a major air offensive against North Vietnam, the air offensive that LeMay and others had foreseen as early as 1964.[48]

The President on 7 May directed the mining of Haiphong and other coastal ports. Code named POCKET MONEY, this mission was the first in a series of actions designed to isolate Hanoi from the rest of Vietnam. On the 10th of May, the President authorized the bombing of most of the targets in North Vietnam that had made up the original JCS 94-target list. The strategy at this point was first to isolate North Vietnam from external support by mining the harbors and destroying the marshalling yards and key choke points along the northeast and northwest railroads, and then to strike all the major supply areas around Haiphong and Hanoi. Restricted zones were established around and within these two cities as in the 1965–1968 campaign; however, the JCS was given much greater freedom in choice of targets, frequency of attacks, and weight of effort; and restricted zones were lifted from time to time. For the first time in the long struggle, airpower was being employed as airmen had advocated. Questions about the effectiveness of a determined application of airpower would soon be answered.

The resumption of bombing in May was known by the code name of LINEBACKER I. Strategy for this campaign differed from the strategy for the original bombing program, ROLLING THUNDER, mainly in that a gradual increase of pressure from the south of North Vietnam to the Hanoi area was no longer the dominating idea. The assumption now was that only the attacks above the 20th parallel would force North Vietnam to realize the futility of trying to conquer South Vietnam by force.

This campaign's objectives could be stated as follows: (1) Restrict resupply of North Vietnam from external sources; (2) destroy internal stockpiles of military supplies and equipment; (3) restrict flow of forces and supplies to the battlefield. [49] Although these three objectives governed our activities in the North during this period, the purpose underlying the entire campaign was to break the enemy's will and ability to continue fighting.

LINEBACKER I and the airpower employed against the ground offensive in South Vietnam apparently had the desired effect, for the President on 23 October halted the bombing above the 20th parallel with the expectation that negotiations would move forward

Within a couple of months after the bombing halt above the 20th parallel, however, the North Vietnamese were again protracting the negotiations, introducing one roadblock after another. It was apparent that they were not seriously interested in a negotiated settlement of the war. As for the suspension of the bombing campaign, the North Vietnamese evidently interpreted it (as they had interpreted earlier suspensions) to be an indication of weakness and lack of resolve. The President determined that a further change in strategy was in order.

On 18 December 1972 he directed an all-out air campaign against North Vietnam's heartland to force a settlement of the war. For the first time, B-52s were used in large numbers to bring the full weight of airpower to bear. What airmen had long advocated as the proper employment of airpower was now the President's strategy—concentrated use of all forms of airpower to strike at the vital power centers, causing maximum disruption in the economic, military, and political life of the country. This air strategy was translated into an 11-day air campaign known as LINEBACKER II which included strikes against point targets by tactical aircraft using laser weapons; neutralization of area targets by B-52s using radar bombing; and suppression of SAMs, AAA, and MIGs by 7th Air

Force and TF–77 fighters. After 11 days, the North Vietnamese sought a cease-fire.

The development of air strategy in World War II, Korea, and Vietnam was a repetitious process. In each case, planners first perceived airpower as a subordinate part of a joint strategy that would employ an extensive ground campaign to end the war on favorable terms. On the other hand, airmen came increasingly to believe that airpower, in its own right, could produce decisive results. The validity of such a view was suggested by results of the Allies' combined bomber offensive in Europe and by the surrender of Japan in the 1940s. Additional evidence came from the skies over Hanoi in December 1972. In a concentrated 11-day test, our air strategy persuaded a determined adversary with a remarkably elaborate air defense system that overt aggression could not be sustained in the presence of unrestricted U.S. airpower.

CHAPTER I

FOOTNOTES

[1] Maxwell D. Taylor, The Uncertain Trumpet (New York: Harper & Brothers, 1959), pp. 47–51.

[2] Robert F. Futrell, Ideas, Concepts, Doctrine: A History of Basic Thinking in the United States Air Force, 1907–1964 (Air University, Maxwell AFB, AL, 1971), p. 152.

[3] United States Air Force Operations in the Korean Conflict, USAF Historical Study No. 71, 25 June–1 November 1950; USAF Historical Study No. 72, 1 November 1950–30 June 1952; USAF Historical Study No. 127, 1 July 1952–27 July 1953; 3 studies (Washington, D.C., Department of the Air Force, 1952–1956), No. 71, p. 1.

[4] Korean Conflict, No. 72, p. 9.

[5] Ibid., p. 43.

[6] Futrell, History of Basic Thinking, p. 152.

[7] Matthew B. Ridgway, The Korean War (Garden City, N.Y.: Doubleday & Company, 1967), p. 258.

[8] William C. Westmoreland, A Soldier Reports (Garden City, N.Y.: Doubleday & Company, 1976), p. 69.

[9] Jacob Van Staaveren, USAF Plans and Policies in South Vietnam, 1961–1963 (Washington, D.C.: Office of Air Force History,June 1965), pp. 14–19.

[10] Townsend Hoopes, The Limits of Intervention (New York; David McKay Company, 1969), p. 13.

[11] The Department of State Bulletin, vol. XLIV, No. 1143, 22 May 1961 (Secretary Rusk's News Conference of 4 May 1961), p. 761.

[12] Francis J. Kelly, Vietnam Studies, U.S. Army Special Forces, 1961–1971, (Washington, D.C.: Department of the Army, 1973), p. 7.

13 Working Paper for CORONA HARVEST Report, USAF Activities in Southeast Asia, 1954-1964, vol 2, book 1 (Maxwell AFB, AL: Air University, January 1973) p. 4-54.

14 Working Paper for CORONA HARVEST Report, Command and Control of Southeast Asia Air Operations, 1 January 1965-31 March 1968, vol 1, book 1, p. 1-1-22.

15 CORONA HARVEST Report, Command and Control, p. 1-1-8.

16 CORONA HARVEST Report, USAF Activities, 1954-1964, pp. 3-26-3-28.

17 Curtis E. LeMay, General, USAF (Ret.) interview held at the Pentagon, Washington, D.C., February 1976. NOTE: General LeMay was the Chief of Staff, United States Air Force, from July 1961 to January 1965. See, also, Jacob Van Staaveren, USAF Plans and Policies in South Vietnam and Laos, 1964 (Washington, D.C.: Office of Air Force History, December 1965), p. 30.

18 CORONA HARVEST Report, Command and Control, book 1, p. 1-1-21 and book 5, part VIII, p. VII-3-2.

19 LeMay, interview, February 1976.

20 Department of Defense Study, United States—Vietnam Relations 1945-1967, book 4 (Washington, D.C.: Government Printing Office, 1971) pp. iv, 7.

21 LeMay, interview, February 1976.

22 Westmoreland, A Soldier Reports, p. 79.

23 United States—Vietnam Relations, book 3, p. a-14.

24 CORONA HARVEST Report, Command and Control, p. 1-1-1.

25 Jacob Van Staaveren, USAF Plans and Policies in Southeast Asia, 1965, pp. 8-22.

26 United States—Vietnam Relations, book 4, p. 77.

27 John P. McConnell, General, USAF (Ret.), interview held at Pentagon, Washington, D.C., March 1976. NOTE: General McConnell was the Chief of Staff, United States Air Force, from February 1965 to July 1969.

28 Working Paper for CORONA HARVEST Report, Out-Country Air Operations, Southeast Asia, 1 January 1965-31 March 1968, book 1, pp. 5-75.

29 McConnell, interview, March 1976.

30 CORONA HARVEST Report, Out-Country Air Operations, p. 147.

31 CORONA HARVEST Report, Command and Control, book II, part III. p. III-3-2.

32 Lyndon B. Johnson, State of the Union Address, Washington, D. C., 10 January 1967.

[33] Summary Report by Preparedness Investigating Subcommittee of the Senate Armed Services Committee, on the Air War Against North Vietnam, 90th Cong., 1st. sess., August 1967, p. 9.

[34] For a detailed discussion of the options considered, see Lyndon B. Johnson, The Vantage Point (New York: Popular Library, 1971), pp. 399–421.

[35] Hearings, Senate Armed Services Committee, 90: 1–1967, S–3–9, part 1, p. 275.

[36] CORONA HARVEST Report, Out-Country Air Operations, p. 156.

[37] Johnson, Vantage Point, pp. 366–368.

[38] Ibid., p. 435.

[39] U.S.G. Sharp and William C. Westmoreland, Report on the War in Vietnam, 1964–1968 (Washington, D.C.: Government Printing Office, 1968), p. 41.

[40] Johnson, Vantage Point, p. 511.

[41] Ibid., p. 368.

[42] CORONA HARVEST Report, Out-Country Air Operations, p. 316.

[43] Westmoreland, Soldier Reports, pp. 183 & 389.

[44] Working Paper for CORONA HARVEST Report, USAF Air Operations Against North Vietnam, 1 July 1971–30 June 1972, p. 9.

[45] William W. Momyer, The Vietnamese Air Force, 1951–1975, An Analysis of Its Role in Combat, USAF Southeast Asia Monograph Series, vol. 3, monograph 4 (Washington, D.C.: GPO, 1977), p. 45.

[46] John A. Doglione, et al, Airpower and the 1972 Spring Invasion, USAF Southeast Asia Monograph Series, Vol 2, Monograph 3, p. 30.

[47] Momyer, Vietnamese Air Force, p. 49.

[48] LeMay, interview, February 1976.

[49] CORONA HARVEST Report, Command and Control, book 1, pp. 1–1–9, 1–1–24.

CHAPTER II

COMMAND AND CONTROL OF AIRPOWER

PRIOR TO VIETNAM WAR

In wars involving two or more services on the same side, command and control of assigned forces has been controversial. This controversy was prevalent in World War II from the early days of North Africa to the invasion of Europe, and it persisted throughout the Korean War. In Vietnam, U.S. politics—because of its greater impact than in other wars, even at the lower levels—aggravated the controversy, making it much more complex and difficult to resolve.

The reason for the controversy is fairly straightforward: The flexibility of airpower and its capacity to concentrate large quantities of firepower in a short time make it a most desirable addition to an army or navy. As a consequence, these two forces have sought the division of airpower, placing it under their control when needed for their own mission.

Airmen, on the other hand, have argued that airpower is a decisive element of war in its own right and that the full effects of airpower can only be achieved when it is centrally controlled and directed against the most vital part of the enemy, whether that part be the industrial base or the military forces deployed to a theater of war. They contend that the fundamentals for directing and using airpower are the same regardless of the strategy for the prosecution of the war. Thus, for airpower to be employed for the greatest good of the combined forces in a theater of war, there must be a command structure to control the assigned airpower coherently and consistenly and to ensure that the airpower is not frittered away by dividing it among army and navy commands.

NORTH AFRICA—THE FOUNDATION

Although World War I provided some indication, the North African campaign of World War II really helped hammer out the doctrine for organizing and employing airpower in theaters of war. Many of the problems associated with these early battles provide a basis for an understanding of the issues that emerged in Vietnam on how to control and employ airpower.

After the invasion of North Africa, it became apparent that the organization of air and ground forces was not successful. Early in the campaign, planners hoped that the Allied armies would be able to reach the Cap Bon peninsula by Christmas of 1942 and trap Rommel between the North African forces and General Bernard Montgomery's advancing 8th Army. But the North African offensive bogged down in the Atlas Mountains, and not until the following spring, after some very tough fighting, were the troops of Colonel General Jurgen von Arnim surrounded, cut off from support from the Italian mainland, and captured.[1]

As the fighting developed in southern Tunisia, the relationship between air and ground forces became a matter of concern to the Combined Chiefs of Staff and General Eisenhower, then serving as the Commander-in-Chief, Allied Forces Northwest Africa. There was no centralized control of either the tactical or strategic air forces. These forces were operating almost independently, with the American Air Force more badly split than the Royal Air Force (RAF).

The doctrine at the time (as set forth in Army Field Manual 1–5) provided that an air support command was attached to an army formation and directed by that ground force commander, who had the more important mission. Airpower, in other words, was adopted to the demands of the ground force commander fighting the battle.

As a result of this doctrine, there was no concerted effort for airpower to gain air superiority over the theater of operation. For example, consider the airpower assigned to the XII Air Support Command (XII ASC, primarily employed in the mission assigned II Corps and the French XIX Corps) and the British RAF 242 Group (located to the north and committed to the support of the British 1st Army). Both of these air forces were trying to provide close air support before obtaining air superiority. Consequently, the German Air Force (GAF) controlled the air in northern and southern Tunisia. Friendly losses were so high that the mission of the air forces and the structure of the command and control system had to change drastically. Ironically—but naturally—not only had Allied airpower failed to achieve air superiority, but they had failed to provide the close air support that the Commanding General of the 1st Army and II Corps had desired. The German fighters, by concentrating against small formations of U.S. and British fighters trying to maintain umbrellas over ground forces throughout the day, made Allied air losses prohibitive.

The German Air Force in North Africa therefore posed a very serious obstacle. If we were to provide our forces air support, we had to destroy the German fighters both on the ground and in the air. Not until we in the XII ASC and the 242 Group had gained air superiority (i.e., when we could conduct missions without undue losses and interference from the enemy) could we concentrate on providing close air support.

40

U.S. P–40s taxiing at Maison Blanche, North Africa, 1943.

Air Chief Marshal Tedder expressed the new doctrine: "Given central-ised control of air forces, this flexibility brings with it an immense power of concentration which is unequalled in any other form of warfare. In other words, if properly used, the flexibility of air force allows it to be highly economical. . . . The important words in my previous sentences are 'if properly used' and 'given centralized control.' " [3]

The need to establish a centralized theater air and ground organization was also vividly brought home during the battle for Kasserine Pass. Rommel made one last bold and desperate push to slice through the rear of the Allied armies to the Mediterranean coast. Had airpower been more concentrated, Rommel would have stopped much sooner.

At this time the reorganization of airpower as approved at the Casablanca Conference in January was finally put into effect. Air Marshal Tedder explains:

> The proposals for the new air command were finally approved by Roosevelt and Churchill on 26 January (1943). An Air Commander-in-Chief for the whole Mediterranean theater would set up his headquarters at Algiers; under him would serve the Air Officers Commanding Northwest Africa, the Middle East, and Malta. He would be surbordinate to the Commander-in-Chief Allied Expeditionary Force in Northwest Africa. . . . In return, the Commander-in-Chief Allied Expeditionary Forces would give every possible help in the North-West Africa Theater for the operation of Mediterranean air forces. [4]

Air Marshal Tedder commanded this new organization. Under him as the Commander of the Northwest African Air Force was General Carl A. Spaatz, whose command consisted of four air forces: the Strategic Air Force, Tactical Air Force, Coastal Air Force, and Troop Carrier Command. With this air organization, Eisenhower had all air elements within a single structure capable of being concentrated on our greatest challenge: to gain control of the air and stop the advance of the German Army.

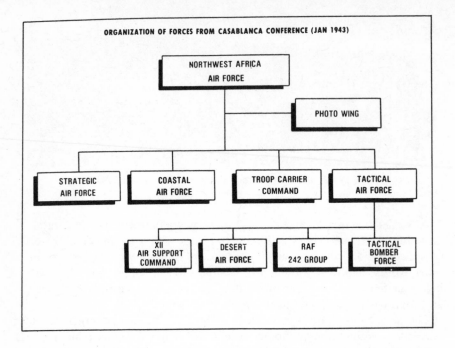

ORGANIZATION OF FORCES FROM CASABLANCA CONFERENCE (JAN 1943)

NORTHWEST AFRICA
AIR FORCE

PHOTO WING

STRATEGIC
AIR FORCE

COASTAL
AIR FORCE

TROOP CARRIER
COMMAND

TACTICAL
AIR FORCE

XII
AIR SUPPORT
COMMAND

DESERT
AIR FORCE

RAF
242 GROUP

TACTICAL
BOMBER
FORCE

The Tactical Air Force was commanded by Air Vice Marshal Sir Arthur Coningham. Coningham, who had commanded the British Desert Air Force and supported Montgomery's 8th Army, was an experienced airman. Tedder writes in his memoirs that "As it transpired, I had no need to be worried about Montgomery's view of the exercise of airpower. He had indeed produced a pamphlet of instructions for the troops which, I told Portal, it would be difficult to improve upon; this was not surprising, in view of the fact that the original document had been prepared by 'Mary' Coningham.''[5]

Coningham understood both the need for controlling the air and the methods for supporting the army. His air force had defeated the GAF on the western desert and had provided the close air support for the breakout at El Alamein. Further, Coningham was familiar with an independent air force: the RAF, unlike the U.S. air forces, had enjoyed equality with ground forces. In the RAF, equality was a fundamental concept which recognized that although ground and air forces have separate missions, those separate missions must blend for the good of the total mission, defeat of the enemy. At a crucial time in North Africa when airpower had been so badly used, the selection of Coningham, a proven commander with extensive combat experience, was indeed wise.

Coningham's headquarters stated as a first order of business that all his forces would concentrate on gaining air superiority.[6] This effort would involve air-to-air combat as well as attacks on all German landing grounds in the area. As soon as they had dominated the German Air Force, they would provide close air support, shifting the 242 Group, Desert Air Force, or XII Air Support Command to wherever the German Army was most vulnerable to concentrated air attacks.*

Concurrent with Coningham's battle for air superiority, General Carl Spaatz, with centralized control of strategic airpower, was directing his forces against the threat that concerned Eisenhower most, the German attempts to reinforce their Afrika Korps. He had the Northwest Strategic Air Force working against airfields in Sicily and Italy so that the Germans would be unable to replace their losses in the battle area. In addition, B–17s, B–25s, B–26s, and Wellingtons were pounding the ports and enemy shipping at Tunis, Bizerte, Palermo and Naples to prevent the movement of supplies and replacements to Rommel and von Arnim. From this campaign airmen derived basic concepts about how best to organize all forces and to employ airpower within a theater of operations.

When the Combined Chiefs of Staff approved the air organization, they also approved the restructuring of the ground forces. General Harold L. Alexander remained in his post as deputy to Eisenhower and also became head of the 18th Army Group on 20 February 1943. Within his Army Group were the British 1st Army, British 8th Army, American II Corps, and French XIX Corps. For the first time, all of the ground forces were under a single theater army component commander. Alexander's counterpart was Spaatz. However, Spaatz delegated most of the detailed coordination on air-ground matters to Coningham's Tactical Air Force, which located its advance headquarters next to Alexander's advance headquarters.

Despite the streamlining improvements made by the Chiefs of Staff, a few organizational problems remained. Alexander's dual role as Deputy Commander-in-Chief of Allied Forces and Commanding General of the 18th Army Group became a matter of concern among airmen. Many of my American and British counterparts believed that the Commander-in-Chief, or Deputy, should not simultaneously be a component commander. We felt that the theater commander should not be immersed in the detailed tactical problems of any one component force. Battlefield situations often demand the full time and attention of a component commander, yet the theater commander's job of determining the strategy for combined air, ground, and sea forces can scarcely afford to be neglected. We felt, too, that a dual-hatted theater commander would tend to be less objective about theater decisions affecting his own component because he would be conscious of having to implement the decisions himself. Since the theater commander is the representative of all the forces, we believed, he should not owe special allegiance to any single component.

---

*Under Coningham's Northwest Tactical Air Force, the XII Air Support Command worked with the American II and French XIX Corps; the Desert Air Force worked with the British 8th Army; and 242 Group with the British 1st Army.

It is especially significant that the North African command structure provided Spaatz, the air component commander, with continuous operational control of all the air elements. The bomber force was actually divided into a tactical bomber force under Coningham's Tactical Air Force and a strategic bomber force under Major General James H. Doolittle. The important point, however, (and this was an issue in Operation OVERLORD and in Vietnam), was that the theater air component commander had these air resources under his direct control. He decided how to use them according to Eisenhower's guidance. Also, the air reconnaissance force for the theater-wide mission came under the direct control of the air component commander, although the Tactical Air Force had its own reconnaissance units for its particular mission. The air component commander, Spaatz, therefore, could gather intelligence for both air and ground battles, and for the strategic campaign against the lines of communication and the rear base areas in Sicily and Italy.

In the Mediterranean, the naval component commander, Admiral Sir Andrew Cunningham, sought control of part of the air resources to protect the fleet when it was within range of German and Italian land-based aviation. [7] (Tedder had been confronted with this issue on numerous occasions during the siege of Tobruk and the campaign on the western desert.) Admiral Cunningham wanted fighters when the Middle East forces were having difficulty in getting supplies through because German airpower dominated the eastern Mediterranean. Convoys to Malta and beyond suffered heavy losses during 1942, and the sea lines of communication were tenuous.

However, Tedder refused to parcel his airpower to the operational control of the Royal Navy. He said that because of conflicting demands for his airpower, he had to employ it from task to task as the nature and intensity of the threat required. He needed to gain air superiority; support the army; defend Alexandria, Cairo, and the desert bases; interdict the land and sea lines of communication that supported Rommel; and protect the fleet, particularly from air attacks. To meet all of those tasks, he had to have centralized control.

Tedder, therefore, couldn't parcel out his airpower to the Navy any more than he could for Montgomery. The only way he could reconcile the needs of both commanders was to shift airpower back and forth. If, for instance, the German Air Force, were to attack a convoy trying to run from Malta to Alexandria, he would shift fighters from the counter air task to cover the movement of the convoy; meanwhile, other parts of the Desert Air Force would strike German fighter bases along the African coast to keep them from attacking the convoy.

Naval aviation, when operating against target systems assigned the tactical or strategic forces, also came under the control of the theater air component commander. During the invasion of Sicily and Italy, the Tactical Air Force commander controlled all airpower used to isolate the objective, provide air defense for the area, and support the troops during their landing and their movement inland.[8] He was responsible for targeting and controlling naval aviation that participated in these campaigns. Even carrier forces not directly involved in the air defense of the fleet and the convoy enroute to the landing area were under his operational control for targeting against airfields and lines of communication leading to the objective area and for covering the landing forces.

The unity of airpower was not only sound in theory, but the theory stood the test of battle and proved to be the most effective method for the command and control of airpower in a theater of operations.

OVERLORD—THE ARGUMENTS ABOUT COMMAND

Even though the North African and Mediterranean experience demonstrated the kind of command structure needed for theater war, preparations for OVERLORD, the forthcoming invasion of Europe, raised some fundamental issues about the command of the tactical air forces and strategic bombers. Churchill and Roosevelt had decided at Quebec in August 1943 that a U.S. commander would be Supreme Commander-in-Chief of the Allied Expeditionary Forces. Before the naming of the commander, many supported General George C. Marshall. Roosevelt, however, felt very keenly the need of keeping Marshall as Chief of Staff of the Army because of the rapid expansion of the Army ground and air force.[9] He considered Marshall almost indispensable to this tremendous task if the forces were to be ready by the invasion date. On the other hand, the British Chiefs of Staff were pressing for a British officer to be the Supreme Commander. After all, they had been in the war from the beginning, had suffered the most casualties and losses on the home front, and until that time had more forces in battle. Perhaps their most telling argument was the number of high British commanders who had been tested in battle compared to the number of American commanders. Only with some misgivings did they accept Churchill's decision that an American would be the Supreme Commander.

With this concession, the British expected to have the top airmen's jobs. Although our Army Air Force was rapidly gaining strength, the British Air Force had more aircraft and personnel in units fighting the enemy. Air Chief Marshal Portal had already been designated by the Combined Chiefs of Staff as their deputy for coordinating the British and American bomber forces for the combined bomber offensive. Although this didn't involve operational command of these forces, it did provide for coordination of targets to be struck, timing of the attacks, and mutual support required from both forces. In a real sense, Portal was the top airman in the Allied air forces even though Spaatz had wide latitude in choosing specific targets within the established priority listing.

At a meeting in Quebec in August 1943, the Combined Chiefs of Staff agreed that there would be a combined command for the air and navy forces, but they did not settle the question on such a command for ground forces. This issue was raised many times throughout the European

45

campaign, particularly following the invasion. Montgomery, who had overall command of the ground forces for the invasion, was not happy about relinquishing command of the American ground forces to General Omar N. Bradley's 12th Army Group.

The selection of the air commander also presented some problems. Portal nominated Air Marshal Leigh-Mallory as the air commander for the invasion, surprising some American airmen who expected Coningham to be nominated for the job. Coningham was the most combat-experienced tactical air force officer in the RAF, having fought across the western desert and into North Africa. Furthermore, he had worked with Montgomery and had considerable experience in working with the American 9th Air Force in the desert campaign and as commander of the Northwest Africa Tactical Air Force. His selection, therefore, appeared most logical. On the other hand, Leigh-Mallory had considerable experience as a fighter group commander during the Battle of Britain and had the job of developing the RAF 2nd Tactical Air Force which would be the counterpart to the 21st Army Group to be commanded by Montgomery. Furthermore, Leigh-Mallory enjoyed the confidence of Portal, which was probably the most significant factor in his nomination. Within his own service, however, Leigh-Mallory evoked considerable differences of opinion, a problem that became even more pronounced among American airmen in the months ahead.

By mid-summer, plans for OVERLORD were taking shape, but the question of the air commander for the invasion was still unsettled. The Americans were opposed to a single air commander, which seemed to be inconsistent with the doctrine developed in the Mediterranean. Actually they were concerned about the implications of giving British airmen all the top jobs, especially since, as the buildup continued, the American air forces in the European war would outnumber the British.[10] American airmen felt that they knew more about strategic bombing than the British, and the Americans feared that if all of the airpower for the invasion were put under Leigh-Mallory, the American bomber forces would be diverted from the air offensive and used on tactical targets.

The bomber force was still growing, and not until the winter of 1943–44 would there be sufficient forces to fly 1,000-plane raids. But if Leigh-Mallory commanded all the forces, the bomber offensive might be prematurely diverted to targets in France directly related to the invasion. American airmen were further concerned that the British would use the bomber force for the British Bomber Command's "city busting" campaign. Even though Major General Ira C. Eaker, Commander, 8th Air Force, had successfully defended daylight bombing at the Casablanca Conference in January, the RAF still might convince Churchill to reopen the question with Roosevelt.

All these issues led the Americans to propose both a tactical and a strategic air command for the invasion.[11] The tactical air commander would have operational control of the American 9th Air Force and the

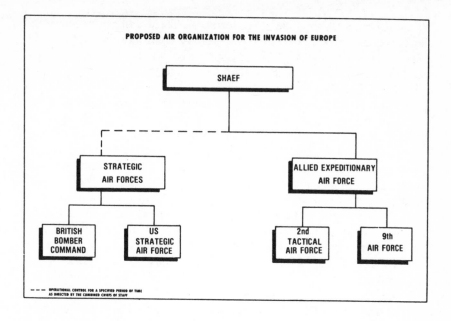

PROPOSED AIR ORGANIZATION FOR THE INVASION OF EUROPE

SHAEF

STRATEGIC AIR FORCES

ALLIED EXPEDITIONARY AIR FORCE

BRITISH BOMBER COMMAND

US STRATEGIC AIR FORCE

2nd TACTICAL AIR FORCE

9th AIR FORCE

— — — OPERATIONAL CONTROL FOR A SPECIFIED PERIOD OF TIME AS DIRECTED BY THE COMBINED CHIEFS OF STAFF

British 2nd Tactical Air Force. These two tactical air forces would work with 12th and 21st Army Groups in the same manner as the Northwest Africa Tactical Air Force worked with 18th Army Group. The air-ground team would, therefore, be designed around this proven concept. The strategic air commander would operate as Doolittle had with his Northwest Africa Strategic Air Force and would report to Eisenhower. This solution would provide an argument for an American to command the strategic air forces for the invasion since a British airman, Leigh-Mallory, would be commanding the tactical air forces. Further, this command arrangement would make it easier to retain the bombers on strategic targets because Eisenhower was more sympathetic than Leigh-Mallory to the strategic bomber offensive.

Leigh-Mallory had originally proposed the turn-over of operational control of 9th Air Force by mid-December since he already had control of the RAF 2nd Tactical Air Force. However, by the time of the Cairo Conference, 5 December 1943, the control of the bombers for support of the invasion was still not settled. With Eisenhower approved as the Supreme Commander, the need for a decision on the air command structure became urgent.

Eisenhower wanted Spaatz to head up American airpower in Europe as he had in North Africa. When Spaatz heard that he would be moving from Africa to take over the American bomber forces, he proposed a United States Strategic Air Force (USSAFE) composed of 8th Air Force in England and 15th Air Force in Italy.[12] USSAFE would be primarily committed to the strategic air offensive, but under emergency conditions it would support Eisenhower in Europe or General Sir Henry Maitland

47

Wilson, Commander-in-Chief, Mediterranean Theater of Operations. Neither of these theater commanders would have control of the Strategic Air Force. The arrangement would continue in which Portal, as deputy for the Combined Chiefs of Staff, would coordinate the targeting and strikes of USSAFE and RAF Bomber Command.

The Combined Chiefs of Staff approved the reorganization of the American strategic bomber forces and the arrangement whereby 15th Air Force in Italy would support an emergency in the Mediterranean theater. Then Spaatz and Harris spoke out against placing the strategic bomber forces under Eisenhower for the invasion, and both adamantly opposed placing their forces under Leigh-Mallory.[13] Spaatz felt the bombers should support the invasion on a case-by-case basis rather than being under Eisenhower for a specified time. Harris held essentially the same view and was even more outspoken than Spaatz on the diversion of RAF Bomber Command from the strategic air offensive. At one point, he informed Portal that he was ready to resign his command, but Portal convinced him that he was needed to carry on the offensive. Portal also said that Churchill and Roosevelt, as a condition for Eisenhower's selection as Supreme Commander, had already agreed to the diversion of the bombers.

Spaatz had appealed independently to General Hap Arnold to prevent the placing of his forces under Eisenhower's operational control for the invasion. The Supreme Headquarters, Allied Expeditionary Forces (SHAEF) planning had proposed that the interdiction campaign start at least three to four months in advance of the invasion, and in Spaatz's opinion this meant that the bombers would be pulled off the strategic air offensive at a decisive time. Spaatz may have been right: 85% of the 2,700,000 tons of bombs dropped on Europe in WWII were dropped after 1 January 1944;[14] Spaatz's Strategic Air Force was rapidly building to reach its full strength in June; and the ability to sustain a thousand-plane raid was at hand. Furthermore, the surplus in the German war economy

B–17s and their fighter escorts enroute to the European mainland.

had been depleted and the capacity of the industrial base to support the war effort was rapidly declining. The Strategic Bombing Survey states, ''If the bombing of Germany had little effect on production prior to July 1944, it is not only because she had idle resources upon which to draw, but because the major weight of the air offensive against her had not been brought to bear.''[15] Hence, any who assess airpower's effectiveness in Europe must consider carefully Spaatz's arguments in favor of extending the strategic air campaign until the last few weeks before the invasion.

In response to Spaatz's arguments, Arnold told much the same story that Portal told Harris. A commitment had been made to Eisenhower to place the bombers under his operational control for the invasion, and Arnold did not intend to raise the issue again with the Combined Chiefs of Staff. Both Spaatz and Harris were at last resigned to the fact that the bombers were going to be diverted. Their question then was who would control their forces.

AIRMAN—DEPUTY THEATER COMMANDER

When Eisenhower was selected as the Supreme Commander, he asked that Tedder be the Deputy Supreme Commander. The association the two had developed in the African campaign was one of mutual respect and confidence. For example, when Lieutenant General George S. Patton had bitterly complained because he wasn't getting air support while his forces were under constant attack from the German Air Force in the El Guettar battle in Tunisia, Eisenhower sent Tedder to Patton's headquarters at Gafsa to iron out the problem.[16] At that point, Coningham and Patton were at an impasse: Patton wanted close air support, but Coningham wanted to sacrifice close air support in favor of attacking the German Air Force directly. So Tedder had a ticklish problem made more difficult by the differences in nationality and service. Still, he handled the

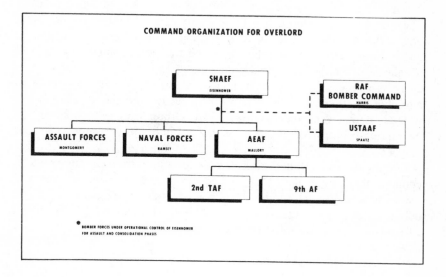

COMMAND ORGANIZATION FOR OVERLORD

disagreement with considerable tact and averted a potentially serious international incident between allies.

Eisenhower's request for Tedder was honored, and Tedder became the Deputy Commander, Supreme Headquarters Allied Expeditionary Force, in 1944. After considerable debate, it was agreed that he would handle the bombers for Eisenhower, and Leigh-Mallory would control only the tactical air forces. Leigh-Mallory's headquarters, however, did get to plan—in large part—the rail interdiction campaign and the campaign against all the *Luftwaffe's* forward staging airfields in France.

The command structure for the invasion was complete, but it was far from clear what the organization would be after the forces were on the continent. As it turned out, Tedder coordinated the American and British bombers during the invasion and after the forces were established ashore. Leigh-Mallory's headquarters accomplished all the detailed planning including the coordination of the fighters with the bombers. For the invasion and during the first months of the war on the continent, there was a single air commander, Leigh-Mallory, for all air operations directly associated with the land campaign. Tedder acted only when necessary to resolve a conflict or to iron out a problem for Spaatz or Harris. Leigh-Mallory actually directed the air campaign except for those forces attacking strategic targets far beyond the ground battle.

COMPONENT COMMANDS—CONTINUING ISSUE

With the establishment of 12th Army Group on 1 August 1944, Eisenhower had to decide whether to create a ground force SHAEF component command to control the operations of Bradley's 12th Army Group and Montgomery's 21st Army Group. The precedent had been established in the North Africa campaign where Alexander, as 18th Army Group commander, functioned as the theater ground force commander controlling and coordinating the efforts of two armies and two corps. If Eisenhower made Montgomery the component commander, he would offend the Americans who were furnishing the most ground forces and, furthermore, were not entirely convinced that Montgomery was a bold, imaginative commander who would rapidly exploit an opportunity for a decisive victory. On the other hand the British would consider it an affront if Montgomery, the hero of El Alamein and the greatest British general of the war, and his 21st Army Group were placed under the operational control of Bradley as the ground force component commander.

Eisenhower was under considerable pressure from Montgomery to continue the arrangement that existed during the invasion, in which all the American ground forces were placed under Montgomery. As the debates developed over the best strategy for breaking out of the Cherbourg peninsula, this command issue was in the forefront. Eisenhower, however, elected to act as his own ground force component commander; and though this arrangement was criticized during subsequent campaigns, he never altered the command structure. The wearing of two hats imposed very heavy demands on Eisenhower, and it appears that political rather than military considerations persuaded him to be both Supreme Commander and ground force component commander. Otherwise, he probably would have made Bradley the ground force component commander and Patton the 12th Army Group commander. These two were the most experienced combat commanders in the American forces, and Eisenhower knew what they could do from their performances in North Africa and Sicily.

When 12th Army Group was activated and Eisenhower elected not to create a ground force component command, American airmen saw no reason why Leigh-Mallory's Allied Expeditionary Air Force (AEAF) should continue as the theater air component command. They especially desired to eliminate the AEAF because it was an obstacle to returning the strategic air forces to the bombing campaign. As long as the AEAF existed, it would continue to exert pressure to employ bombers in extensive support of the ground campaign, allowing only an occasional use of the bombers for major strategic offensives. The ground campaign's need for heavy bombers was not continuous, however, and lateral coordination among commanders could meet the occasional needs without AEAF's organizational machinery.

The most telling argument for eliminating the AEAF was that the U.S. tactical bombers and fighters in the 9th Air Force already worked closely with Bradley's 12th U.S. Army Group; AEAF wasn't needed to coordinate between them. The 9th Air Force commander, General Hoyt S. Vandenberg, could and did coordinate with his British counterpart, Air Marshal Coningham, Commander of 2nd Tactical Air Force; and there was adequate direction from Eisenhower's headquarters on the individual responsibilities of the British and American tactical air force commanders. Also, Tedder, as Eisenhower's deputy, already had the responsibility for coordinating the efforts of the U.S. 9th Air Force and the British 2nd Tactical Air Force and for arranging support from Spaatz's U.S. Strategic Air Force and Harris' Bomber Command. Since Tedder wasn't an air deputy but Deputy Commander-in-Chief, SHAEF, he had the full authority and staff support afforded the Commander-in-Chief.

When the AEAF was dissolved on 15 October 1944,[17] nothing resembling a theater air component command was left. The need for such a command, however, was most apparent: Detailed coordination of the tactical and strategic air forces demanded a component command whose staff was primarily concerned with such matters hour by hour. This level of detailed planning was left to SHAEF. Unfortunately, the SHAEF staff wasn't equipped to handle both the long range strategic planning and the tremendous job of tactical planning for all air and ground forces.

Thus Eisenhower's decision to sidestep the problem of choosing either an American or British ground force component commander resulted indirectly in the unfortunate lack of an air component command. Even

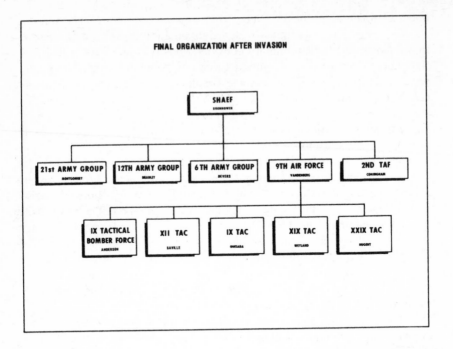

FINAL ORGANIZATION AFTER INVASION

SHAEF
EISENHOWER

| 21st ARMY GROUP | 12TH ARMY GROUP | 6TH ARMY GROUP | 9TH AIR FORCE | 2ND TAF |
| MONTGOMERY | BRADLEY | DEVERS | VANDENBERG | CONINGHAM |

| IX TACTICAL BOMBER FORCE | XII TAC | IX TAC | XIX TAC | XXIX TAC |
| ANDERSON | SAVILLE | QUESADA | WEYLAND | NUGENT |

though Leigh-Mallory was not the most acceptable commander to either Coningham or Vandenberg, the AEAF would probably have been retained if Eisenhower had designated a ground force component commander. By electing to be his own ground force commander, Eisenhower strengthened the argument that an air component wasn't necessary either since Tedder could function for the air forces in the same manner as Eisenhower did for the ground forces. For a number of reasons, then, Eisenhower's air organization was not a model for the future.

As they do in every war, personalities played a major part in the way the air and ground organizations developed. Many airmen concluded from the Mediterranean and European experience, however, that in a theater of war an air component commander should continuously control the theater air forces. The advantages of this doctrine seemed clear for tactical and naval air forces, but the issue was somewhat debatable in the case of strategic air forces. The question of control of strategic air forces when employed in a theater air campaign became even more debatable as nuclear weapons became the dominant element of our defense policies in the fifties.

KOREA—STRUCTURE OF A THEATER OF OPERATIONS

In Korea, command and control again became a major problem. General MacArthur, as the United Nations Commander, controlled all of the Allied forces; for the command of U.S. forces (for which he reported to the Joint Chiefs of Staff), his title was Commander-in-Chief, Far East.

The Far East Command (FECOM) was a unified command reporting to the JCS.

Since it was a unified command, staff representation from the components was to be equitably divided so that each of the services would have appropriate authority and rank in the decision process. A unified staff, since it does represent all of the services, is expected to function with a minimum of service parochialism although it is never feasible or even desirable to erase all the bias of an officer who has spent 20 or more years in his service. Careful selection of officers and balancing of rank and positions within the staff, however, produce a minimum of service bias, and that bias is tempered by the common mission of the theater headquarters. When a headquarters that is supposed to control multiservice forces is not structured with a balanced staff, inter-service problems tend to become magnified since there is inadequate consideration of at least one service's view at the outset. A balanced staff will tend to resolve or at least minimize these problems before viewpoints become hardened.

There was continuing difficulty within the Far East Command structure because of MacArthur's failure to establish an army component command. He reserved to himself the roles of the Far East Commander and the Commander of Army Forces, Far East. On the other hand, the Air Force and Navy established component commands designated the Far East Air Force and Naval Forces Far East. These component headquarters had staffs manned to direct the forces throughout the area of MacArthur's responsibility.

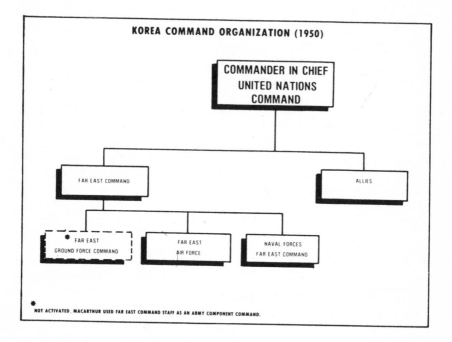

KOREA COMMAND ORGANIZATION (1950)

COMMANDER IN CHIEF
UNITED NATIONS
COMMAND

FAR EAST COMMAND

ALLIES

* FAR EAST
GROUND FORCE COMMAND

FAR EAST
AIR FORCE

NAVAL FORCES
FAR EAST COMMAND

* NOT ACTIVATED. MACARTHUR USED FAR EAST COMMAND STAFF AS AN ARMY COMPONENT COMMAND.

With MacArthur as his own Army component commander, the Far East Command staff was weighted excessively with Army personnel since it had to do the work of a component command as well as that of theater staff. This imbalance resulted in the staff getting into problems that should have been worked out by the field commands. Air Force targeting, which should have been done by the air component commander, with Far East Command generally confining itself to policy and adjustments of priorities, is a good example of the problems created by an improperly organized staff.

TARGETING—JURISDICTIONAL QUESTION

Lieutenant General George E. Stratemeyer, the Far East Air Force (FEAF) commander, urged in 1950 that FEAF should plan the targeting of air missions for MacArthur's Far East Command. He contended that FEAF was the only agency with the professional ability to determine the best air targets and the best way of destroying them. Despite Stratemeyer's arguments, though, MacArthur's chief of staff elected to establish a General Headquarters (GHQ) Target Group.[18] This group, made up of officers within the Far East Command staff, lacked the experience and depth of knowledge for targeting an air force. Compared to the European operation in World War II in which a joint British and American staff recommended targets for the U.S. Strategic Air Force and RAF Bomber Command, the FECOM effort was inadequate.

After this inauspicious beginning, though, FECOM activated a GHQ Target Selection Committee with a much higher level of representation, including the Vice Commander for Operations of the Far East Air Force and a senior representative of the Naval Forces Far East. This committee continued throughout the war although the Far East Air Force assumed most of the tasks for targeting USAF and carrier-based aircraft.

By 1952, FEAF's own targeting committee was composed of representatives from 5th Air Force, Far East Bomber Command, and Naval Forces Far East. The FEAF committee met every two weeks and made recommendations to the FEAF commander. After his approval, the Far East Commander considered the recommendations which then became basic guidance for the development of the air campaign plan. Thus by the summer of 1952 the FEAF commander was performing the targeting functions appropriate for an air component commander.

OPERATIONAL CONTROL WiTHIN FEAF

On 8 October 1950, Stratemeyer requested Far East Command to assign him operational control of all the air units engaged in the war over North and South Korea. (Operational control meant coordinating air activities with the activities of other forces and specifying the amount of forces to be employed, the type of munitions, the time on and off target, and the controlling agencies.) He believed successful integration of the efforts of all assigned air forces could only be accomplished by a single air commander directly responsible to MacArthur.

Stratemeyer had already made considerable progress in gaining command and operational control of units deployed from Strategic Air Command (SAC) and Tactical Air Command (TAC). SAC, at the direction of the Joint Chiefs of Staff, assigned three B–29 groups to the operational control of Stratemeyer. This new Far East Bomber Command was on a command level with the fighters of the 5th Air Force, and both organizations reported to FEAF.

FEAF eventually chose all targets through its targeting committee, as I have explained, then developed air campaign plans and sent them to 5th Air Force and Far East Bomber Command for implementation. Fifth Air Force assumed responsibility for coordinating the fighter escort. Route planning and approaches to the target, though, were joint concerns of 5th Air Force and Far East Bomber Command. After considering enemy fighter defenses, 5th Air Force usually accepted Bomber Command's proposed routes. Disagreements, if any, were resolved by FEAF. Bomber Command also generally chose how the target would be struck if the target were not associated with Allied ground activities. When bombers (other than B–29s) were used for close air support, however, 5th Air Force was primarily responsible for the details of the missions. The methods used for directing fighters and B–26 light bombers for close air support were not used for B–29s because of the B–29s' enormous firepower. For them, special control arrangements helped avoid injuring friendly ground forces. The 5th Air Force combat operations officer in the name of the 5th Air Force commander had final control over these

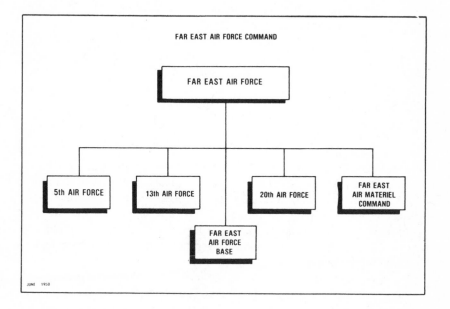

FAR EAST AIR FORCE COMMAND

FAR EAST AIR FORCE

5th AIR FORCE   13th AIR FORCE   20th AIR FORCE   FAR EAST AIR MATERIEL COMMAND

FAR EAST AIR FORCE BASE

JUNE 1950

missions, and a Joint Operations Center with representatives from the Army and Navy coordinated the strikes at the major decision level.

CONTROL OF STRATEGIC AIR FORCES IN A THEATER OF OPERATIONS

Placing the bombers under the operational control of the air component commander was a return to the arrangement that succeeded in North Africa. In Europe, however, the bombers were not under operational control of the air component commander (Leigh-Mallory) for the Normandy invasion. For close air support after the invasion and after the Allied Expeditionary Air Force was inactivated, the bombers came under the control of 9th or 2nd Tactical Air Force for specific missions in support of ground operations. For the most part, however, strategic bombers in Europe remained outside the control of the air component commander. The rationale, as we recall, was that the primary mission of the theater was to destroy Germany's industry, economy, and will to fight, a mission essentially beyond the theater mission of defeating the German forces deployed in the field.

The situation in Korea seemed to call for for a command and control arrangement more like the one used in North Africa because North Korea did not have an industrial base comparable to Germany's. The targets in Korea were fewer and required less force to destroy or neutralize; most of them were associated with airfields, lines of communication, and the logistical structure directly supporting the Chinese and North Korean forces. These types of targets would generally have been assigned to tactical air forces in Europe. In my opinion, it is fundamentally sound under such conditions to place the strategic forces under the operational control of the tactical air force or air component commander for as long as the air campaign may last.

The FEAF commander used the B–29s in Korea against airfields along the Yalu, against interdiction targets, against industrial facilities, and at times for close air support. Only on rare occasions were the Joint Chiefs of Staff involved with the targets assigned these forces, and then usually only to prescribe broad policy for an entire target system.

With this kind of guidance, the FEAF commander had considerable freedom. He could plan campaigns to cover several months of operations with little fear that major changes would be dictated by Washington. The only restrictions established by Washington were on targets along the Yalu, particularly airfields, bridges, power plants, and logistical centers. These restrictions, however, were not applied to targets within North Korea. A more general restriction intended to limit the war was the prohibition of attacks against China. "Hot pursuit" under some conditions was authorized, but attacks against aircraft taking off from bases across the Yalu were not.

The important prohibition of flights into China notwithstanding, both Stratemeyer and his successor, Lieutenant General Otto P. Weyland, were largely successful in their urging that all of the airpower in the Korean War should be controlled by the air component commander. FEAF controlled the bombers, as requested, and also the Allied fighters. (They employed Allied fighters just as they did those from the United States except for fighter assignments along the Yalu. Because relatively low-performance Allied fighters couldn't cope with the MIG–15s, USAF F–86s were given the Yalu patrols.) Except in China, Stratemeyer and Weyland could send Allied airpower wherever it could most damage the enemy.

CONTROL OF NAVAL AVIATION

Stratemeyer had a more difficult time, however, when he insisted that the principle of centralized control of airpower in a theater also applied to naval aviation. Even though naval aviation would help gain and maintain air superiority, interdict the battlefield, and provide close air support, Naval Forces Far East (NAVFE) opposed placing this force under the operational control of the air component commander.

An earlier attempt to eliminate this sort of problem had resulted in the Key West Agreement of 1948* which stipulated that air campaigns would be the business of the air commander and sea campaigns would be the business of the naval commander. Consistent with this Joint Chiefs of Staff agreement, but more fundamentally consistent with the overpowering experience in the air campaigns of World War II, Stratemeyer and Weyland argued that FEAF should control naval aviation since it was clearly an augmentation of the forces assigned to FEAF for the theater air mission.[19]

NAVFE's arguments were essentially the same as those used by Admiral Cunningham in the Mediterranean in World War II. The Navy argued that their primary mission was to gain and maintain control of the sea and to secure the sea lines of communication. If they were to carry out this mission, their forces could not be restricted to the control of a theater commander, but had to be free to engage opposing naval forces.

The Navy maintained that although its forces would support the theater commander, he should not control them—an enormously important distinction because Joint Chiefs of Staff publications allow supporting force commanders great freedom. The JCS defines the relationship of "supported/supporting" as follows: "The commander of the supported force indicates in detail to the supporting commander the support missions he wishes to have fulfilled and provides such information as is necessary. . . . The commander of the supporting force. . . . takes such action to fulfill them within his capabilities. . . . The supporting commander prescribes the tactics, method and procedures to be employed by his

---

*Secretary of Defense James V. Forrestal assembled the Joint Chiefs of Staff at Key West, Florida, on 11 March 1948, to decide "who will do what." The Key West Agreement is the basis of the current roles and missions of the services.

forces."[20] While this definition provides a limited basis for integrating the naval air effort, it doesn't go far enough for the air component commander to construct a campaign plan in which the targets are shifted as priorities change. The support arrangement is essentially tailored for a highly planned operation of a few days. In a brief operation the support relationship may effectively harmonize the efforts of two or more forces. However, large operations extending over a long time require the more dependable, authoritative relationship of operational control or command. Thus with naval forces committed to the continuing air campaign in Korea, and with no threat from an opposing fleet, FEAF's argument for operational control made sense.

In response to Stratemeyer's request for operational control of all aircraft operating in the Korean campaign, the FECOM chief of staff replied on 8 July 1950 that the Commander, FEAF, would have command or operational control of all aircraft in the execution of the FEAF mission as assigned by the Commander-in-Chief, Far East (CINCFE). He went on to say that the Commander, NAVFE, would have command or operational control of all aircraft in the execution of the NAVFE mission as assigned by CINCFE.[21] If the reply had stopped at this point, the issue would probably have been settled. FEAF was responsible for the air campaign, and all air units having a capability to participate in that campaign would have come under the operarational control of the air component commander. Since there was no naval campaign of note, all naval air was in fact committed to the campaign.

However, in the same directive the Chief of Staff stated that when both NAVFE and FEAF were assigned missions in Korea, "coordination control, a Commander-in-Chief, Far East, prerogative," was delegated to the Commander, FEAF. This portion of the directive reopened and obscured the question. Weren't the air resources of NAVFE, when committed for strikes in either North or South Korea, operating in an area of responsibility already established as the prerogative of the air component commander? And what did the term *coordination control* mean? The FEAF commander construed it to be another way of expressing "operational control," while NAVFE assumed it referred to the aforementioned support relationship. Further, for the first two years of the war the Commander, NAVFE, considered his forces to be in support of FECOM, not FEAF. Thus NAVFE looked to the targeting committee at FECOM to provide the targets and the time to strike them. After receiving these targets from FECOM, NAVFE did coordinate with FEAF and 5th Air Force, but only a naval liaison officer in the 5th Air Force Joint Operations Center had contact with Task Force–77 (TF–77), the carrier force off the east coast of Korea.

NAVFE initially requested that all of its targets come from exclusive naval air areas of operation. These areas were to be on the east coast of Korea, which was closer to TF–77; therefore, they would provide carrier aircraft longer time in the target area. FEAF objected to assigning an area of operation to NAVFE on the grounds that all air forces had to be employed wherever they could confront the enemy air force and the main lines of communication.[22] Naval aviation had to be employed along the Yalu and against the airfields around Pyongyang since these were the places where the enemy was trying to establish control of the air. The east coast had some important rail lines, but since the main LOCs were through the west coast and central region, the interdiction campaign had to be concentrated there. Thus the contribution of naval aviation to the air campaign would be appreciably reduced if naval air were confined to an area of operation along the east coast.

NAVFE further argued that for close air support, naval aviation should be used on a specific sector of the front; again, because of proximity to the carriers, the area should be on the eastern part of the front. FEAF again contended that we should not shackle our own airpower with such arbitrary geographical limitations. Airpower should not be committed to a specific part of the front unless the divisions in that area were heavily engaged, offensively or defensively. Airpower should be used across the front wherever the ground action justified a diversion of forces from the air superiority or interdiction campaigns. Weyland further argued that air support was much more effective when it was massed than when it was spread out in small increments.

Even though Far East Command didn't modify its original directive on coordination control, FEAF and NAVFE came to an arrangement by mid-1952 which, in fact, recognized that FEAF was the controlling authority for all air operations. By that time, most of the detailed air operations were concentrated in the Joint Operations Center (JOC) of 5th Air Force. TF–77 placed a naval section in the JOC, and the combat operations officer assigned missions to TF–77, through the naval section, just as he did to Air Force units. At the same time that TF–77 established a naval section in the JOC, it also assigned an officer to the FEAF targeting committee which had assumed most of the responsibility for selecting and recommending targets. Also, by 1952 carriers were providing close air support using the control procedures long established for 5th Air Force and Allied aircraft. At last the full weight of Air Force and Navy airpower was being applied in a concerted manner.

CONTROL OF MARINE AVIATION IN A THEATER OF OPERATIONS

Control of Marine aviation, when the Marines are committed to a land campaign, has also been a difficult and emotional problem. Marine aviation has been justified on the basis of its ability to support an amphibious operation, which the Marines are assigned as a primary mission. Since amphibious forces are without the artillery support normally organic to an Army division constituted for sustained land warfare, Marine landing forces are dependent upon naval gunfire, carrier based air, Marine air, and Air Force air (if within range) for fire support. After the forces hit the beach, Marine air augments the limited organic artillery. Then, since the Army is responsible for the conduct of prompt

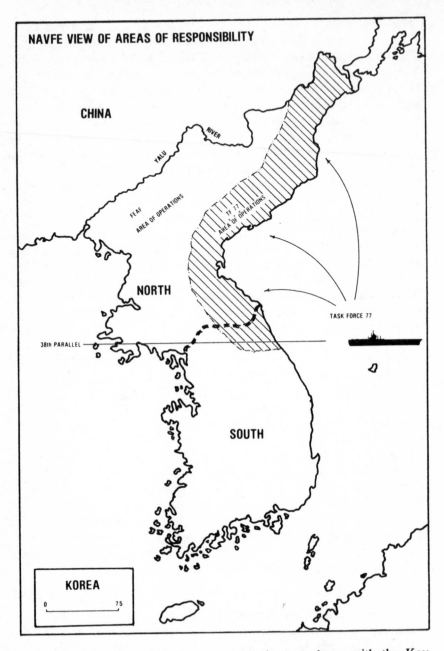

NAVFE VIEW OF AREAS OF RESPONSIBILITY

CHINA

NORTH

SOUTH

38th PARALLEL

TASK FORCE 77

YALU

RIVER

FEAF AREA OF OPERATIONS

TF 77 AREA OF OPERATIONS

KOREA

0   75

and sustained combat operations on land (in accordance with the Key West Agreement of 1948),[23] Army forces replace Marines after the objective area is secure and the Marines either withdraw or become a part of the Army forces.

Marine airpower is thus basically tailored to the needs of the landing force, including some fighters for local air defense. For interdicting the

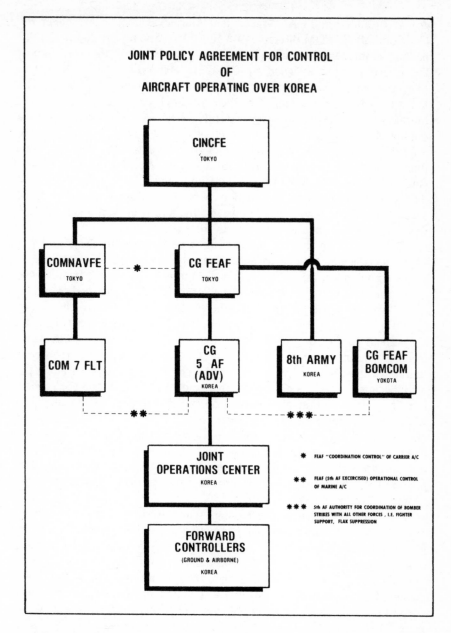

JOINT POLICY AGREEMENT FOR CONTROL
OF
AIRCRAFT OPERATING OVER KOREA

CINCFE
TOKYO

COMNAVFE
TOKYO

CG FEAF
TOKYO

COM 7 FLT

CG
5 AF
(ADV)
KOREA

8th ARMY
KOREA

CG FEAF
BOMCOM
YOKOTA

JOINT
OPERATIONS CENTER
KOREA

FORWARD
CONTROLLERS
(GROUND & AIRBORNE)
KOREA

* FEAF "COORDINATION CONTROL" OF CARRIER A/C

** FEAF (5th AF EXCERCISED) OPERATIONAL CONTROL OF MARINE A/C

*** 5th AF AUTHORITY FOR COORDINATION OF BOMBER STRIKES WITH ALL OTHER FORCES, I.E. FIGHTER SUPPORT, FLAK SUPPRESSION

landing area and gaining control of the air, the Marines are dependent upon carrier-based air and land-based air. Thus, these high priority missions are outside the basic responsibility of Marine aviation, which is close air support.

After the introduction of the Marines into Korea, FEAF maintained that Marine aviation should come under the operational control of 5th Air Force since Marine air had the task of providing close air support to 8th

61

Army. In addition, FEAF argued that Marine aviation came under the Chief of Staff FECOM directive of 8 July 1950, which provided for FEAF having command or operational control of all aircraft in the execution of the FEAF mission assigned by CINCFE. According to this directive, Marine aviation should come under the operational control of FEAF. The Marines, however, wanted all of their air employed in direct support of Marine ground forces, or more explicitly, they, like the Navy, wanted a section of the 8th Army front assigned as an exclusive area of operation.

FEAF opposed this employment of Marine aviation using the same fundamental arguments Tedder had used in World War II to avoid parceling Allied airpower to the Royal Navy. Fifth Air Force did concede that whenever the tactical situation permitted, Marine aviation would support Marine ground forces; however, the airpower would still be used where it could do the most damage to the enemy.

With the Inchon landing, Marine air was assigned to the Army's X Corps for the amphibious assault. General Edward M. Almond, the X Corps commander, argued for retaining Marine aviation as organic to the Corps for close air support. He further proposed the continued operation of X Corps as separate from 8th Army.[24] But with X Corps assigned to 8th Army after exploiting the invasion, Marine aviation returned to the operational control of 5th Air Force and was used across the 8th Army front or on interdiction according to the tactical situation. However, Marine fighters did not participate in the counter air campaign along the Yalu because of the superior performance of the MIG–15 and the need to have a greater portion of the force on the interdiction campaign. Also, with a stable front there were fewer requirements for close air support and consequently more need for the Marine air units in the interdiction campaign. As a result of the integration of Marine air operations with 5th Air Force operations, centralized control of all the airpower assigned to the Far East theater of operation provided the flexibility that it did in the campaigns of World War II.

With the conclusion of the Korean War, airpower had again demonstrated the need for a command structure that didn't arbitrarily divide forces between mission areas. The command structure had to be capable of using airpower in a variety of tasks simultaneously or in sequence. The fundamental point, though, was that the theater air component commander had to control all the airpower in the theater so that he could support ground, naval, or air operations—wherever the enemy was weak.

CHAPTER II

FOOTNOTES

[1] Wesley F. Craven and James L. Cate, eds. The Army Air Forces In World War II, vol. 2: Europe: Torch to Pointblank (Chicago: The University of Chicago Press, 1949), pp. 200–205.

[2] Ibid., p. 137.

[3] The Lord Tedder, Air Power In War, The Lees Knowles Lectures for 1947 (London: Hodder and Stroughton, n.d.,), p. 89.

[4] The Lord Tedder, With Prejudice, the War Memoirs of Marshal of the Royal Air Force (Boston: Little, Brown and Company, 1966), p. 393.

[5] Ibid., p. 396.

[6] Craven and Cate, Torch to Pointblank, p. 496.

[7] Tedder, With Prejudice, p. 114.

[8] Craven and Cate, Torch to Pointblank, p. 493.

[9] Ibid., p. 748.

[10] Ibid., pp. 735–739.

[11] Ibid., pp. 738–742.

[12] Ibid., p. 741.

[13] Ibid., pp. 735–740.

[14] The United States Strategic Bombing Survey, Overall Report (European War) (Washington, D.C., 1945), pp. 1, 6.

[15] Ibid., p. 71.

[16] Tedder, With Prejudice, pp. 410–411.

[17] Craven and Cate, The Army Air Forces in World War II, vol. 3: Argument to V-E Day, p. 622.

[18] United States Air Force Operations In the Korean Conflict, USAF Historical Study No. 71, 25 June–1 November 1950 (Washington, D.C.: Department of the Air Force, 1952), pp. 11–13.

[19] Korean Conflict, Study No. 71, p. 11.

[20] Unified Action Armed Forces (UNAAF), Joint Chiefs of Staff Publication 2 (Washington, D.C., 1974), p. 56.

[21] Korean Conflict, Study No. 71, p. 12.

[22] Korean Conflict, Study No. 72, (1 November 1950—30 June 1952), pp. 77–78.

[23] Armed Forces Bulletin #1, Department of the Air Force, Washington, D.C., 21 May 48, p. 5. (This bulletin reproduced the Secretary of Defense staff paper on the functions of the Armed Forces and the JCS which were determined at the Key West Conference, Florida, March 11–14, 1948.)

[24] Korean Conflict, Study No. 72, p. 75.

CHAPTER III

COMMAND AND CONTROL OF AIRPOWER IN THE VIETNAM WAR

With the onset of the Vietnam War, the command structure for a theater air component command was fairly well established. The experiences of World War II and Korea had been translated into publications, and field exercises and maneuvers to some extent followed these doctrinal principles. As Strike Command (STRICOM,* to be redesignated Readiness Command or REDCOM on 1 January 1972) became more active in the conduct of joint force training exercises such as SWIFT STRIKE, DESERT STRIKE, and GOLDFIRE early in the 1960s, however, a trend developed that deviated from the lessons of World War II and Korea.

STRICOM exercises featured considerable concentration of authority within the Joint Staff for the detailed direction of subordinate forces. This assigning of detailed operational responsibilities to the Joint Staff elevated the level of control of component forces while at the same time reducing the tactical competency at that level of control. Most of us involved in tactical air operations in World War II and Korea felt that the theater staff should not be involved in the detailed tactical direction of forces. These responsibilities should be left with the component forces where there is the highest competency in dealing with daily tactical operations.

A Joint Task Force headquarters, even though it is far from being a theater headquarters in size and scope of operation, contains all the activities of such a headquarters and does provide the basis for expansion into a theater headquarters if the scope of operations increases. Thus by increasing Joint Staff controls, the STRICOM field exercises tended to reduce the authority of the component commanders. And the impact of

---

* A Unified Command established under the Joint Chiefs of Staff in 1961 to conduct joint exercises, to deploy forces based in the United States to established theaters of operations or contingencies, and to recommend joint doctrine. The Navy and Marines have no forces assigned to REDCOM.

these exercises was inevitably felt in contingency planning in which Joint Force deployments were projected to have roughly the same command structures employed in the field exercises.

The Joint Force headquarters employed in the STRICOM exercises were never large enough to include provisions for all the various air missions that an air component commander would direct in an active theater of war. Further, marine and naval aviation were not brought into play because STRICOM included no Navy or Marine forces. Thus the same difficulties that plagued the command structure in World War II and Korea prevailed in the STRICOM field exercises. Even though these command issues were eventually settled in combat, I doubt that they have really been resolved, since individual service doctrine was not altered to conform to the realities of combat.

Unfortunately, there is never enough money in peacetime for large exercises in which the forces and command structure can be evaluated in conditions like those of a real theater of operations where an entire tactical air force of some 1,200 aircraft and an army of six divisions are put into the field. Such exercises would provide a better understanding of how marine and naval aviation are controlled by the theater air component commander in meeting mission requirements. Because of budgetary restraints, though, current field exercises do not achieve this objective. Although small exercises are useful for some purposes, they may make it more difficult in some ways for the forces to adjust to a theater command organization when they enter combat. Thus command problems which troubled us in World War II and again in Korea survived the STRICOM exercises to appear yet again in Vietnam, where they were compounded anew by national and international political concerns.

COMMAND STRUCTURE—THE BEGINNING

The beginning command structure for the Vietnam War came from the Military Advisory Group (MAG). This advisory group was established on 17 September 1950 when the French granted a degree of autonomy to Vietnam, Laos, and Cambodia within the French union.[1] With the MAG the United States began direct military aid to the forces of these three countries through the French. The role of U.S. advisors was very limited, and both the Air Force and Army sections of the MAG wanted to become more directly involved in the use of American equipment. After the fall of Dien Bien Phu in 1954, the role of the MAG began to change profoundly. On 1 November 1955, the MAG was redesignated the Military Assistance Advisory Group, Vietnam (MAAG). This title reflected the change in relationship between France and the United States on the training of the Vietnamese. In the same year, the French granted autonomy to the Vietnamese armed forces, which paved the way for the U.S. to provide direct assistance.

From 1955, the United States was directly involved in the organizing and training of Vietnamese units. The Vietnamese Air Force (VNAF) was a very small element of their armed forces, and its organization had to be built from the bottom up. What organization the VNAF possessed was inherited from the French and was considerably different from the USAF arrangement for fighting a theater war. With the introduction of FARM GATE* T-28s and B-26s, it became increasingly apparent that we had to have a USAF command structure to control our flying units. Even though FARM GATE units were restricted to combat training missions, it was inevitable that such missions would involve actual combat, and in fact combat provided the best conditions for the training.[2] Furthermore, FARM GATE was providing more and more close air support and air cover for convoys which were being ambushed with increasing frequency. The VNAF simply couldn't meet all these requirements, and mission directives to FARM GATE provided for its employment in emergency situations that the VNAF was unable to handle.

2ND AIR DIVISION ADVON—THE BEGINNING OF AN AIR COMPONENT

On 15 November 1961, 13th Air Force in the Philippines activated an advanced echelon (ADVON) headquarters of its 2nd Air Division at Tan Son Nhut airfield on the outskirts of Saigon.[3] Detachments at Danang and Nha Trang were created under the ADVON to control the expanding air force. Senior officials chose to designate the headquarters an ADVON rather than an air division headquarters because U.S. policy at the time was that American forces in Vietnam were there only to train the South Vietnamese. The title of air division headquarters seemed to exaggerate the extent of combat operations by USAF units.

With the introduction of a tactical air control system, additional aircraft for reconnaissance and airlift, and a liaison squadron, the inadequacies of the small ADVON headquarters soon became apparent. Further, senior U.S. officials were considering expanding the MAAG in light of their decisions to introduce Special Forces to train Army of Vietnam (ARVN) rangers and to separate advisory and training functions from the administration and logistical functions of the MAAG. Technically, expanding the MAAG would not necessarily mean expanding the ADVON headquarters because the 2nd Air Division ADVON was not identified as the air section of the MAAG. The ADVON was in a peculiar command position: It was an element of the 13th Air Force, but 13th Air Force had no responsibility for its activities in Vietnam. During this initial period the MAAG, in fact, was the controlling agency for the combat activities of the 2nd Air Division ADVON; the commander of the ADVON reported to the MAAG chief on the activities of his assigned forces. Although this was not a clear command arrangement, its very obscurity helped underscore the need to restructure the MAAG to configure it for the

---

* See page 11 for discussion.

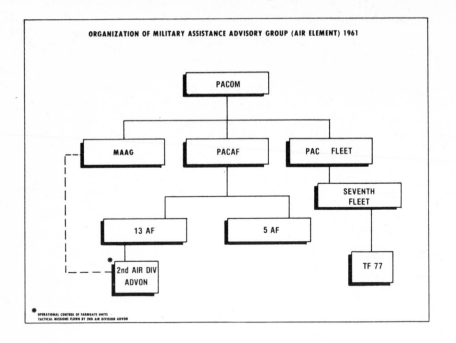

ORGANIZATION OF MILITARY ASSISTANCE ADVISORY GROUP (AIR ELEMENT) 1961

```
                        ┌──────────┐
                        │  PACOM   │
                        └──────────┘
                             │
      ┌──────────────────────┼──────────────────────┐
 ┌──────────┐          ┌──────────┐          ┌──────────┐
 │  MAAG    │          │  PACAF   │          │ PAC FLEET│
 └──────────┘          └──────────┘          └──────────┘
   │                         │                     │
   │                         │              ┌──────────┐
   │                         │              │ SEVENTH  │
   │              ┌──────────┴──────┐       │  FLEET   │
   │         ┌──────────┐      ┌──────────┐ └──────────┘
   │         │  13 AF   │      │   5 AF   │      │
   │         └──────────┘      └──────────┘      │
   │              │                         ┌──────────┐
   │         ┌──────────┐                   │  TF 77   │
   └─────────│ 2nd AIR DIV                  └──────────┘
             │  ADVON   │
             └──────────┘
```

* OPERATIONAL CONTROL OF FARMGATE UNITS
TACTICAL MISSIONS FLOWN BY 2ND AIR DIVISION ADVON

control of combat operations and to lay the foundation for future expansion.

As a result of the Vietnam visit of Taylor* and Rostow* in October 1961, President Kennedy decided to expand the training and advisory forces in Vietnam.[4] There was to be a significant increase in the size of the Air Force: The airlift force would be increased; the RANCH HAND (defoliation unit) detachment would be expanded; and additional control facilities and more forward air controllers with liaison aircraft would be added. General Taylor further recommended, and the President agreed, that the MAAG should be reorganized and expanded to control the rapidly growing United States commitment. On 8 February 1962, Military Assistance Command Vietnam (MACV) was formed with General Paul D. Harkins as Commander, U.S. Military Assistance Command Vietnam (COMUSMACV). With the establishment of MACV, advisory activities were formally separated from training and operational activities. In essence, MACV was an operational headquarters with the nucleus of a staff that could direct expanded combat operations.

Most of the planning for a war in the Far East had assumed that China would intervene, and the projected combat command structures were also based on that assumption. Whether CINCPAC would exercise control through a theater sub-unified command, though, was an unanswered

---

* At this time, Rostow was serving as Deputy Special Assistant to the President for National Security Affairs, and General Taylor had the position of "Military Representative" to the President.

question. The Army and Air Force believed that a theater unified command would be needed and that it should report to the Joint Chiefs of Staff as the Far East Command did in the Korean War. Under the theater unified command would be an Army, Navy, and Air Force component, each capable of expanding to match the scope of the conflict. Most senior planners believed that any small war in the Far East could well expand into a major war and that the command structures projected in our contingency plans should be able to accommodate such an expansion. In anticipation of the need for a rapidly expandable theater unified command, Korat, Thailand, had been selected as the headquarters for Commander, United States Southeast Asia (COMUSSEA).

But senior Navy spokesmen opposed the idea of a separate theater command reporting to the Joint Chiefs of Staff. They supported CINC-PAC in his view that the initial organization in Southeast Asia should be a sub-unified command under CINCPAC and that a later evaluation of the level of combat should be used to determine whether a separate theater command reporting to the Joint Chiefs of Staff were needed. The Navy proposal called for continuing the existing command structure which gave CINCPAC, as a unified commander, control of all forces in the Pacific.

MACV—A SUB-UNIFIED COMMAND

MACV began its life as a sub-unified command under CINCPAC in 1962 with debate continuing about what its structure would be in the future. The composition of the MACV staff, however, was a matter of

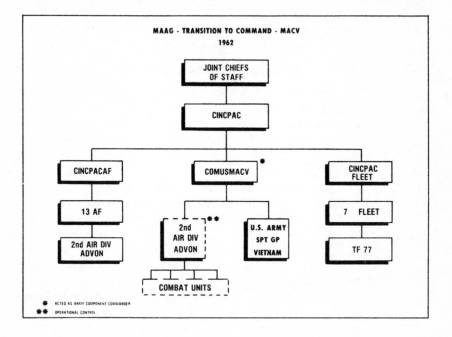

MAAG - TRANSITION TO COMMAND - MACV
1962

immediate concern to the Joint Chiefs. Airmen argued that the staff should be balanced in accordance with the principles of a tri-service command and should provide for equitable representation and equal rank in the key staff positions. They argued, then, for the organization that had operated successfully in the North African theater headquarters, the Mediterranean theater headquarters, and the European theater headquarters of World War II. In Vietnam, airmen were concerned that the MACV staff would be weighted in favor of Army officers as the Far East Command had been. Such an imbalance had created a lack of understanding of airpower in the Korean War, and if the staff in MACV were to be dominated by the same imbalance, the result would be the same.

In fact, Army spokesmen were arguing in 1962 that counterinsurgency was primarily a land war and that the Army was responsible for such wars. Hence the command structure should reflect this relationship, and assigned airpower should come under the control of the Army commander responsible for the campaign. Harkins was one of those who felt that the war in South Vietnam was essentially a ground war and that the command structure of an Army specified command would be most appropriate. Although he did not express this view in a formal proposal to CINCPAC until 1964,[5] his thinking had a decisive influence on the composition of the initial MACV staff.

Of the primary staff agencies, airmen held the J–2 and J–5 positions, while Army officers held the J–1, J–3, and J–4 positions. The head of the Combat Operations Center, a key position, was given to a Marine officer, as well as that of Chief of Staff. Thus from the outset the MACV staff was heavily weighted in favor of ground officers, but even at this early date airpower had become fundamental to all combat and civic action operations. Strong airpower would be essential to the South Vietnamese if they were ever to stand off the enemy alone.

PACAF VIEW OF COMMAND ORGANIZATION

While others were trying to devise a workable command structure for MACV, CINCPACAF, the air component commander under CINCPAC, was working to establish an appropriate air command structure for our forces in South Vietnam and Thailand. General Jacob E. Smart, CINCPACAF, and the Air Staff had held the view that the main threat to U.S. interests in the Far East was China and that PACAF's command structure should be designed to meet that threat. Consequently, Smart believed that the air command structure in Southeast Asia should provide for direct control from PACAF headquarters to the numbered air force (in this case 13th Air Force) for the execution of the air campaign. The air forces assigned to South Vietnam should be limited to those absolutely needed by COMUSMACV to accomplish his mission.[6] If many air units were assigned to MACV solely for in-country use, the forces available to CINCPACAF for the conduct of any larger air campaign would be limited unacceptably, since forces assigned to MACV could only be withdrawn for missions outside of South Vietnam with the approval of CINCPAC. Smart assumed that if combat became widespread in Southeast Asia it would be difficult to get Harkins to release forces, and that Admiral Harry D. Felt, CINCPAC, would be understandably reluctant to overrule Harkins in such matters.

It was these concerns about our wider responsibilities in the Far East that shaped the early thinking of General LeMay (then Chief of Staff of the Air Force), General Smart, and the key members of the Air Staff. Smart was probably the most expressive on the need to keep the main elements of the Pacific Air Force under a single air command structure. He pointed to the structure of CINCPACFLT in which naval forces were held under a Navy command structure and placed in support of a theater commander for specific missions. This would be the most appropriate method of providing air support for MACV. That is, most of the Air Force units in Southeast Asia would continue to be assigned to 13th Air Force. When COMUSMACV needed more air effort than his assigned air units could deliver, all or part of 13th Air Force would be placed in support of MACV for a specified period of time. In this manner, Smart contended, PACAF could support MACV and also preserve its flexibility for dealing with broader conflict or even war with China.

JTF–116—A NEW ELEMENT

As MACV was beginning to expand, the situation in Laos was deteriorating, and our hopes for North Vietnamese compliance with the 1954 cease-fire agreement were rapidly disappearing. In order to display U.S. concerns about Laos and the status of the cease-fire agreement, the President decided to send a Joint Force to Thailand. Composed of Marine, Army, and Air Force units, Joint Task Force 116 deployed on 12 May 1962.[7] With the deployment of JTF–116, new questions about command relationships had to be answered: What should be the relationship of JTF–116 with MACV? What headquarters should be responsible for JTF–116 air operations outside of South Vietnam?

The Air Force's situation was especially troubling because its forces in Southeast Asia were fragmented among three commands. The units in South Vietnam were under the command of the 2nd Air Division ADVON; the units in Thailand were under the command of 13th Air Force; and those recently deployed from Tactical Air Command belonged to Joint Task Force 116. Obviously, this arrangement forfeited all the important advantages of centralized control of airpower.

General Harkins' recommendation for a command arrangement was that MACV should control all U.S. forces in Vietnam and Thailand. This arrangement would be consistent with previous planning for the contingency of an expanded war in which all U.S. forces in Southeast Asia would be controlled from a single headquarters at Korat. But Admiral Felt didn't approve the idea of a single command under Harkins; he

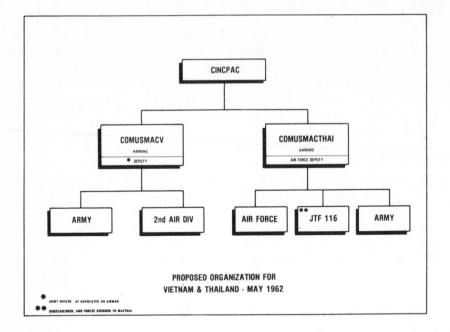

CINCPAC

COMUSMACV
HARKINS
* DEPUTY

COMUSMACTHAI
HARKINS
AIR FORCE DEPUTY

ARMY

2nd AIR DIV

AIR FORCE

** JTF 116

ARMY

PROPOSED ORGANIZATION FOR
VIETNAM & THAILAND - MAY 1962

* ARMY OFFICER AF ADVOCATED AN AIRMAN

** DISESTABLISHED, AND FORCES ASSIGNED TO MACTHAI

preferred to maintain separate headquarters in Vietnam and Thailand for the forces based in those countries.

Felt proposed to the Joint Chiefs that Harkins be given two deputies, one for Thailand and one for South Vietnam. Harkins would command both MACV and the Military Assistance Command Thailand (MAC-THAI). Harkins' deputy for MACTHAI would be an Air Force lieutenant general who would have, in turn, an Army major general as his deputy. After some debate, the Joint Chiefs accepted the proposal, and JTF–116 was deactivated on the 13th of May 1962 with its forces being assigned to MACTHAI.

Shortly afterward the Air Force, too, was compelled to face up to the need for a better command structure in Southeast Asia. After some sharp differences of opinion among his subordinates, LeMay decided that the 2nd Air Division ADVON should become a reinforced air division. On 8 October 1962, 2nd Air Division was activated, and all the Air Force units that had deployed as a part of JTF–116 were placed under its command. [8] These tactical fighter and airlift units were on temporary assignment from Tactical Air Command, but it was understood that other units would replace them periodically to sustain a highly visible military presence in Thailand until the situation in Laos improved.

2ND AIR DIVISION—EXPANDED AUTHORITY

The 2nd Air Division commander, from the outset, was expected to perform two roles: air component commander for MACV, and forward commander for 13th Air Force, in the latter of which he was responsible for all USAF operations in Southeast Asia. COMUSMACV would have operational control through 2nd Air Division of USAF units flying in South Vietnam. But the 2nd Air Division commander would be performing as a forward commander for 13th Air Force when he controlled air operations in Laos or anywhere else outside South Vietnam. If the war should expand against China, 13th Air Force would become the air component for COMUSSEA, and the 2nd Air Division commander's role as a forward commander would dissolve. With this unorthodox command arrangement, PACAF preserved its control over most of the air units in Southeast Asia.

Although the arrangement seemed improbable, it was designed with an eye to the future. For there were those in 1962 who doubted that the insurgency could be brought under control in South Vietnam by ground action and who strongly suspected that the use of airpower against the North Vietnamese homeland would eventually be necessary. If that should happen, it would be essential to have the air campaign controlled by airmen through a central, theater-wide command structure. MACV seemed unlikely to provide a suitable structure for such a campaign, given MACV's thoroughgoing commitment to treating the conflict as an insurgency which had to be settled in-country and on the ground. Thus PACAF and the Air Staff were most anxious that control of the main elements of the Air Force in the Pacific remain under the CINCPAC component command structure and not under MACV as a sub-unified command.

Early in 1962, General LeMay recognized that the insurgency in South Vietnam was demanding a more imaginative employment of airpower. On a visit to South Vietnam on 23 April 1962, he talked with General Harkins about the need to make airpower more responsive.[9] The command system was too cumbersome; the tactical air control system (TACS) was not being allowed to operate as efficiently as it had during World War II and the Korean War. Requests for air cover and for strikes against the ambush forces operating along most of the major roads were being processed much too slowly. LeMay emphasized that airpower could make such ambushes very costly if it were called in quickly enough. Proper use of the TACS and a more direct method of processing requests would eliminate much of the delay. LeMay also pointed out that more airmen were needed on the MACV staff to help improve the effectiveness of FARM GATE units and to create a better understanding among Vietnamese officers and Army advisors of how to use airpower's capabilities. Airpower was available, and it could offer significant protection to convoys, support for civilian irregular defense groups, and close air support for major ground operations; but to do these things, it would have to be centrally and efficiently controlled by an Air Operations Center (AOC) using the facilities of the already established TACS.

Harkins agreed that the Air Operations Center should coordinate all air activities including helicopters, but he decided that operational control of the helicopters should remain under the corps advisors. The AOC of 2nd Air Division would be responsible for the overall coordination of air assault operations. The Army would locate an Army air element within the AOC to facilitate the coordination and to expedite requests for immediate air missions to cover convoys and to respond to an enemy ambush or attack against an isolated garrison.

AIRMAN AS DEPUTY MACV PROPOSED

From the experience in the European war of having an airman as Deputy Commander-in-Chief of SHAEF, LeMay believed that making an airman the Deputy Commander of MACV would lead to a clearer understanding and a better employment of airpower in the expanding war. LeMay proposed to Harkins that he designate an Air Force lieutenant general as his deputy, similar to the Tedder-Eisenhower relationship. The Deputy Commander would provide the command direction often missing when an air deputy is restricted to advising and coordinating. With the Deputy Commander an airman, a better balance of experience would prevail resulting in a more effective use of forces. Furthermore, the Commander and Deputy Commander, MACV, were positions requiring broad tactical experience; it was immaterial whether COMUSMACV was a soldier or an airman as long as he could direct the forces assigned him by the Joint Chiefs of Staff. Whatever the military service of COMUS-MACV, his deputy should be of another branch having large forces assigned. Since Harkins, as COMUSMACV, was a soldier, his deputy should be an airman because of the assigned airpower and the projected buildup in it.

Harkins was opposed to assigning an airman as his deputy.[10] He again stated that the war was primarily a ground war and that his deputy should be a soldier. He considered airpower a significant factor in campaign strategy but saw the war as a counterinsurgency conflict in which the Army had the dominant interest. The situation did not require an airman as deputy commander; neither was there a need for an airman in the position of air deputy at this time. Harkins believed his command structure satisfactory—Major General Rollen H. Anthis, the 2nd Air Division commander, was his air component commander. As such, Anthis had direct daily access to Harkins. Calling Anthis a deputy commander or an air deputy would not produce a better understanding or closer working relationship.

Since General Anthis was air component commander, airmen thought MACV should be organized as a theater of operations with an Army component command in which COMUSMACV would not be the component commander. Army planners, remembering MacArthur's command organization in the Korean War, thought an Army component would be redundant; the MACV staff could function in both capacities. This arrangement also existed in the Vietnamese command structure, and it would be easier working with the Vietnamese if MACV were similarly organized. Nevertheless, Harkins established an Army component command in August of 1963.[11] He served as the component commander with Major General Joseph W. Stilwell as his deputy. Stilwell was primarily responsible for U.S. Army Support Command activities involving administration and logistics. Harkins reserved all operational matters for the MACV staff which functioned in a dual role as theater and Army component commander's staff.

The Joint Chiefs of Staff continued to discuss the question of a Deputy Commander, MACV. They were divided on the issue, and were still considering it when Secretary of Defense McNamara visited South Vietnam in December 1963. McNamara discussed the deputy question with Harkins and agreed that a soldier should be appointed as Harkins' deputy, effective 27 January 1964.[12]

Early in 1964 Harkins proposed that Anthis be given additional duty as the Deputy Commander, MACV, for Air Operations (or Air Deputy) although only a few months earlier Harkins had rejected the idea of a Deputy Commander for Air. The Air Staff did not support his request for an Air Deputy; in fact, LeMay had rejected the proposal prior to December 1963. In rejecting the 1964 proposal, LeMay stood firm on his earlier position that an airman should be appointed as Deputy Commander, MACV, not just Deputy for Air.[13] Harkins was due to leave Vietnam in June, and Westmoreland, the current deputy, was in line for his job. Thus the deputy position would soon be vacant again.

MACV AS UNIFIED COMMAND UNDER JCS—A PROPOSAL

The issue of making MACV a unified command reporting to the JCS came up again early in 1964. There were some cogent arguments why MACV should report to the JCS and not to CINCPAC, with the most important point being that basic decisions were made by the JCS; it would be a more effective arrangement for them to go directly to MACV which would carry out those decisions. Furthermore, experience in World War II and Korea indicated the need to place control close to the scene of action—a fundamental principle of command. CINCPAC was only a reviewing authority and could only slow down the decision process between Washington and MACV. He had to go to the JCS for guidance since strategic and political matters were beyond the scope of his authority. And even if matters were within the scope of his authority, he would still have to go to the JCS for guidance because of the political overtones of most of the military decisions. From these considerations, it was argued that MACV should be separated from CINCPAC and placed under the JCS as a unified command.

Support for the idea of MACV being a unified command under the JCS came also from some elements within the State Department. Admiral Felt, however, was opposed to the proposal because it could split South Vietnam from the rest of Asia. In his view, it was necessary that one headquarters be responsible for the entire Pacific theater, with sub-

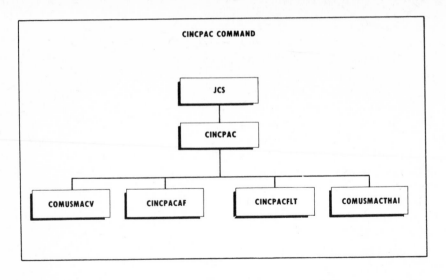

CINCPAC COMMAND

JCS

CINCPAC

COMUSMACV CINCPACAF CINCPACFLT COMUSMACTHAI

commands established as necessary to handle operations in particular areas. He further argued that CINCPAC needed control over all Pacific forces so he could use them as he saw fit to meet any threat. The Chinese threat remained dominant, and CINCPAC wanted to be able to direct forces to counter that threat without debating their use with the JCS. Initially, Harkins and the JCS agreed with Felt that MACV should remain as a sub-unified commander under CINCPAC.

DEPUTY COMMANDER FOR AIR OPERATIONS

Then in May 1964, Harkins submitted a proposal to make MACV a specified command reporting to the JCS. The difference between a specified and a unified command is fundamental. A specified command recognizes the dominance of one service in military operations. Other forces come under its direction, and the executive staff is weighted in favor of the dominant service. On the other hand, a unified command represents a multi-service activity. The interests of the participating services are equally divided, with a balanced staff of representatives from the participating services. The mission of the unified command recognizes no one service as dominant, but requires the integration of all participating services in an equitable relationship.

Harkins, in submitting his proposal for a specified command with the Army as executive agent, based his argument on the war being a counterinsurgency operation, the primary responsibility of the Army. This was the same argument that had been advanced in early 1961 on the control of air units assigned to South Vietnam. But Admiral U. S. Grant Sharp (the new CINCPAC as of 1 July 1964), opposed the establishment of a specified command and stated that the unified effort needed strengthening, not diluting. The JCS agreed with Sharp, and the issue of a specified command never again arose. The issue of a theater unified command, however, remained active.

The air organization of MACV remained a concern for the Air Staff. They were worried that much of the airpower assigned to CINCPACAF would be placed under the operational control of 2nd Air Division as the air component of MACV. By maintaining 2nd Air Division under 13th Air Force, PACAF retained the option of employing forces outside South Vietnam. Once air forces were assigned to MACV, however, the Commander of 2nd Air Division would have to persuade COMUSMACV to release them. If COMUSMACV objected, which was likely since most unified commanders are not particularly happy about releasing their forces to another command, PACAF would have to ask CINCPAC to release the forces. This could be a delicate request unless CINCPAC were concerned about the same areas as CINCPACAF.

The assignment of Westmoreland as COMUSMACV on 20 June 1964 forced the issue on the question of an airman as his deputy. LeMay had not given up on the idea; he felt airpower was crucial to the success of our policy in Vietnam. The JCS again considered the question of an airman as Deputy COMUSMACV, but were as divided as when the question was previously considered. To resolve the split, McNamara sided with the Chairman, General Earle G. Wheeler; and Lieutenant General John L. Throckmorton, U.S. Army, was appointed Deputy on 2 August 1964.

Before long, Westmoreland again raised the question of an Air Deputy. Uncertain how the North Vietnamese would react to the strikes above the DMZ, he argued against bombing North Vietnam in late 1964 because he didn't feel he had enough troops to stop an invasion by the remaining regular divisions of the North Vietnamese Army.[14] If such an invasion did come, however, airpower would be the critical factor in stopping it. Thus in September 1964, Westmoreland proposed that the commander of 2nd Air Division be appointed the Deputy Commander, MACV, for Air Operations as an additional duty. Sharp agreed with the proposal in light of the continuing buildup of air activity within South Vietnam. But the Air Staff maintained their original position on an Air Deputy for MACV. Their feeling persisted that the air organization within South Vietnam should be limited and that 2nd Air Division was adequate. The main organization for operations in Laos and North Vietnam should be 13th Air Force with an advanced headquarters in Thailand located at Korat or Udorn. Westmoreland's proposal for an Air Deputy in September 1964 was not approved although almost a year later the position would be created.

Thus at the end of 1964 the organization for the conduct of the war was still not settled. At the higher levels, discussions continued on the merits of establishing a separate theater versus those of maintaining MACV as a sub-unified command of CINCPAC. With increasing air activity along the lines of communication in Laos, and PIERCE ARROW air strikes in

403-892 O - 83 - 7

North Vietnam, though, clarification of air command arrangements was obviously needed.

ROLE OF CINCPAC COMPONENT COMMANDS

Admiral Sharp was opposed to any basic change in the PACOM command structure. He initially believed the air war in North Vietnam and Laos should be fought by his two component commanders, CINC-PACAF and CINCPACFLT, while the war in South Vietnam should be fought with forces assigned to MACV but supported by PACAF and PACFLT forces located outside South Vietnam. He believed this organization provided flexibility for concentrating his forces in the Pacific against the Chinese should that contingency develop. Further it gave him overall direction of the air war, which was a divided responsibility between CINCPACAF and CINCPACFLT. When he was given authority for strikes in North Vietnam and Laos, Sharp determined which missions would be assigned to PACAF and which to PACFLT. Each component commander then detailed the missions to his subordinate commands. PACAF directed 13th Air Force to carry out the specified strikes, and 13th Air Force in turn directed its subordinate command, 2nd Air Division, to execute them. PACFLT was given specific targets which were relayed to the 7th Fleet. Seventh Fleet then assigned the missions to a subordinate task force, Task Force 77 (TF–77) located in the Gulf of Tonkin above the 17th parallel.

The commanders of 2nd Air Division and TF–77 coordinated their

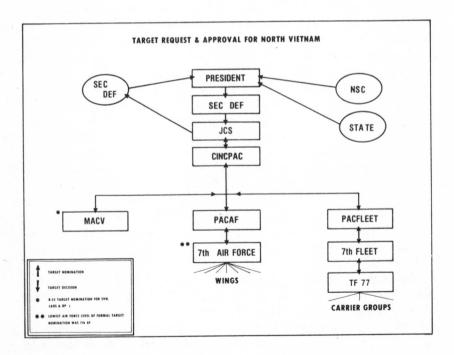

TARGET REQUEST & APPROVAL FOR NORTH VIETNAM

strikes in accordance with CINCPAC's designation of either PACAF or PACFLT as the coordinating authority for a specific strike. Although no formal PACOM procedure existed for such coordination, it is a principle of all military operations that commanders have an inherent responsibility to coordinate their operations with lateral or adjacent commanders when such operations affect the other's forces. Although the 2nd Air Division commander and the commander of TF–77 coordinated on these initial strikes, with the establishment of ROLLING THUNDER in March 1965 it was apparent the command arrangement for the conduct of the air war was inadequate. Many airmen advocated establishing a single air commander for the command and control of all air operations—Air Force, Navy, and Marine.

The control of naval aviation in an air campaign was a problem during the Korean War, and now it again needed to be resolved. There was a distinct difference, however, between the command problems in Vietnam and those in Korea. In the earlier war, Far East Command was given total responsibility and had no subordinate sub-unified commands. Far East Command approached, at least after 1952, the command structure of World War II in the Mediterranean and European theaters. In the 1960s, though, PACOM existed as a theater above a theater (MACV) with the total U.S. war effort divided between two headquarters. To further complicate the command problem, PACOM, as the theater command headquarters, was some 7,000 miles from the scene of the air and ground battles. By contrast, during World War II and Korea, theater headquarters were only a few hundred miles from combat. Not only was the command structure in Vietnam a problem, but with headquarters so distant from combatant forces, the decision process became even more difficult. Nevertheless, CINCPAC continued to believe a theater command in Southeast Asia reporting directly to the Joint Chiefs of Staff would be a mistake; and Air Force, Navy, and Marine representatives to the JCS supported his view.

Within the complex PACOM–MACV arrangement, if Air Force and Navy airpower were to be centrally controlled for ROLLING THUNDER, PACAF as the air component of PACOM was surely the appropriate controlling organization. Thus CINCPACAF pressed for operational control of naval air when employed in North Vietnam and Laos to include the selection of targets, time on and off target, control agencies, coordination, and weight of effort. (This degree of control conforms to the definition of operational control used in World War II and Korea.) However, CINCPAC did not see the necessity of placing carrier forces in the Gulf of Tonkin under the operational control of CINCPACAF.

General John P. McConnell, who succeeded LeMay as Chief of Staff on 1 February 1965, was principally concerned about the air command structure below the PACAF level.[15] He didn't believe there would be any major changes in the PACOM structure that would affect the Southeast

Asia theater. But McConnell did want an adequate theater command structure for the rapid buildup of USAF airpower in South Vietnam in response to the President's decision to expand all U.S. forces there. Base construction had to be hurried to take care of the wings arriving almost monthly; in a few short months the fighter force based in South Vietnam would number close to 400 aircraft. Faced with such expansion, McConnell urged reopening the issue of an airman as deputy to MACV.

Second Air Division was not large enough to assume the additional task of building an air force in combat. An air division is usually a small operational headquarters of 20 to 30 people, and as a sub-unit of a numbered air force, it is dependent upon that air force for administration, logistics, and other support. Thirteenth Air Force, higher headquarters for 2nd Air Division, was having difficulty supervising the large construction program in Thailand which was growing at the same time as the expansion in South Vietnam. Consequently, 2nd Air Division had to assume logistical and engineering responsibilities in addition to its primary operational function.

Major General Joe H. Moore, Commander of 2nd Air Division since 31 January 1964, had recommended expansion into a numbered air force reporting directly to CINCPACAF.[16] But his air division was still subordinate to 13th Air Force for strikes in Laos and North Vietnam, and to MACV for operations in South Vietnam. Moore was in a most difficult position: 13th Air Force and MACV often presented him with conflicting demands, and 2nd Air Division, though inadequately staffed, had responsibility for building bases and logistical facilities. PACAF alleviated some of the construction problems by temporarily assigning people to the 2nd Air Division staff from units in the Pacific and the United States.

AIR DEPUTY POSITION ESTABLISHED

In mid-1965, the Deputy Commander MACV issue was reopened by the pending reassignment of Lieutenant General Throckmorton. McConnell proposed that a three-star airman replace Throckmorton for the same reasons advanced by LeMay. But this proposal was no more successful than previous ones. Westmoreland believed as Harkins that the major task in South Vietnam was the ground battle and that he needed a soldier as his deputy to help share the burden. Airpower was important, but the main task was that of the soldier "finding and fixing" the enemy, then bringing in airpower to help artillery destroy him. Based on this strategy, air power was a supporting element rather than a dictating consideration. Thus Westmoreland argued that Throckmorton's replacement should be another soldier just as when he was deputy to Harkins in 1964.

---

* According to Westmoreland's Report on the War in Vietnam (1968), p. 279, Throckmorton was replaced by Heintges on 5 November 1965. Heintges was replaced by Abrams on 1 June 1967, two years after the events being discussed here.

Westmoreland's counterproposal was, again, that a three-star airman be the Deputy for Air Operations and simultaneously commander of 2nd Air Division. Unsuccessful in getting approval for the Deputy Commander, MACV position, McConnell now agreed with Westmoreland, and the JCS approved the Air Deputy position on 25 June 1965.

CONTROL OF ARMY HELICOPTERS & MARINE AIRCRAFT

Moore, as the Air Deputy, requested control of Army helicopters just as Anthis had done in 1962. In Moore's opinion, proper control would require locating an Army aviation element in the Tactical Air Control Center (TACC) where joint planning would include helicopter assault operations. Further, the tactical air control system would control helicopters the same as other aircraft; the Combat Operations Officer at the TACC would determine if any mission could proceed based on the amount of enemy ground fire and upon other air operations that had developed as a result of unforeseen enemy action. If a diversion or delay were necessary, the Operations Officer would consult with the senior Army representative at the TACC. If there were a difference of opinion, the 2nd Air Division commander, acting as either air component commander or Air Deputy, could present the matter to COMUSMACV for resolution. The closeness of MACV headquarters to 2nd Air Division headquarters would expedite the decision process.

In spite of arguments advanced by Moore and his staff, however, the directive establishing the Air Deputy position excluded any reference to his controlling Army helicopters. This absence of control was a problem throughout the war, for the large number of aircraft sorties and the absolute necessity to counter enemy ground fire during helicopter assaults demanded unified planning and control. In fact the demands for air support are greater during a helicopter assault than for a traditional airborne operation. In an airborne assault the force is traveling at a much higher penetration speed with minimum exposure, and it has a higher degree of survivability compared to a helicopter assault. Both types of operations require close coordination between all military forces, and for this reason an airman in previous wars had control of such operations until the troops landed; after landing, the ground force commander took over. The air commander in earlier wars decided whether defenses in the target area would permit the assault to proceed, but this was not to be the case in helicopter assault operations in South Vietnam.

The Air Deputy's authority was further restricted by MACV Directive 95–4 on 6 May 1965. Among other things, this directive prescribed the relationship of Marine air and its control.

> Marine Corps aviation resources are organic to III MAF and are commanded and directed in support of tactical operations as designated by the Commanding General III MAF. The Marine Corps tactical air control system will exercise positive control over all USMC aircraft in support of Marine Corps operations

and over other aircraft as may be in support of such operations. In the event COMUSMACV declares a major emergency, 2nd Air Division will assume operational control of certain air resources designated by COMUSMACV.[17]

Under this directive, airpower was further fragmented by the establishing of all elements of two separate tactical air forces in the theater, one controlled by the theater air component commander and the other by the equivalent of a corps commander. This fragmentation grew unworkable as the war progressed, and by the time of Khe Sanh (January 1968), it was apparent Marine airpower had to be controlled by the air component commander as it had been in Korea. In the years before this change occurred, each of the air component commanders constantly raised the issue with COMUSMACV

FURTHER DEFINITION OF 2ND AIR DIVISION RESPONSIBILITIES

In November 1965, McConnell directed that 2nd Air Division be separated from 13th Air Force. It would report to PACAF on air operations outside South Vietnam and on all Air Force service matters within South Vietnam. (Air operations in South Vietnam were the prerogative of COMUSMACV as the sub-unified commander.) General Hunter Harris, Jr., who replaced Smart as Commander of PACAF, wanted to keep 2nd Air Division under 13th Air Force since such a structure would have kept 13th Air Force in control of all air operations outside South Vietnam, but the expansion of 2nd Air Division (it was rapidly approaching the size of 5th Air Force in the Korean War) made such a relationship with 13th Air Force impractical. But now, what headquarters would control airpower located in Thailand?

When MACV was organized in 1964, COMUSMACV was also designated COMUSMACTHAI and was made responsible for planning United States military participation in SEATO. However, the Thai military expressed concern over this arrangement. Their concern was probably that the dual-hatted commander would link them directly to the actual fighting in South Vietnam. Further, the Thais may have feared they would exercise less influence in the U.S. military structure if the dual command arrangement prevailed. At the request of the Thai government, the two positions were separated on 30 April 1965.

The Thai government requested, further, that a commander based in Thailand direct all USAF units based in their country. Presumably the same reasons advanced for not having COMUSMACV command forces based in Thailand motivated this similar request. The U.S. Ambassador, however, had a significant role in determining the command of U.S. airpower based in Thailand. If the air commander remained in South Vietnam, the Ambassador's influence would be reduced; if the commander were based in Thailand, however, the Ambassador would have direct control since he was the senior U.S. official in the country.

Various proposals treated the air command structure for controlling Thailand based units. It had become accepted fact, however, that the control of air operations in South Vietnam would differ from the control of those in North Vietnam. Sharp was adamant in his positon—his headquarters would control the air war against North Vietnam while MACV would control airpower employed in South Vietnam. Confronted with the conflicting needs to have all airpower under a single air commander and to meet the request of the Thai government, McConnell proposed a unique arrangement: He would appoint one airman as deputy commander for 2nd Air Division and for 13th Air Force in Thailand.[18] This unique arrangement satisfied the Thais' request to have the forces based in Thailand under a commander located there; the arrangement also provided centralized direction of the total air effort under the commander of 2nd Air Division.

The Deputy Commander, 2nd Air Division/13th Air Force, had logistical and administrative responsibility for all units located in Thailand, and he reported to the 13th Air Force commander on both these matters. However, the commander of 2nd Air Division in Saigon did not pass operational control of the forces in Thailand to the deputy. Moore, like all subsequent commanders of 7th Air Force, elected to control all of these forces from his headquarters. This was the most efficient way to use them for missions in North Vietnam, Laos, and South Vietnam. Although Thailand based units could not be used for missions in South Vietnam except during emergencies, South Vietnam based aircraft could be used in all three areas. Thus there was greater flexibility in retaining control at 2nd Air Division headquarters.

It was soon evident that the 2nd Air Division staff was not of sufficient size to direct the units coming into South Vietnam. Normally, six or more wings need a tactical air force structure for operations. This provides a command and control system and other support activities for about a thousand airplanes. Second Air Division reached the size of a numbered air force in mid-1966; however, considering the airfield construction program alone, 2nd Air Division needed the staff of a numbered air force as early as 1964.

On 14 March 1966, 2nd Air Division was inactivated, and the 7th Air Force of World War II fame was reestablished to direct the air war in North and South Vietnam. Activating 7th Air Force and assigning an air deputy to COMUSMACV completed the top air command structure for the rest of the war except for the final phase-out of MACV headquarters in Spring 1972. The 2nd Air Division/13th Air Force position was redesignated Deputy Commander, 7th Air Force/13th Air Force.

The deputy position in Thailand was a source of increasing concern since the 7th Air Force commander had not given the deputy operational control of any forces. It was not feasible to give control of forces to the deputy commander without increasing his staff to supervise the execution of the mission. But the increased staff would have entailed a considerable

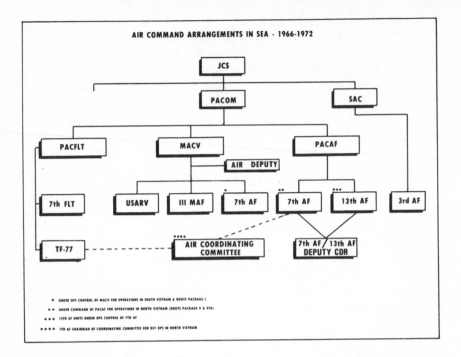

AIR COMMAND ARRANGEMENTS IN SEA - 1966-1972

● UNDER OPS CONTROL OF MACV FOR OPERATIONS IN SOUTH VIETNAM & ROUTE PACKAGE I.
●● UNDER COMMAND OF PACAF FOR OPERATIONS IN NORTH VIETNAM (ROUTE PACKAGE V & VIA)
●●● 13TH AF UNITS UNDER OPS CONTROL OF 7TH AF
●●●● 7TH AF CHAIRMAN OF COORDINATING COMMITTEE FOR KEY OPS IN NORTH VIETNAM

duplication of resources, which were limited by the maximum number of people who were authorized to be based in South Vietnam and Thailand. More fundamental, however, was the need to control all air operations from a single point. The conflicting requirements and the shifting of forces to continually cover competing demands from MACV, PACAF, and the Ambassador in Laos necessitated centralized control in 7th Air Force. And with about 1100 combat aircraft, the force was well within the span of control of a numbered air force headquarters.

CONTROL ARRANGEMENTS FOR OPERATIONS IN LAOS

The Deputy Commander, 7th/13th Air Force, was primary liaison for the 7th Air Force commander in dealing with the Ambassadors in Thailand and Laos. Since the Ambassador in Thailand was responsible for all activities of U.S. forces based there, he requested a daily report of missions flown by units in Thailand. The Deputy Commander, 7th/13th Air Force, met with the Ambassador frequently and kept him briefed on potential missions and those already underway. The Ambassador then used these reports to keep the Thai government informed. If matters required special attention, the 7th Air Force commander met personally with the Ambassador. Also, as 7th Air Force commander from 1966–1968, I provided a monthly resume of missions. However, the Ambassador in Thailand exercised no control over the operations of the force. His responsibilities were to keep the Thai government informed on the air war and to obtain facilities needed for basing our forces.

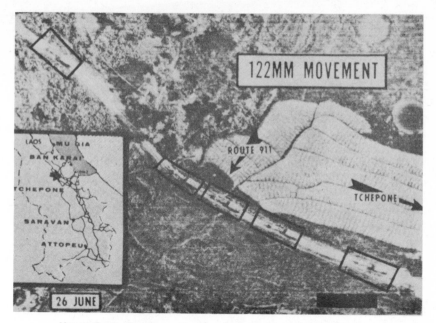

Heavy North Vietnamese traffic along the Ho Chi Minh Trail in Laos.

Command relationships with the Ambassador in Laos were complex and difficult. The Ambassador, as the senior United States official, was responsible for all U.S. military activities; consequently, all air operations came under the detailed surveillance and control of the embassy. In effect the embassy air attaché functioned as an air commander since he could determine 7th Air Force employment through the authority of the Ambassador.

COMUSMACV had responsibility for air operations in southern Laos (the two principal areas here were called TIGER HOUND and STEEL TIGER), and under the unified command principle, he further delegated this responsibility to the 7th Air Force commander. Since there were no U.S. troops in the area, except for occasional long range patrols to gather intelligence, most of the activities were air missions: interdiction, reconnaissance, air superiority sorties. Command of the forces followed prescribed military channels, and the Ambassador in Laos could approve or disapprove certain targets. Approval was a continuing problem because senior Laotian Commander, Vang Pao, at the embassy's direction, maintained road patrols along the Ho Chi Minh network. The 7th Air Force commanders thought these teams produced questionable information about enemy movements along the trail; yet concern for the safety of the teams, whose locations could never be pinpointed, denied large areas to our strike forces. In fact our planners urged that the teams be withdrawn so any target in the area controlled by North Vietnamese could be cleared for attack. Because of these teams and fear of hitting

85

friendly villages, the Ambassador insisted upon limiting attacks to 200 yards either side of the road network unless strike forces were under the control of a forward air controller (FAC), unless the Laotians approved the strike.

These restrictions finally led to the assignment of a Laotian to the airborne command, control, and communications (ABCCC) aircraft. He represented the military region commander's headquarters and had the authority to clear targets for attack. Thus a typical strike in Laos started with a forward air controller in an O–2 or OV–10 aircraft locating a target and requesting strike approval from the ABCCC. The Combat Operations Officer, with the Laotian officer's approval, would then release the target to the FAC. This process took only a few minutes and followed the pattern for a normal mission in which strike aircraft reported to the ABCCC which passed them to the FAC's control. If the Laotian officer didn't believe he could clear a target because of its proximity to a village, the ABCCC radioed either the nearest Air Operations Center colocated with a military region headquarters or the Air Operations Center at Vientiane for approval. The process took time, and if North Vietnamese forces were moving near a village, the delay could be too long. However, most of the targets were within the approval authority of the Laotian officer.

BARREL ROLL (the operations in northern Laos) created many command problems not experienced in STEEL TIGER or TIGER HOUND. The latter two areas had no significant ground activities except along the Bolovens Plateau, and control of strike forces involved relatively little coordination. The 7th Air Force commander, as the air component commander, exercised operational control of all airpower employed in these areas whether Navy, Marine, or Army. Thus, the command and control system operated to a great extent in the same manner as it did in South Vietnam.

In BARREL ROLL, Vang Pao's forces operated under direct control of the embassy. The U.S. air attaché in Laos played a major role on the embassy staff in selecting targets and proposing the size of forces employed daily. In essence, the activity sealed off a geographical area, and airpower was fragmented for that area. The embassy constantly sought the dedication of certain amounts of airpower to the support of Vang Pao's forces. For a number of reasons, 7th Air Force vigorously resisted assigning such control to the embassy. The most significant reason was that fragmenting the force prevented its use where it would provide the greatest potential for decisive action. The embassy also wanted a wing of propeller aircraft at Nakon Phanom under its operational control, but Moore and I (as Seventh Air Force commanders in 1965 and 1966) contended that enemy defenses would preclude employing propeller aircraft against the targets the embassy proposed. We were sure that it would be necessary to employ high performance jet aircraft.

Aircraft use was a continuing problem with the embassy, a problem that abated only when the North Vietnamese increased their forces in the Plain of Jars and began a systematic assault on Vang Pao's troops and his base at Long Tieng. By this time the sortie requirements were so many that a single wing dedicated to Vang Pao's support would obviously have been inadequate. Since B–52s and all fighters used in North Vietnam were being employed in Laos in increasing quantities by the middle of 1967, BARREL ROLL needed the same arrangement that existed in South Vietnam for control of airpower. The embassy's special staff were not the ones to control sophisticated air operations required to hold a country with inadequate ground forces or to stop an enemy who introduced more troops and weapons.

Still, the Ambassador never felt that enough airpower was being devoted to the war in Laos, and he raised the issue through diplomatic and military channels on several occasions. His argument was one frequently heard from an organization wanting sole control of airpower to support its missions. But if airpower had been divided as the Ambassador proposed, there would have been insufficient forces for the other missions in South Vietnam, southern Laos, and North Vietnam. CINCPAC considered the war in North Vietnam a priority commitment; COMUS-MACV considered his mission in South Vietnam dominant; and the Ambassador in Laos was convinced the preservation of the status quo in Laos deserved extensive airpower. The only way all the conflicting requirements could be satisfied was through 7th Air Force's centralized control of airpower. The 7th Air Force commander finally had to decide upon each request based upon the criticality of the situation and the amount and type of airpower available.

Each time Moore, I, or a later commander decided to reassign air support from one area to another, we provoked an energetic response from the losing activity. JCS pressure on CINCPAC to eliminate a target released for attack carried priority, often resulting in reduced support for BARREL ROLL or for interdiction strikes in STEEL TIGER or TIGER HOUND. Priorities shifted frequently, and whenever a situation developed constituting a grave threat to the security of forces or facilities, we diverted airpower to stabilize the situation. Invariably our decisions displeased other headquarters because they, removed from the scene of action, were bound to assess the situation somewhat differently.

With the bombing halt of 1968, our efforts focused increasingly on the war in Laos. Our control of airpower then resembled the control in South Vietnam although it never reached the same high degree of efficiency and effectiveness. The embassy in Laos, however, placed fewer restraints on targets since the North Vietnamese advance had to be stopped. As restraints lifted, the rather elementary AOCs could direct airpower to the control of FACs and long-range navigation (LORAN) bombing much more efficiently. Excellent bombing by F–111s, F–4s, and A–7s made the defense of Long Tieng, Vang Pao's headquarters, possible. The enemy's

130mm artillery pieces located on high terrain above Long Tieng could have been removed by Vang Pao's infantry, perhaps, but such an up-hill ground attack would have been extremely costly. However, our airpower kept the enemy gun positions under constant attack using all-weather bombing techniques since clouds obscured the peaks much of the time. We processed target requests and controlled strikes in this campaign with the same general procedures employed in South Vietnam.

CONTROL OF NAVY STRIKE AIRCRAFT DIVERTED FROM NORTH VIETNAM

Arrangements for the control of strikes in North Vietnam differed importantly from the arrangements in Laos and South Vietnam. Navy aircraft employed in South Vietnam and Laos came under my control as 7th Air Force commander, which permitted me to select the time on and off target, and the controlling agency. Most of the "diverts" from North Vietnam into Laos were made by the Navy; for them the ABCCC assigned targets and FACs according to its general attack plan. Often the scant notice received before a strike diversion required the ABCCC to make rapid decisions and adjustments. Thus if the weather forecast for North Vietnam was marginal, I always had contingency targets given to the ABCCC.

Diversions into South Vietnam were less frequent, but when they happened, most aircraft were used in one of the two northernmost Military Regions (MR–I and MR–II). Both target areas stretched the range limits of Navy aircraft, so the Navy sorties that diverted into these areas sometimes had to be supplied air refueling. Also, it was difficult to accommodate diversions into South Vietnam because of the nature of the fixed targets. Many targets directly associated with ground force contacts required strikes to occur at a precise time. Of course the ABCCC didn't know when Navy aircraft would be diverted, so at best there would be only a very short time to arrange the strike. Since Navy sorties couldn't be as effective in South Vietnam as they were against selected targets along the Ho Chi Minh Trail network in Laos, most Navy "diverts" were sent to Laos.

As set forth in MACV Directive 95–4, I, as the Deputy Commander for Air, could assume control of all airpower supporting MACV at the direction of COMUSMACV. During the Tet offensive early in 1968, Westmoreland requested CINCPAC to authorize the commitment of carrier air to my control. CINCPAC approved what amounted to an extended diversion of TF–77 carrier air from the bombing campaign in North Vietnam. The TF–77 commander committed daily sorties for my use, and we controlled these aircraft using the same procedures established for all other aircraft under 7th Air Force jurisdiction. The arrangement worked.

I should mention here my concern in 1968 about diverting either Air Force or Navy fighters to targets in South Vietnam, thereby taking the

U.S. Navy A4–C "Skyhawk" drops 250lb "iron bombs" on a suspected Viet Cong stronghold in South Vietnam.

pressure off North Vietnam at a time when, psychologically and militarily, an intensive air campaign in the north was needed to help counter the Tet offensive. Under the circumstances, I was most anxious not to hold forces for the war in the south unless absolutely essential.

CONTROLLING THE STRIKES UP NORTH

Perhaps the toughest question facing Admiral Sharp (CINCPAC) early in 1965 dealt with who would control strikes in North Vietnam and how they would be coordinated. The ad hoc arrangement for FLAMING DART (February 1965), in which coordination was delegated on a mission-by-mission basis, was totally inadequate for ROLLING THUN-DER (beginning on 2 March). A formal command arrangement had been established for forces in South Vietnam and Laos, but the question of air operations in North Vietnam was unsettled.

The Air Force, based on experience with a similar situation in Korea, advocated placing carrier air under the control of CINCPACAF as the theater air component commander. PACAF would then delegate control to 2nd Air Division (7th Air Force) in Saigon, and the unity of airpower would be preserved. As in the Korean War and World War II, strikes in North Vietnam had to be closely integrated, which could be assured by having all airpower under the air component commander.

However, the Commander-in-Chief of the Pacific Fleet (CINC-PACFLT) contended that naval airpower was an inherent part of the fleet, and its mission and could not be separated. MacArthur had used the term "coordinating authority" to harmonize FEAF and NAVFE air forces in Korea; the same arrangement seemed suitable for the Vietnam situation. Therefore it was appropriate to designate one of the components the "coordinating authority" with that authority limited to such things as exchanging information on strike plans, requesting support for a particular operation, and establishing procedures to prevent conflicting activities. Essentially, CINCPACFLT wanted TF–77 in the Tonkin Gulf to have the same relationship to 2nd Air Division/7th Air Force that it had initially with 5th Air Force in the Korean War. Senior air officials feared that this arrangement would create the same problems it had created in the Korean War; it was not the command relationship needed to adequately direct both forces to a common objective.

COORDINATING AUTHORITY

CINCPAC agreed with CINCPACFLT. In March 1965, PACAF was designated the coordinating authority for ROLLING THUNDER,[19] but the directive clearly stated that such authority did not involve the operational control of the carrier forces. The charter also established the ROLLING THUNDER Armed Reconnaissance Coordinating Committee (later changed to the ROLLING THUNDER Coordinating Committee) to coordinate and resolve items of mutual interest to the Navy and Air Force. The committee was to eliminate overlapping areas of interest, reduce duplication of effort against targets in North Vietnam, and promote an effective ROLLING THUNDER program.

CINCPAC would assign targets released by the JCS to PACAF and PACFLT. PACAF would then ensure that strike forces didn't conflict with one another in approaching attacking, and withdrawing from the target. CINCPAC expected PACAF to delegate coordinating authority to 2nd Air Division, and Moore did receive that authority.

Moore's experiences with 5th Air Force in the Korean War convinced him that a commander needed something other than coordinating authority to conduct a theater air campaign properly, but he faced up to the problem of how to convert the arrangement into some workable relationship with TF–77. Missions were becoming more complex, so both forces needed specific guidelines to carry out their assignments. Yet, it was not feasible to exchange plans for every mission with the many variables and communication problems that inevitably develop during intensive operations.

Moore formed a working committee with TF–77 to formulate a proposal for controlling the two forces within the guidelines of CINCPAC's directive. The initial proposal by 2nd Air Division was to establish a time-sharing arrangement of three-hour intervals for striking North Vietnamese LOCs south of the 20th parallel. Various segments of the routes would be assigned to either TF–77 or 2nd Air Division for three-hour periods. Assignments would be planned a week in advance to cover all lines of communication and would permit one force to operate in the other force's area should the latter not elect to use its three-hour period.

TF–77 didn't like the arrangement because the range limitations of Navy strike forces wouldn't allow them to reach distant targets. Without air refueling, they couldn't attack the passes and roads along North Vietnam's western border. As a counter proposal, TF–77 recommended North Vietnam be divided on a north-south axis with TF–77 responsible for the coastal area. However, most of the targets and lines of communication were within 25 to 30 miles of the Tonkin Gulf, except for the passes leading into Laos which were bottlenecks in the road network. At the head of these passes, the North Vietnamese stored large quantities of war material awaiting shipment further south. Except for these supplies, almost all targets below the 20th parallel were along the coastal highway (Highway #1) and near the many small ports used to stage supplies for movement south. Although the north-south division would have helped the Navy's range problem, it would not have promoted the effective use of forces assigned to 2nd Air Division; areas without significant targets would have had high coverage while areas with many targets would not have received enough effort.

After considering different methods for coordinating air support, the 2nd Air Division/TF–77 ROLLING THUNDER Coordinating Committee proposed to divide North Vietnam into a series of route packages beginning at the DMZ.[20] There were six route packages with the sixth divided into two parts, VI A (Air Force) and VI B (Navy). Dividing North Vietnam into route packages gave the Air Force three areas and the Navy four; however, the Air Force had a much larger and more hazardous area to cover than did TF–77. The basic point, though, wasn't exposure over enemy territory but that the division of airpower into geographical areas was improper.

ROUTE PACKAGES—A CONTROL ARRANGEMENT

The Committee assigned Route Package I to 2nd Air Division; it covered an area from the DMZ to just above the 18th parallel. MACV had initially considered this area an extension of the ground battle zone, and our operations in Route Package I were directed by MACV rather than by PACAF. If this logic had prevailed when the route packages were

U.S. Navy strike aircraft were launched from TF–77 carriers cruising in the Gulf of Tonkin.

being designed, the same argument might have led the Committee to place all route packages under MACV since the entire air campaign against the LOCs was meant to affect the battle in South Vietnam. In any case, the assignment of Route Packages I and II would be reexamined later.

TF–77 controlled Route Package II, covering an area from the 18th parallel to just below the 19th parallel, and from the coast to the Laotian border. The most significant target in Package II was the Vinh area and the logistical activities surrounding it. Coastal shipping and traffic on the coastal highway were also major target systems. However, all of the major passes leading from North Vietnam into southern Laos were on the southern edge of Route Package I; there were no passes in Route Package II.

The Navy also controlled Route Package III, the largest geographical area but with less significant targets except for the coastal rail and road network. Barthelemy Pass was a major target, however, since most of the supplies supporting the Pathet Lao and North Vietnamese forces in the Plain of Jars moved over Route Seven. Seventh Air Force covered this pass and most of the movements leading into it, while the Navy covered the eastern end of Route Seven.

Except for Route Package VIB, Package IV was the most active area assigned the Navy, and few of the targets in Package IV were restricted.

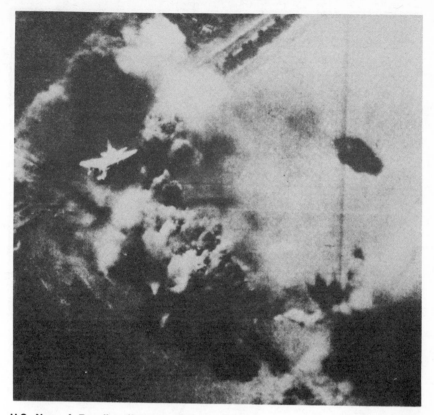

U.S. Navy A–7 pulls off target after destroying a span of the Hai Dong (west) railway bridge.

The most important targets were the rail and road networks and the bridge at Thanh Hoa. In addition, Nam Dinh was a major rail yard and marshalling area for logistics. At the time Bai Thuong was the only all-weather airfield in the area, and enemy fighters used it as a staging field for patrols south.

The Air Force was responsible for Route Package V—twice the size of any other area. It contained most of the railroads in the northwest and the LOCs supporting North Vietnamese forces in northern Laos. Package V was bounded on the east by a line along the 150°30′ longitude, on the west by the Laotian border, on the north by the Chinese border, and on the south by an imaginary extension of the northeast rail line until it intersected the Laotian border.

By far the most important of all route packages was Route Package VI. Most of the targets were in this area, and enemy defenses there were the strongest. Package VI was divided between the Air Force and Navy along the northeast rail line. Using the railroad as a dividing line gave the least chance for error were pilots from either 7th Air Force or TF–77 to stray

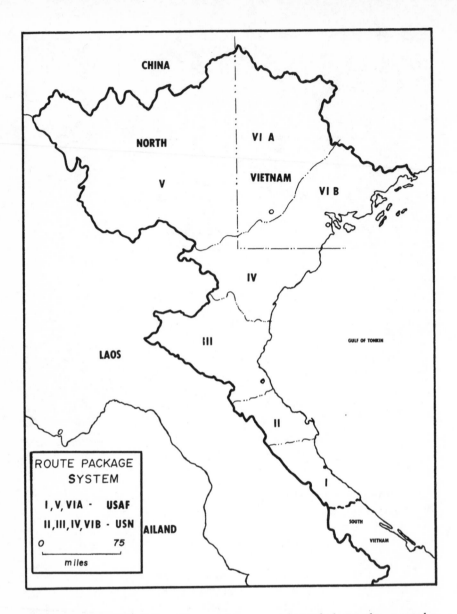

CHINA

NORTH

VI A

V

VIETNAM

VI B

IV

III

II

I

LAOS

GULF OF TONKIN

SOUTH
VIETNAM

ROUTE PACKAGE
SYSTEM

I, V, VIA · USAF
II, III, IV, VIB · USN

0   75

miles

AILAND

and be assumed hostile. The package was bounded on the west by 150°30′ longitude and on the south by a line just to the north of Nam Dinh.

Most of the original 94 targets on the JCS list were in Route Package VI. Some targets were actually so close together there was no clear way of separating them. The Paul Doumer bridge, for example, was only a couple of miles from the Hanoi Railcar Repair Shop and five miles from the Canal Des Rapides bridge. With fighters traveling at speeds over 500 knots, it made little difference which targets were assigned since exposure to the defenses was already maximum. Thirty miles beyond Hanoi, however, the defenses thinned out and the number of important targets decreased.

Dividing North Vietnam into route packages compartmentalized our airpower and reduced its capabilities. One result was that 7th Air Force diverted too many sorties into Route Package I when weather prevented strikes in Route Package V or VI and the ABCCC was fully committed with aircraft along the LOCs in Laos. On the other hand, TF–77 had an inadequate number of aircraft for 24-hour coverage of its assigned route packages. The situation resembled the one in World War II which led Air Marshal Tedder to write, "There were so many cooks that wanted to stir the bombers' broth that, had there been no centralised control, no head cook with a firm hand, there would have been a very real danger of flexibility in itself resulting, not in concentration and economy of effort, but in dispersal and waste of effort. . . . Air warfare cannot be separated into little packets; it knows no boundaries on land or sea other than those imposed by the radius of action of the aircraft; it is a unity and demands unity of command."[21]

CINCPAC allocated Navy targets to the Commander of TF–77 through CINCPACFLT and the Commander of 7th Fleet. Carrier task group commanders selected their daily targets from the ones they received from TF–77. Each of the four route packages was divided into two sectors with a carrier responsible for one sector in each package. Within these sectors, task group commanders had maximum freedom in their daily planning except for those targets controlled by the JCS and released for attack only when approved by the President.

The route package system was valuable to TF–77 in cycling its forces and providing localized control in a single area. However, the route package system was fundamentally wrong for the best application of all U.S. airpower. Certainly the Navy needed a solution to its range problem, but a fragmented command structure is not the best way to accommodate a mixed force of fighters, bombers, and attack aircraft with varying ranges. In fact, theater air component commanders in previous wars adjusted force employment plans for widely differing aircraft capabilities. In the European campaign of World War II, for example, 9th Air Force controlled P-47s, P-51s, P-61s, B-25s, and B-26s. A single consolidated plan provided for their employment according to differences in range, speed, and payload. The net effect was a unified action that shifted attacks as required against airfields, vehicles, marshalling yards, troop concentrations, and enemy ground forces.

The route package system was a compromise approach to a tough command and control decision, an approach which, however understandable, inevitably prevented a unified, concentrated air effort. Within 7th Air Force and TF–77, aircrew ability to carry out assignments against heavily defended targets was outstanding. So the disagreement wasn't over the training and capabilities of crews, but over how best to control two air forces from two different services. The same issue arose in the Korean War, and my present fear is that our continuing failure to settle this issue may be exceedingly costly in some future conflict such as, for instance, a NATO war. Any arrangement arbitrarily assigning air forces to exclusive areas of operation will significantly reduce airpower's unique ability to quickly concentrate overwhelming firepower wherever it is needed most.

In 1966, when it appeared the coordination authority agreement would not be revised to permit my control of carrier air, I proposed establishing in the Gulf of Tonkin an ABCCC to control all strike forces against lines of communication in Route Packages I through IV. This ABCCC would use the same techniques used in Laos for the control of strikes on the Laotian road network. The system was virtually perfected after almost two years of operation and it had worked under some difficult conditions. When carrier forces were diverted to Laos, the ABCCC handled as many as 300 unplanned sorties a day in addition to the 200–250 sorties normally under its control. When Navy diverts were available, the ABCCC called up additional FACs to augment existing FACs in various parts of the network. Couldn't this flexible system improve our efforts in the lower route packages as well?

My proposal, which was not adopted, would have had the ABCCC located over the Gulf in contact with TF–77 and the 7th Air Force command center at all times. Naval air representatives would have augmented the Air Force staff of the ABCCC. Prior to ABCCC takeoff, its on-board combat commander would have received the daily mission plans of 7th Air Force and TF–77 or the individual carrier task group commanders. With these missions before him, he could have reassigned incoming strike aircraft to different route packages should better targets have appeared. The ability of an on-scene commander to shift airpower over the four route packages would have permitted us to destroy more time-critical targets such as trucks, trains, and boats. If a strike uncovered a supply area or other lucrative target, the ABCCC commander would have reported to the 7th Air Force and TF–77 operations centers that diversions would be made and would continue as long as the new target was more valuable than the preplanned targets. If I judged the diversions too disruptive to the planned interdiction effort for the day, I could have denied the diversions, limited their number, or launched additional forces under ABCCC control. The responsibility would have been mine for the overall decision affecting both 7th Air Force and TF–77 strike forces.

Strike aircraft taking off from our bases in Thailand and from the carriers would have routinely checked in with the ABCCC. The ABCCC could either reaffirm their targets or assigned other targets. At the same time, the ABCCC would have notified high speed FACs (because of defenses, low speed FACs couldn't work the LOCs in North Vietnam) that strike aircraft would report to their control. From then on, the mission would have resembled any other FAC-controlled strike.

This arrangement for force control below the 20th parallel would not have been suitable for attacks in Route Packages V, VIA, and VIB, however, because of the strong defenses in the northern areas. Almost all missions into these areas required extensive preplanning to minimize exposure and to achieve a high degree of coordination among supporting forces. The more complex the defenses, the greater was the need for preplanning.

In September 1966, I requested operational control of all strikes against targets in North Vietnam. I believed the ROLLING THUNDER Coordinating Committee was not promoting mutual support. Although the committee was acting under its charter, it couldn't direct either force to modify its operations or to schedule strikes at different times.

I should point out that even if my request had been approved, which it wasn't, TF–77 would have had relatively little flexibility in launching and recovering strikes because of the time required by the carrier task groups to position themselves and turn into the wind for recovery and launch. The carrier force worked on a twelve-hour operational cycle, and their schedules could not be modified quickly. On the other hand, we in 7th Air Force had almost no limitations on launch or recovery times. A single self-imposed limitation came from the fact that daylight was needed for the rescue of downed crews. Thus we launched afternoon strikes early enough to permit a few hours of daylight after our forces left the target area. We were especially careful about this timing for strikes near Hanoi where some of our aircraft would probably go down.

The difficulties with the coordinating authority and route package system surfaced when the JCS released for attack a target like Phuc Yen, as they did in October 1967. Although planning for such an attack was complete, defenses kept changing, and the number of aircraft on the field varied. Surprise was an important consideration in this first attack if we were to destroy the five or six IL–28 Beagle light bombers. These were the only enemy aircraft which, without staging, posed a threat to Danang and other logistical and operational facilities, so our planners monitored the location of these aircraft closely. Earlier, the North Vietnamese had moved the IL–28 into China whenever a raid developed. But when the airfield was not released for destruction during the first half of 1967, the North Vietnamese evidently decided to leave their IL–28s in place, confident that we would not attack for fear of escalating the war.

When the JCS released Phuc Yen for attack on 24 October, both 7th Air Force and TF–77 received only a few hours advanced notice. Although our strike forces had been in Route Package VI A that morning and were just coming back, I decided to change the afternoon mission for a strike on Phuc Yen, and I notified my commanders just before noon. This mission necessitated changing the bomb loads and briefing the pilots on the attack since they were not familiar with the target. Under normal circumstances a pilot would have a couple of days to study such a target in greater detail. The airfield was heavily defended, so precise timing and exact identification of targets within the airfield complex were most important.

As 7th Air Force redirected its attack force, the complex process of coordination with TF–77 was underway. Detailed arrangements were necessary since congestion over the target might lead to collisions, and clusters of aircraft would certainly provide excellent targets for North Vietnamese gunners. Thus, concentrating the strikes for maximum effect was desirable, but having strike forces holding on the periphery of the target area waiting their turn was not. Also, we needed to distribute the targets so that all of them would be hit if for any reason one force didn't penetrate and others did. The second force in an attack like this one should reinforce the strikes of the first force and cover the high value targets a second time.

TF–77's role was critical, and of course the Navy commanders and pilots cooperated fully; but the important point is that our complex "coordination" relationship was obviously inadequate. Although we hit Phuc Yen successfully on 24 and 25 October 1967,[22] the command structure did not give me sufficient authority to guarantee that I could respond immediately with the full weight of Air Force and Navy airpower in any similar situation. I could respond immediately with Air Force airpower, and I could coordinate with the Navy. Coordinating authority is simply inadequate when operations must be changed rapidly and when intricate details must be quickly resolved.

Throughout the remainder of the 1965–1968 air offensive and in 1972, 7th Air Force periodically raised the issue of command arrangements for air operations in North Vietnam. But Admiral Sharp (CINCPAC) remained convinced that coordinating authority was the best arrangement,* and Sharp's successor, Admiral John S. McCain, Jr., took the same position. The issue was not significant between 1968 and 1972 because there were few protective reaction strikes in North Vietnam, and the 7th Air Force commander had operational control of all air efforts in South Vietnam, Laos, and Cambodia.

In May 1972, with the decision to begin bombing North Vietnam, the route package issue was raised again. General Lucius B. Clay, CINCPA-CAF, didn't think route packaging and coordinating authority were satisfactory methods for controlling the air effort.[23] He expressed the same view held by all the commanders of 7th Air Force—operational control of naval air by the air component commander was the only sound arrangement. The combat level was not an appropriate place for committee decisions. Admiral Noel Gayler, who replaced McCain as CINCPAC on 6 October, stated that "the route package boundaries as outlined in the basic LINEBACKER/BLUE TREE operations order would continue to remain in effect. However . . . to improve efficient use of resources and to attain mass application of force where indicated, the geographical

---

* See previous reference on page 90 for Sharp's view of coordinating committee.

area which includes the NE/NW rail line and Hanoi environs will be designated an integrated strike zone. This is the most vital area in North Vietnam. To bring the necessary weight to bear, CINCPACAF and CINCPACFLT will schedule strike missions into one another's geographical area." [24] The war only lasted a couple of more months, and force employment didn't change enough to permit an evaluation of the proposed "integrated strike zone." Had the war continued beyond December, it's likely Gayler would have been compelled to discard the route package system because of the difficult command and control problems that developed in the final 11-day air offensive with the B–52s.

COMMAND OF BOMBER FORCE—CONTINUING PROBLEM

Command and control of the B–52s was a continuing problem throughout the war. When they were introduced into South Vietnam in 1965, the Air Force was most anxious that the bombers not come under the operational control of MACV or his Air Deputy. Air Force leaders believed that the air war would be fought outside South Vietnam; therefore, the bombers should be kept outside the command structure of MACV.

While operational control of the bombers was withheld from COMUS-MACV, it was also withheld from CINCPAC, the theater commander. The Air Force argument prevailed that because the B–52s also had a nuclear mission for general war, they should remain under the control of SAC, a specified command reporting to the Joint Chiefs of Staff. Of course CINCPAC also reported to the JCS, and any of SAC's forces under the operational control of CINCPAC could be withdrawn at any time. But the Air Force argued that in an emergency valuable time could be lost in debate over pulling the forces from CINCPAC's control. No commander voluntarily gives up forces, particularly in a war with the burden for success or failure on his shoulders, but any delay in SAC's regaining control of the forces from CINCPAC could be critical in a general nuclear war.

COMUSMACV seemed little concerned with having operational control of the bombers as long as he had a say in selecting the targets. Since the aircraft would be used in South Vietnam, at least initially, the Air Force had no objection to COMUSMACV as a sub-unified commander having the authority to nominate targets. He could not approve strikes, but he would be the principal source of target selection.

The 2nd Air Division/7th Air Force commander and all the ground force commanders nominated targets to MACV. MACV consolidated the list and established an order of priority before sending it to CINCPAC. CINCPAC reviewed the list but rarely made any significant changes before forwarding it to the Joint Chiefs of Staff. The JCS thoroughly reviewed the list before submitting it to the Secretary of Defense, who coordinated with the Secretary of State before submitting it to the President. This was a long and involved process, considering the nature

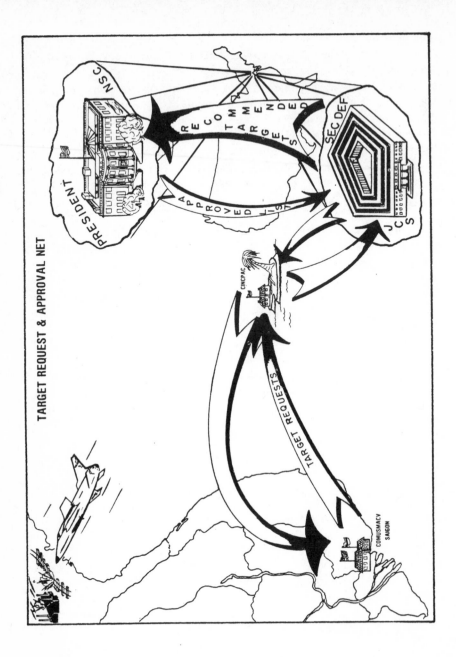

TARGET REQUEST & APPROVAL NET

NSC

PRESIDENT

SEC DEF

RECOMMENDED TARGET LIST

APPROVED LIST

JCS

CINCPAC

TARGET REQUESTS

COMUSMACV
SAIGON

of targets in South Vietnam. The coordination process was simply not effective for employing bombers against suspected enemy areas. The drawbacks were soon recognized, though, and by early 1966 the system changed to permit JCS approval of targets within South Vietnam. Soon B–52s were being used selectively against targets on the LOCs in Laos, with approval for those strikes coming from the Secretary of Defense after coordination with the President.

SAC established a liaison section in MACV headquarters to coordinate air strikes requested by MACV. This liaison section reported to the Deputy Chief of Staff for Operations at SAC headquarters, but dealt mostly with SAC's 8th Air Force headquarters in Guam. Eighth Air Force commanded all the B–52s, tankers, and strategic reconnaissance aircraft in Southeast Asia. As 7th Air Force commander, I had no control over the targeting, timing, or attack profile of the bombers. Basically, I provided advisory assistance through the tactical air control system, pre-strike and post-strike reconnaissance of the target area, fighter cover in the vicinity of the DMZ and in Laos, and ECM support with Wild Weasels and EB–66s. Seventh Air Force adjusted its operations so all these supporting arrangements were accommodated to the attack times MACV and the SAC liaison section agreed on. Most of the time coordination of our own strikes with the bomber strikes was not a problem since the B–52s bombed from over 30,000 feet while tactical air operations were usually under 10,000 feet.

When the B–52s were used against the LOCs in Laos during the early and middle months of 1966, 7th Air Force was the primary agency for selecting targets. These targets were blended into the interdiction campaign plan for a continuity in attacks from bombers and fighters. Since 7th Air Force's use of B–52s in this role competed with MACV's proposed use of the bombers for targets in South Vietnam, and since COMUSMACV was responsible for the LOCs in Laos from STEEL TIGER south, he made the final decision on whether interdiction strikes or strikes inside South Vietnam would have priority. At the weekly strategy meetings, the 7th Air Force commander proposed the number of sorties to be flown by the B–52s in the interdiction campaign. COMUS-MACV then decided how to apportion B–52 strikes between the out-country interdiction program and in-country attacks. Between these weekly meetings, of course. COMUSMACV occasionally diverted addi-tional strikes into South Vietnam for unanticipated high priority targets.

In July of 1966, I proposed to Westmoreland a change in the command arrangement for B–52s.[25] Experiences in North Africa, Europe, and Korea had underscored the importance of giving the tactical air com-mander control of the bombers committed to his mission. And it seemed logical that the B–52s should be placed under the operational control of the 7th Air Force commander who was responsible for the total air effort in South Vietnam and Laos and for the coordination of strikes in North Vietnam.

In September 1966, I again raised the issue with Westmoreland. I didn't propose to assume MACV's final authority for B–52 targeting, but I did propose that 7th Air Force compile the targets, plan the missions, and control their execution. If the nominated targets exceeded the allocated B–52 sorties, MACV would still establish priorities.

After some discussion and further clarification of the proposal, Westmoreland agreed that the B–52s would come under the control of 7th Air Force and that the SAC liaison section would be attached to 7th Air Force headquarters. It was understood MACV would continue to control the final target and priority list submitted to CINCPAC and the JCS. Since MACV had no responsibility for targets in North Vietnam other than those in Route Package I, 7th Air Force would continue to nominate targets in the other route packages to CINCPACAF who would forward them to CINCPAC.

After Westmoreland had agreed in principle to placing B–52s under control of 7th Air Force, he proposed to discuss the assignment of Route Package II with CINCPAC. In his opinion Route Package II, like Package I, was an extension of the battle in South Vietnam, and therefore his air deputy should be responsible for the control of air operations in this area. Sharp, however, did not react favorably to the proposal. He felt MACV had enough responsibilities in South Vietnam, Laos, and Route Package I. Further, TF–77 was covering the area with adequate strikes, and Sharp could see nothing to be gained by altering the assignment of route packages. PACAF supported Sharp's position, and the matter rested. However, the targeting problem was not resolved, and it became a critical issue with the resumption of bombing below the 20th parallel in May of 1972.

McConnell, then USAF Chief of Staff, agreed we needed a better organization to coordinate the B–52 effort, but he was not prepared to place the B–52s under the control of the MACV air component commander. He did agree, however, that a small SAC advanced headquarters responsible for the targeting, planning, and control of B–52 strikes could be attached to 7th Air Force. MACV was not the proper level for planning and conducting such operations, although COMUS-MACV should continue to establish target priorities for areas under his jurisdiction.

General Joseph J. Nazzaro, Commander of SAC, voicing the opinions of McConnell and Harris, proposed that the SAC liaison section located at MACV headquarters be merged with a new SAC ADVON, and that the ADVON be attached to the Air Deputy, MACV, in his role as the component commander. Again I stated this arrangement didn't solve the problem; the real question was operational control, and a SAC ADVON was not the solution. I further pointed out that the Air Deputy had no staff and possessed no operational functions except that of advising. Seventh Air Force was really the air component command, and the SAC ADVON should be assigned to it.

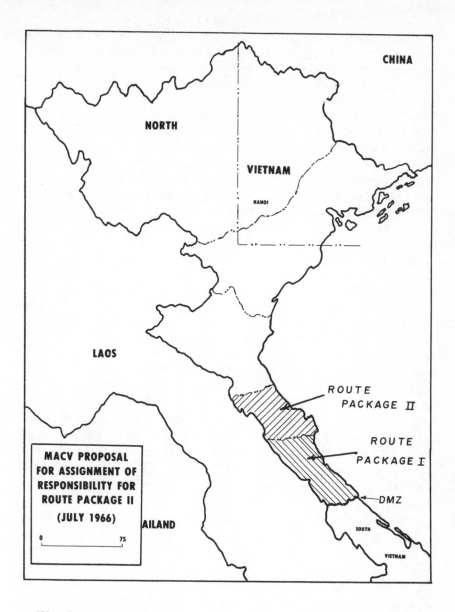

MACV PROPOSAL
FOR ASSIGNMENT OF
RESPONSIBILITY FOR
ROUTE PACKAGE II
(JULY 1966)

0 _____ 75

CHINA

NORTH

VIETNAM

HANOI

LAOS

ROUTE
PACKAGE II

ROUTE
PACKAGE I

DMZ

SOUTH

VIETNAM

AILAND

We airmen couldn't agree on the operational control of the B–52s. Although subsequent commanders of 7th Air Force raised the issue, particularly during the 11-day offensive in 1972, B–52s stayed under SAC's control for the remainder of the war. The SAC ADVON was attached to the Air Deputy MACV, on 10 January 1967, but it actually functioned as part of the 7th Air Force headquarters for the reasons I've discussed. SAC and the Air Staff believed MACV or the theater headquarters level was the proper place for policy and planning on B–52 operations—the same concept that 7th Fleet held on naval support, and

103

that the Marines held in coordinating their air support with 7th Air Force. In contrast, 7th Air Force commanders believed the Air Deputy position to be redundant and held that 7th Air Force was the real air component of MACV. Hence, the control of all airpower, including B–52 operations, should be vested in that organization.

In May 1972 when the President decided to resume bombing below the 20th parallel, the control of B–52s posed a complex problem because of the assignment of route packages among MACV, 7th Air Force, and TF–77.[26] Until this time, 7th Air Force had been largely controlling B–52s through the SAC ADVON even though no formal directive gave 7th Air Force this control. Realities of the situation—the increasing use of B–52s outside South Vietnam and the removal of U.S. ground troops—had left 7th Air Force the dominating headquarters and made it necessary for 7th Air Force to control the bomber strikes.

With divided responsibilities for targeting in the route packages, there was no single agency except PACOM with the authority to adjust priorities. For Route Packages II, III, and IV, TF–77 nominated B–52 targets through 7th Fleet and PACFLT to PACOM. In the case of Route Package I, 7th Air Force nominated targets through MACV to PACOM. In both cases, CINCPAC made the final determination of priorities before forwarding the list to the JCS. This long and involved process was too slow to meet the time requirements for targets released by the JCS.

During this time, COMUSMACV returned to the earlier view that the lower route packages were really an extension of the ground battle area since they contained the logistical base of enemy forces fighting in South Vietnam. Although MACV had been turned down on the assignment of Route Package II to its jurisdiction, it now requested that the Air Deputy as the representative of COMUSMACV control all B–52 operations below the 20th parallel.[27] This request, made in September 1972, was also disapproved.

In the meantime, CINCPAC requested the JCS grant him authority to approve B–52 targets below the 20th parallel. Obviously the JCS needed to delegate control of targeting to the command level responsible for the day-to-day fighting of the war. I believe this control should have gone to the 7th Air Force commander, since the Air Force was charged with the mission of conducting the air campaign and interdicting the ground battlefield. Nevertheless, the JCS approved CINCPAC's request on 4 October 1972, but only for 10 days. The ten days covered a period of intensive bombing against LOCs and other targets in Route Packages I, II, III and IV. When CINCPAC received approval authority, he delegated B–52 targeting to his two component commands as he had done with the JCS targets. He delegated responsibility in Route Packages II, III, and IV to CINCPACFLT, who in turn delegated responsibilities to the commander of TF–77. Thus TF–77 became the prime agency for selecting targets and coordinating strikes by the B–52s in these three route packages. Furthermore, since PACAF had no responsibility for Route Package I, MACV continued to target this area.

MACV took exception to the assignment of targeting responsibilities to TF–77 for reasons previously cited, recommending that all B–52 targeting be centralized under MACV as was done for strikes in South Vietnam.[28] The issue actually involved a great deal more than targeting since the B–52s flying over these areas required extensive fighter cover, ECM support, and reconnaissance forces from 7th Air Force. The real question was whether a single air commander should control the forces of TF–77 and SAC. Missions in the high threat areas required the carefully integrated planning characteristic of 7th Air Force strikes near Hanoi. However CINCPAC decided not to change the assignment of geographical areas, expressing the view that the Coordinating Committee, chaired by 7th Air Force, was the proper agency to coordinate the efforts of the B–52s, 7th Air Force, and TF–77. He also proposed that the Coordinating Committee include representatives from MACV and the SAC ADVON. With this enlarged membership, the Committee would have representatives from all air activities. SAC supported CINCPAC in this proposal.

The ROLLING THUNDER Coordinating Committee had not been an effective instrument for controlling air operations during the 1965–1968 bombing campaign. It was now being placed in the more difficult position of trying to resolve the conflicting demands of three different forces in a highly complex operation involving SAM suppression, extensive ECM activities, fighter screens, protection of reconnaissance platforms, and air refueling of a large portion of the force. The limited number of strikes in October indicated to 7th Air Force that a single air commander, not a committee, was the only real solution to the problem; the ROLLING THUNDER Coordinating Committee could not do the job. However, bombing above the 20th parallel was halted on 23 October, and no significant changes were made in command and control arrangements prior to the resumption of the Hanoi bombing on 18 December 1972.

With the breakdown in negotiations, the President decided to conduct a major air offensive against the greater Hanoi area. The planning for this campaign was tightly guarded, and few military people were even aware that such plans existed. Staffs that normally would have done the planning for such missions were not cleared. Security instructions were so severe, in fact, that it was difficult to assemble the minimum number of people to ensure all aspects of the mission were covered.[29]

The initial instructions from the JCS on the 15th of December called for a three-day campaign with planning for prolonged operations. The JCS approved 31 targets, most of them in Route Package VI A, for attack. The campaign would employ B–52 strikes throughout the night and tactical air strikes during the day, thus placing a heavy load on 7th Air Force to support around-the-clock operations.

Although CINCPAC was responsible for all air operations in North Vietnam, SAC had made plans for such a campaign in case it should be

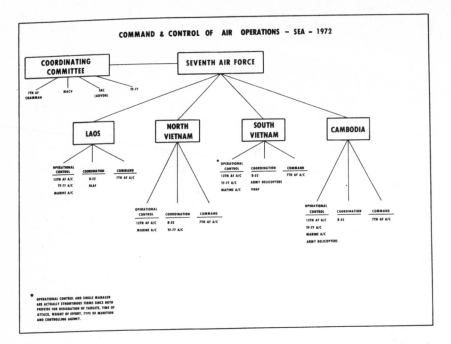

COMMAND & CONTROL OF AIR OPERATIONS – SEA – 1972

dii ected to execute an attack on short notice. As a specified command, it had direct communication with the JCS and used it to plan the targets for the B–52s. The JCS target list was disseminated to CINCPAC, SAC, MACV, PACAF, PACFLT, 7th Air Force, and TF–77. Based on the location of the targets, either 7th Air Force or TF–77 would be responsible for particular strikes as long as CINCPAC remained firm on the route package assignment. But which targets would the B–52s strike, and who would select those targets? If the route package structure prevailed, 7th Air Force would select the targets in VI A, and TF–77 in VI B. SAC, however, opposed target selection by these two commands. Complicating matters further, CINCPAC had indicated in an earlier message that SAC would determine the sorties and targets for the B–52s. Obviously an agreement on targets was needed to eliminate duplication of day and night strikes.

Although the problem had existed since divided responsibilities were established in 1965 for the air war, the need for a single air authority was never clearer. The arrangement for the first three days of the campaign gave SAC the dominant voice in the selection of targets for the B–52s. Once these targets had been informally coordinated with the JCS, General John C. Meyer, Commander of SAC, discussed support arrangements with Gayler and General John W. Vogt, Jr., Deputy Commander, MACV, and commander of 7th Air Force. Vogt, as Chairman of the Coordinating Group, was responsible for detailed coordination with representatives from SAC ADVON, TF–77, and MACV. Vogt was not satisfied with the arrangement because SAC presented the targets so late that there was inadequate time for the detailed planning of fighter cover, ECM, and Wild Weasel support. He felt SAC should notify him at least 18 hours before a planned mission. Further, Vogt felt his forces were being spread too thin by trying to support both day and night operations, and he understood that his priority was to support day laser strikes near Hanoi.[30]

After the first three days of the offensive, the initial coordination problems were resolved although target selection remained split among three levels of command. Gayler, on 21 December, modified his earlier view on the targeting of B–52s and now took the position that his headquarters and SAC headquarters would jointly determine the targets in accordance with guidance from the JCS. Once SAC and PACOM agreed on the projected targets, and after JCS approval, the Coordinating Committee at Saigon would work out the details of the mission and plan 7th Air Force and TF–77 support.

Gayler further revised the targeting procedure after the Christmas standdown. After reviewing the coordination problem and the many messages on details of conducting strikes in the Hanoi-Haiphong area, Gayler issued a new directive restating the authority of CINCPAC to conduct the air campaign in North Vietnam. He now took the position that all requests for target validation from SAC, 7th Air Force, and TF–77 should be sent to his headquarters. His headquarters would then approve or disapprove the targets, and those that were beyond his authority would be sent to the JCS for approval. Once the list was approved, all mission details would be worked out by the Coordinating Committee. Gayler didn't believe his headquarters was the proper level of command to work the daily mission which involved many operational details on specific target selection, ECM support, and suppression of North Vietnamese air defenses. These matters had to be dealt with by people closer to the combat theater.[31]

This then was the final arrangement for the control of forces during the 11-day offensive. It was a return to the original procedure established for 2nd Air Division and TF–77 in the fall of 1965. Although the arrangement worked and coordination was achieved, the fundamental issue of the unity of airpower was not clarified.

SUMMARY

Throughout the three wars, World War II, Korea, and Vietnam, the command and control of airpower has been a major issue. Airpower has great flexibility to perform many tasks in war, and its ability to respond with varying levels of firepower to a variety of targets has led Army and Navy commanders to seek control of airpower as part of their forces. But to give in to these understandable wishes of surface commanders is to destroy the very thing that gives airpower its strength—the ability to focus quickly upon whatever situation has the most potential for victory or for defeat. Airmen know the centralized control of airpower in a theater of war can best serve armies and navies; to fragment airpower is to court defeat. In North Africa, Europe, Korea, and Vietnam this principle has been proven time and again. As Air Marshal Tedder writes, "Air warfare cannot be separated into little packets; it knows no boundaries on land or sea other than those imposed by the radius of action of the aircraft; it is a unity and demands unity of command." [32]

CHAPTER III

FOOTNOTES

[1] George S. Eckhardt, Command and Control, 1950–1969, U.S. Army Vietnam Studies (Washington, D.C.: Government Printing Office, 1974), p. 7.

[2] William C. Westmoreland, A Soldier Reports (Garden City, N.Y.: Doubleday & Company, 1976), pp. 110–111.

[3] Official History of the 2nd ADVON, 15 Nov 1961–8 Oct 1962, Historical Division, 2nd Air Division, vol. 1, p. 17.

[4] Lyndon B. Johnson, The Vantage Point (New York: Popular Library, 1971), pp. 55–58.

[5] Eckhardt, Command and Control, p. 42.

[6] Working Paper for CORONA HARVEST Report, Command and Control of Southeast Asia Air Operations, 1 Jan 1965–31 Mar 1968, book 1 (Maxwell AFB, Al: Department of the Air Force, 1971), p. II–1–34.

[7] CORONA HARVEST Report, USAF Activities in Southeast Asia, 1954–1964, book 1, p. 4–32.

[8] Official History of the 2nd ADVON, 15 Nov 1961–8 Oct 1962, Historical Division, 2nd Air Division, vol. 1, p. 17.

[9] CORONA HARVEST Report, USAF Activities in Southeast Asia, 1954–1964, book 1, p. 4–44.

[10] Ibid. See also, Westmoreland, Soldier, p. 75.

[11] Eckhardt, Command and Control, p. 36.

[12] Westmoreland, Soldier, p. 41.

[13] Curtis E. LeMay, General, USAF (Ret.), interview held at Pentagon, Wash. D.C., February 1976. See also, CORONA HARVEST Report, USAF Activities in Southeast Asia, 1954–1964, book 1, p. 4–44.

403-892 O – 83 – 9

[14] Westmoreland, Reports, pp. 105, 109–112.

[15] John P. McConnell, General, USAF (Ret.), interview held at Pentagon, Wash. D.C., March 1976.

[16] CORONA HARVEST Report, Command and Control, book 1, p. II–1–34.

[17] MACV Directive 95–4, U.S. Air Operations in RVN, 28 June 1966, para 3.e.

[18] McConnell, interview, Mar 1976. See also, CORONA HARVEST Report, Command and Control, book 1. p. II–1–34.

[19] Working Paper for CORONA HARVEST Report, Out-Country Air Operations, Southeast Asia, 1 Jan 1965–31 Mar 1968, book 1, p. 50.

[20] Ibid., p. 53.

[21] The Lord Tedder, Air Power in War, The Lees Knowles Lectures for 1947 (London: Hodder and Stroughton, n.d.), p. 91.

[22] CORONA HARVEST Report, Out-Country Air Operations, book 1, p. 156.

[23] CORONA HARVEST Report, USAF Air Operations in Southeast Asia, 1 Jul 1972–15 Aug 1973, vol. 2 (Hickam AFB, Hawaii: Department of the Air Force, HQ Pacific Air Force, 1973), p. IV–38.

[24] Ibid., p. IV–34.

[25] CORONA HARVEST Report, Command and Control, book 5, p. VIII–6–3.

[26] CORONA HARVEST Report, USAF Air Operations, SEA, vol. 2, p. IV–24.

[27] Ibid., p. IV–25. Also, John W. Vogt, Jr., General, USAF, (Ret.), interview.

[28] Ibid., p. IV–27.

[29] John W. Vogt, Jr., and Glen W. Martin, General, USAF, (Ret.), interview.

[30] Ibid.

[31] CORONA HARVEST Report, USAF Air Operations, SEA, vol. 2, p. IV–320.

[32] Tedder, Air Power, p. 91.

CHAPTER IV

THE COUNTER AIR BATTLE
(AIR SUPERIORITY)

The first task of air power is to gain and maintain air superiority. Air superiority is essential to sustained air, ground, and sea operations. As defined in JCS Publication 1, it is "that degree of dominance in the air battle of one force over another which permits the conduct of operations by the former and its related land, sea and air forces at a given time and place without prohibitive interference by the opposing force."[1]. Air Chief Marshal Tedder, General Eisenhower's deputy, said after the war, "We were to find out in the hard school of war that without air supremacy, or as we now say, 'air superiority,' sea power could no longer be exercised; and without air superiority, air power itself could not be exercised. . . . But the outstanding lesson of the late war was that air superiority is the prerequisite to all war winning operations, whether at sea, on land, or in the air."[2] For a very short operation, an air commander may be willing to accept relatively high losses by conducting other missions before having attained at least local air superiority. However, for sustained operations air superiority is essential.

AIR SUPERIORITY, WORLD WAR II—8TH & 9TH AIR FORCES

To achieve air superiority in the combat theaters of World War II, we brought all elements of the enemy air forces under attack. A portion of our tactical air force was devoted to offensive air sweeps; another part was devoted to the destruction of airfields; and another part was continually devoted to the destruction of anti-aircraft installations. By attacking all elements of the enemy air force, we achieved a position of superiority which allowed us to turn the full strength of the Air Force to helping the ground forces. If too much of our air effort had been diverted to the ground battle before control of the air was established, our losses, first in the air and then on the ground, would have increased sharply.

In North Africa we destroyed the German fighter force on the ground and in the air. Our attacks had to be repeated continually since new fighters could be brought into the theater and airfields could be repaired in a matter of twelve to eighteen hours. Thus, the first task of our tactical air force was to go after the entire structure of the German Air Force in North Africa and keep it under attack.

In North Africa, as in other theaters of World War II, there were no limitations on attacks against the enemy air forces. No matter where the fighters were, no matter where the airfields were, and no matter where the anti-aircraft guns were, all were subject to attack; and the enemy knew it. There was no real escape from our airpower. By pulling back from combat they could reduce losses temporarily, but when that happened, we gained local air superiority without a fight.

Prior to the Normandy invasion, General Eisenhower realized that air superiority would have to be obtained if the enormous amphibious operation were to succeed.[3] Air Chief Marshal Tedder says that Eisenhower "might have added that it was not simply a question of air superiority but of air power, which can be exercised in its full form only after air superiority has been gained."[4] Because of his great concern for the invasion, Eisenhower wanted the full weight of the U.S. Strategic Air Force and RAF Bomber Command devoted to neutralizing the German airfields in France and the transportation system that would be used by the Panzer divisions in their attempt to push the Allied forces back into the Channel.

General Carl Spaatz, Commander of the U.S. Strategic Air Force, recommended against too early diversion of the bombers to these targets. He pointed out that the best way to ensure air superiority for the invasion was to continue the attacks against synthetic fuel plants as long as possible, drawing the German fighters up to defend these plants. His argument was founded on the knowledge that the *Luftwaffe* was badly short of fuel to fly the fighters that were still operational and to train replacement pilots.* Thus, Spaatz reasoned, the fighters wouldn't come up to defend the airfields and rail system in France since they were not considered vital. On the other hand, the GAF would have to come up in force to protect the synthetic oil plants since the loss of these plants would ground them; and with this loss, the war, for all practical purposes, would be over. Spaatz proposed that our bombers be kept on the synthetic oil and aircraft production facilities and the airfields deep in the German homeland until a few weeks before the invasion. Then the total effort of 8th Air Force and RAF Bomber Command would be shifted to the airfields and lines of communication within France.

---

*It was not learned until after the war that the average German fighter pilot was receiving only 35 hours of specialized training as compared to at least 85 hours in 1940.[5] This lack of training was a direct result of the shortage of fuel.

The combined efforts of the Allied air forces during the 90 days prior to the invasion were concentrated on targets in France. Destruction of the GAF was the primary objective. Our fighters struck airfields that might be used by the Germans as advanced staging bases for attacks on the Allied landing force. Medium and heavy bombers were scheduled night and day against the railroad bridges, marshalling yards, and staging areas that could be used by the German armored forces to mass for attack against our beachhead. We estimated that this all-out effort would weaken the GAF to the point that it could mount only 700 sorties a day against the beachhead. On the day of the invasion, however, the GAF flew only about 200 sorties, and the effect of these sorties was negligible. Of the 160 German fighters in commission in France, only 60 were able to take to the air. Allied planes, on the other hand, flew some 14,000 sorties on the first day of the invasion. Not a single aircraft was lost to German fighters. As Lord Tedder writes, "despite operation 'Pointblank,' German production of fighter aircraft rose steadily, indeed swiftly, in 1944. After the war Speer (Albert Speer, German Minister of Munitions) was asked to explain how it was that the *Luftwaffe* nevertheless grew weaker. He replied: "The answer to that was simple—the Allies destroyed the aircraft as soon as they were made."[6] The total neutralization of the German Air Force before the Normandy landing on 6 June 1944 was a superb performance, a classical case of gaining and maintaining air superiority.

During the last few months of the European war, the Germans introduced the ME–163, a single engine rocket aircraft, and the ME–262, a twin engine jet. It is interesting to speculate on what might have been the outcome of the combined bomber offensive if the Germans had converted more of their production to these jet fighters and had devoted less effort to the V–1 and V–2 missiles. Certainly our P–51s and P–47s would have had great difficulties combating large formations of these fighters, for the ME–262's top speed was at least 100 knots faster than that of the P–51. Also, with their better performance at high altitudes around 25,000 feet, these fighters would have been able to engage and disengage the B–17s and B–24s with greater ease. From the point of view of the *Luftwaffe,* however, ME–262 production was a perfect example of a remedy that was too little, too late. By the time ME–262s took to the air, the *Luftwaffe* was overwhelmed. Consequently, 8th Air Force was able to claim 146 ME–262s destroyed against the loss of 10 fighters and 52 bombers during the last ten months of the war.[7]

AIR SUPERIORITY—KOREA—5TH AIR FORCE

Seven years later in another part of the world, the need for air superiority was again apparent. During the Korean War, the mission of 5th Air Force was to gain and maintain air superiority, interdict the battlefield, and provide close air support to 8th Army. Just as achieving air superiority was the first concern in World War II, it also became the top priority mission in the Korean War. In fact, air superiority was

Sequence shows MIG–15 pilot leaving his aircraft after it has been disabled by an F–86.

perhaps even more important in Korea because of the superiority in numbers of the Chinese ground forces over the ground forces of the United Nations Command. [8] The ground war could have been a disaster if the Allied air forces had not been able to control the air. This control permitted better than 39% of the daily sorties to be turned to interdicting the battlefield, preventing the Chinese from being able to mount a sustained offensive.

Our 5th Air Force contained the North Korean Air Force (NKAF). Of course the NKAF was not all Korean, but basically Chinese with Russian and Polish pilots as well. Further, there is substantial reason to believe that most of the fighter squadrons actively engaging the F–86's were Soviet squadrons being rotated through the front at about six-week intervals. [9]

General Otto Weyland, Commander of FEAF, stated that the first

Close-up of F–86 flight patrolling "MIG Alley" in northwest Korea.

priority of his air force was to keep the air force in North Korea neutralized so that the NKAF could not attack Allied ground forces.[10] There were 75 airfields in North Korea that could have supported MIG–15s. During the course of the war, these airfields were suppressed by the combined efforts of 5th Air Force and FEAF Bomber Command.

The enemy's sanctuary in China greatly compounded our problems in maintaining air superiority, of course, for we could neither destroy the MIGs on the ground at their Chinese bases nor follow them into Chinese airspace to destroy them in the air. We dealt with the situation primarily through the use of fighter sweeps and screens. Fighter sweeps were commonly used, as they had been in World War II, to entice the enemy to come up for battle. These sweeps were made in areas such as "MIG Alley" where the probability of engagement was high. The frequency and size of the sweeps depended on the availability of our fighters, the probability of enemy reaction, and the supporting effect such flights would have on other operations. Other F–86 patrols along the Yalu screened the fighter-bombers conducting attacks against the rail network and other targets associated with the enemy's logistical system. The F–86s, by interposing themselves between the fighter-bombers and the MIG–15s based in the Antung area, allowed the F–84s, F–80s, and F–51s to carry out their missions with almost complete security.

Engagements between the MIGs and our F–86s were frequent: In December 1952, over 3,997 MIGs were observed, 1,849 engaged, and 27 destroyed.* But though the MIG–15s were able to penetrate the F–86

---

*The overall exchange rate of enemy to friendly losses for the Korean War was 10 to 1.[11] Almost all of these kills were without benefit of on-board radar even though the F–86 had a forward ranging radar. Most of the pilots used a fixed sight setting and got their kills from the 6 o'clock position with the six fifty caliber machine guns. Most of the fighter kills in World War II were also from the 6 o'clock position.

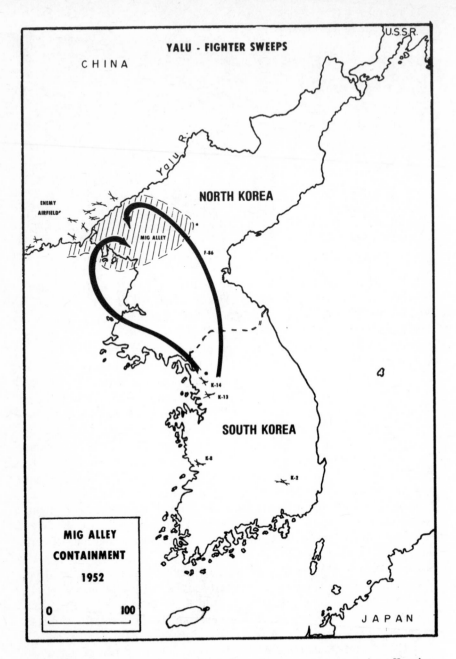

YALU - FIGHTER SWEEPS

CHINA

U.S.S.R.

Yalu R.

ENEMY
AIRFIELD⁼

NORTH KOREA

MIG ALLEY

F-86

K-14
K-13

SOUTH KOREA

K-8

K-2

MIG ALLEY
CONTAINMENT
1952

0   100

JAPAN

screen now and then, it was only during the second communist offensive
in the spring of 1952 that they posed a significant threat to the fighter-
bombers. Even then, only a few MIG–15s were successful in attacking
the F–84s and F–80s.

As a result of the F–86 patrols and the attacks on airfields in North
Korea, then, the NKAF was not able to mount any significant air attacks

against the fighter-bombers, nor were there any significant attacks against United Nations' ground forces. Additionally, no airfields of 5th Air Force were struck except by a small biplane nicknamed "Bedcheck Charlie." Attesting to the effectiveness of the screening-out of the NKAF, Lt General Nam Il, chief North Korean delegate to the negotiations at Kaesong, stated in August of 1951, "I would like to tell you frankly that in fact without direct support of your tactical aerial bombing alone your ground forces would have been completely unable to hold their present positions. It is owing to your strategic air effort of indiscriminate bombing of our area, rather than to your tactical air effort of direct support to the front line, that your ground forces are able to maintain barely and temporarily their present positions."[12] Although we might point out that strategic bombing of railroad lines, bridges, marshalling yards, and power plants is hardly "indiscriminate," the critical importance of our air superiority was obviously a point on which both sides at the negotiating table could agree.

At the conclusion of the Korean War, the missions that were maintaining our air superiority were those that had succeeded in World War II. In both wars it was necessary to hit the enemy air force on the ground and in the air. Our attacks had to be kept up day-in and day-out so that the enemy air force never had a chance to recover. The success of these missions dedicated to air superiority gave us the freedom to employ our airpower in the other missions needed to bring hostilities to an end.

AIR SUPERIORITY—NORTH VIETNAM—A NEW DIMENSION

In the air campaign against North Vietnam, air superiority permitted us to conduct the interdiction campaign, provide close support to ground forces in South Vietnam and Laos, and protect the vital logistical and ·

"Gaggle" of F–86s approaches "MIG Alley" in search of MIG–15s.

population centers in South Vietnam.* Operations in Southeast Asia demonstrated again that air superiority is not a condition that can be achieved once and for all. It must be won continually as long as the enemy has any aircraft, missiles, or guns left. His entire air defense system must be attacked repeatedly if we are to use the enemy's airspace freely.

THE NORTH VIETNAMESE AIR DEFENSE SYSTEM

The air defense system** in North Vietnam was a thoroughly integrated combination of radars, AAA, SAMs, and MIGs. It was Soviet in design and operation. During the early days of 1965, it was in an embryonic state and could have been destroyed with no significant losses to our force. However, the policy of the U.S. at the time was not to destroy this system because such an action might be considered an escalation of the conflict. General Westmoreland recorded his response to this situation in his book A Soldier Reports: "Some of McNaughton's [John T. Mc-Naughton, Assistant Secretary of Defense for International Security Affairs, 1964–1967] views, in particular, were incredible. On a visit to Saigon at a time when my air commander, Joe Moore, and I were trying to get authority to bomb SAM–2 (a Soviet-made missile) sites under construction in North Vietnam, McNaughton ridiculed the need. 'You don't think the North Vietnamese are going to use them!' he scoffed to General Moore. 'Putting them in is just a political ploy by the Russians to appease Hanoi.' " Westmoreland continues, "It was all a matter of signals, said the clever civilian theorists in Washington. We won't bomb the SAM sites, which signals the North Vietnamese not to use them. Had it not been so serious, it would have been amusing." [13] Because of our restraint, the system was able to expand without any significant interference until the spring of 1966, at which time systematic attacks were permitted against elements of the system. We were never allowed to attack the entire system.

NORTH VIETNAMESE RADAR

The Soviet air defense system relies upon many more radars than does the U.S. system. There were about 200 radars in the North Vietnamese air defense system with three major ground control intercept (CGI) sites: Bac Mai, Phuc Yen and Kep. Bac Mai and Phuc Yen normally controlled most of the air defense missions, although for missions staged into the southern portion of North Vietnam, a subordinate control unit was

---

*The concentration of logistics in South Vietnam at Cam Ranh Bay, Danang, Qui Nhon and Saigon made these installations exceedingly vulnerable to air attack. Consequently, preventing the North Vietnamese from bringing these areas under air attack was one of the especially important missions of 7th Air Force.
**air defense system: A combination of fighters, surface-to-air missiles, anti-aircraft artillery, radars, and a command and control organization thoroughly integrated to destroy aircraft and other vehicles penetrating a given air space.

established at Vinh. This site was not capable of handling a large number of aircraft.

The NVN radar system was effective in detecting our flights, in vectoring MIGs for attack, and in coordinating SAM and AAA engagements. With so many radars located in such a small area, it was impossible for us to jam all of them at once. In fact, we didn't try; we jammed only those specific GCI radars at any particular time that were vectoring interceptors toward our aircraft. Their radar coverage included so many redundancies that it was almost always sufficient to provide good GCI control during an engagement regardless of our countermeasures.

After 1965, the MIGs were under GCI control from takeoff until landing. And the control was excruciatingly positive throughout the mission. Controllers vectored the MIGs into position for attack with surprising detailed instructions, even to the point of telling an individual pilot when to arm his weapon and when he was "cleared for attack." If the situation didn't look favorable, the controller would direct the pilot out of the area of the potential engagement. North Vietnamese radar control was so thorough and so detailed that MIG pilots had very few opportunities to exercise their own initiative in deciding whether to engage or not.

ANTI-AIRCRAFT ARTILLERY (AAA)

By the summer of 1966, the AAA defenses had become formidable. We estimated at that time that there were some 7,000 guns of all caliber in North Vietnam[14] About 3,500 of these were located in Route Packages V, VI A, and VI B. The largest concentration was around Hanoi and Haiphong. During the early part of 1966, improvements were made in the North Vietnamese control system, resulting in better integration of 57mm and 85mm gun defenses with SAMs and fighters.

The AAA defenses within 30 miles of Hanoi and 10 miles of Haiphong were comparable to those found in World War II around key industrial areas and in the Korean War around the airfields along the Yalu and near Pyongyang. Many experienced pilots said the Hanoi flak was the heaviest in the history of aerial warfare, and it may well have been. In fact, though, our heaviest losses in the Hanoi area rarely exceeded four percent with an overall loss rate for attack sorties from 1965–1968 of 4.10.[15] By comparison, bomber losses to anti-aircraft fire and fighters while we were attacking targets in the Ruhr valley in August of 1943 were near ten percent.[16] Our reduced loss rate notwithstanding, the point defenses around the Doumer bridge, Hanoi Railcar Repair Shop, Thai Nguyen steel mill, Viet Tri thermal power plant, and many others were as tough as one could possibly imagine.

There were pockets of AAA around Thanh Hoa and Vinh that resembled those around Hanoi and Haiphong. However, since Thanh Hoa and Vinh are

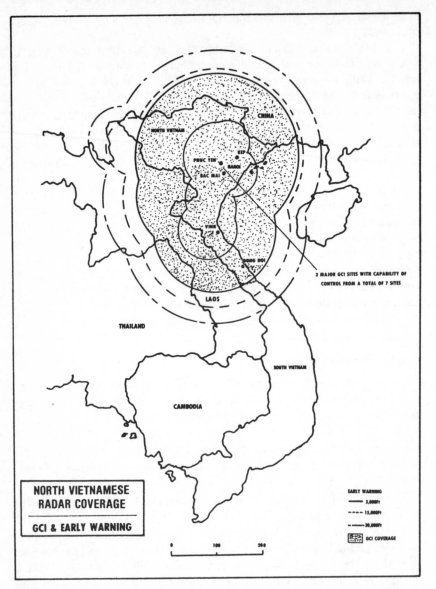

2 MAJOR GCI SITES WITH CAPABILITY OF
CONTROL FROM A TOTAL OF 7 SITES

NORTH VIETNAMESE
RADAR COVERAGE

GCI & EARLY WARNING

EARLY WARNING
——— 3,000Ft
– – – 15,000Ft
– ·· –– 20,000Ft
GCI COVERAGE

0   100   200

located along the coast, strike aircraft had no prolonged gauntlet of fire to run when approaching and departing these targets. Most of the targets in these areas were struck by the Navy; to get to them, carrier pilots had to penetrate an average of less than fifteen to twenty-five miles of defended airspace. On the other hand, the targets in the Hanoi delta assigned to the Air Force required the penetration of over 100 miles of defended airspace.

Although total time of exposure to AAA was important, the few seconds spent directly over the target were the time when the 37mm and

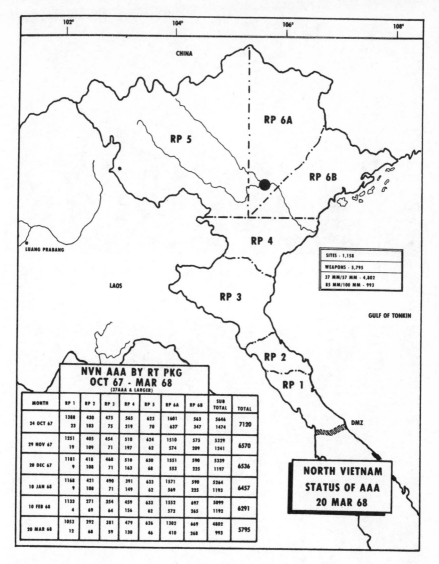

NVN AAA BY RT PKG OCT 67 - MAR 68
(37AAA & LARGER)

MONTH	RP 1	RP 2	RP 3	RP 4	RP 5	RP 6A	RP 6B	SUB TOTAL	TOTAL
24 OCT 67	1388 / 23	430 / 103	475 / 75	565 / 219	623 / 70	1601 / 637	563 / 347	5646 / 1474	7120
29 NOV 67	1251 / 19	405 / 109	454 / 71	510 / 197	624 / 62	1510 / 574	575 / 209	5329 / 1241	6570
20 DEC 67	1181 / 9	418 / 108	468 / 71	510 / 163	630 / 68	1551 / 553	590 / 225	5339 / 1197	6536
10 JAN 68	1168 / 9	421 / 108	490 / 71	391 / 149	633 / 62	1571 / 569	590 / 225	5264 / 1193	6457
10 FEB 68	1133 / 4	271 / 69	354 / 64	459 / 156	633 / 62	1552 / 572	697 / 265	5099 / 1192	6291
20 MAR 68	1053 / 12	292 / 68	381 / 59	479 / 130	626 / 46	1302 / 410	669 / 268	4802 / 993	5795

SITES - 1,158
WEAPONS - 5,795
37 MM/57 MM - 4,802
85 MM/100 MM - 993

NORTH VIETNAM STATUS OF AAA 20 MAR 68

57mm guns took their heaviest toll. Our strike forces delivering conventional ordnance (in North Vietnam we didn't use "smart" laser-guided bombs capable of being delivered accurately from a much higher altitude until May 1972) had to start their dive toward the target at about 12,000 feet and pull out above 4,500 feet. During the moments of stable flight between roll-in and pull-out, our aircraft were the most predictable and therefore the most vulnerable. And it was during these moments that the enemy would open fire with everything they had. Because of the value of the targets, but also because of this vulnerability of our aircraft during attack, most targets of any significance were heavily defended with interlocking AAA.

North Vietnamese 57mm anti-aircraft battery firing at USAF jet. This photograph was taken in 1965.

More sophisticated gun emplacements. This photo was taken in 1967.

AAA defenses didn't change much after 1968. Our crews in the 1972 air offensive faced about the same quality of AAA, although they did encounter more radar-directed fire from 85mm and 100mm guns. "Smart bombs" permitted us to strike from higher altitude in 1972, and our higher altitude of operation brought more of these bigger guns into action.

Since there were no SAMs in Korea, it isn't surprising that AAA accounted for most of our aircraft losses there. But in Vietnam, too, about 68% of our losses were to anti-aircraft fire. AAA did the major damage to our strike forces and caused the most problems in selecting routes of penetration and egress. These facts seem highly improbable today as I look back over contemporary accounts of the war, for most of the reporters who discussed air operations in North Vietnam were understandably less interested in AAA than in the newer threat, SAMs.

SAMs

The first loss of an aircraft (an Air Force F–4C) to a SAM occurred on 24 July 1965.[17] After that date, SAM defenses expanded rapidly. During 1966 and 1967, these defenses continued to grow in response to our increasing air effort. In all of 1965, the enemy fired 180 SAMs and destroyed 11 aircraft.[18] This compares with the 11-day offensive in 1972 when more than 1,000 SAMs were fired with the resulting loss of 15 B–52s and three other aircraft.

The SAMs were initially deployed in a 30-to 40-mile circle centered on Hanoi. Extensions of the initial SAM network stretched along the northeast railroad and to some extent the northwest railroad. A Chinese AAA regiment also covered most of the northwest rail line with particularly heavy defenses around Yen Bai airfield, but there is no evidence that they were involved in the SAM system. Apparently the SAMs at all locations were operated almost entirely by the Russians and North Vietnamese, with the Russians acting as technical advisors most of the time.

The number of SAM sites remained fairly constant after 1967; there were about 200. But the equipment could be moved from one location to another to reduce vulnerability to air attack. Thus despite our intense reconnaissance activity, it was practically impossible to determine precisely where the SAMs would be in advance of any given mission. Our response, as I shall discuss later, was to send special flights ahead of strike forces to cause the SAM radars to come on the air in preparation for a launch. When the SAM radars came on, the strike flights, which were about five minutes behind, could determine the location of the SAMs and take evasive action.

No mobile SAMs were deployed throughout the war, the mobile SAMs being the SA–4s and 6s which are mounted on tracked vehicles, but SAMs were deployed on transporters into Laos and Route Package I. Usually, these deployments included only one or two launchers. The missile would be fired from a concealed position, and then its launcher

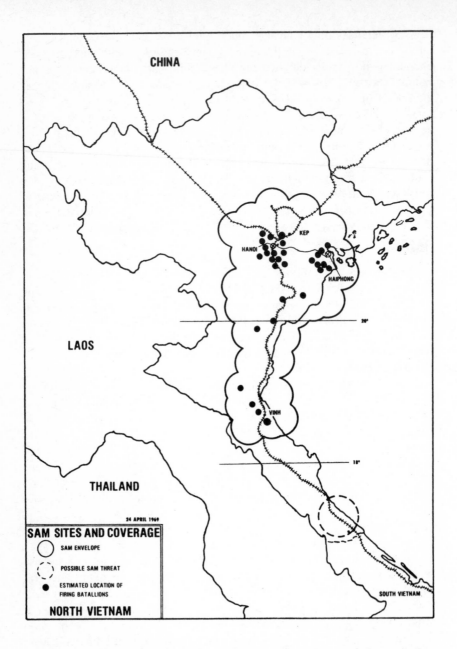

CHINA

KEP

HANOI

HAIPHONG

LAOS

VINH

20°

18°

THAILAND

24 APRIL 1969

SAM SITES AND COVERAGE

◯ SAM ENVELOPE

◌ POSSIBLE SAM THREAT

● ESTIMATED LOCATION OF FIRING BATALLIONS

SOUTH VIETNAM

NORTH VIETNAM

would be shifted immediately to a new location. Most of these deployments were made during 1967 and 1968 when the North Vietnamese tried repeatedly to shoot down a B–52. Fortunately, not a single B–52 was lost during this period to SAMs.

Near Hanoi, the enemy had about twenty to thirty active SAM battalions, each having four to six launchers. This represented a high of about 180 launchers during 1967 and an estimated 200 or more during the

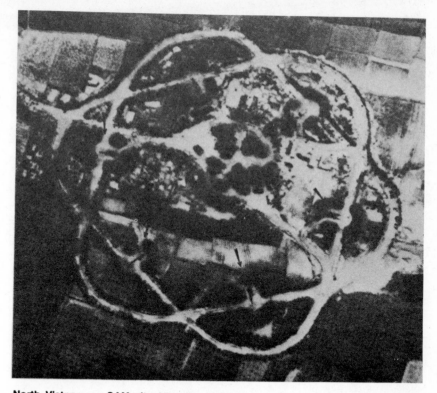

North Vietnamese SAM site 25 miles northwest of Hanoi. Note five of six SA–2s poised on their launchers.

1972 offensive. The 30 battalions maintained about 100 ready missiles-on-launchers at any given time. The total inventory of missiles in North Vietnam was about four to five hundred, with about two hundred missiles at launch sites and another two to three hundred in the supply system. This rather limited inventory was inadequate at times, as we could discern from the sharp decrease in the firing rate after three or four days of intensive operations. The inventory problem was especially evident in December 1967 when an unexpected period of good weather allowed us to fly an all-out effort, and in the 11-day offensive in December of 1972.

COUNTERMEASURES

During World War II, bombers used Electronic Countermeasures (ECM) and chaff extensively in jamming German GCI and fire control radars. Most of the effort, however, was devoted to jamming the AAA radars. The 88mm AAA was the most potent weapon of the time, and by jamming its tracking and acquisition radar we could force the gunner to depend on a visual sighting device which was not very effective for the altitudes (25,000 to 29,000 feet) at which the B–17s and B–24s were flying.

125

Early North Vietnamese SAM site with its distinctive road network.

Analysts estimated that our attrition rate was reduced by at least 25% by these countermeasures.[19]

But the anti-aircraft threat in World War II and Korea was not considered severe enough to warrant ECM equipment for fighters. In addition, most fighter pilots believed that their aircraft had sufficient maneuverability to avoid the most menacing concentrations of anti-aircraft fire. They preferred not to trade performance for ECM equipment. For these reasons there was little advancement in fighter ECM technology for many years.

It was not until the fall of 1961 that a research and development program was undertaken to obtain ECM for fighters. SAC had extensive ECM equipment, but even though SAMs had been in the Soviet air defense system for a number of years, we fighter people were slow to accept the fact that it would take more than maneuverability and speed to defeat a SAM defense system. Initially, the decision was made to put ECM equipment in a pod that could be carried on an external station; we would then be able to choose whether to carry the pod or not depending upon the threat.

Development of the first operational pods was slow, and when we finally received them, we encountered more difficulties. Pods were introduced into the war in 1966, but they worked so intermittently that operational commanders had little confidence in them. As the SAM threat continued to increase, a new sense of urgency was created to get the pods corrected. In November 1966, pods were reintroduced into combat and were soon recognized as the most important new development in enhancing the fighting potential of 7th Air Force.[20] Although the SAM threat remained serious throughout the war, the pods gave us our first effective means of managing the threat.

We began experimenting at once to find the flight formation that would give us the best ECM protection against SAMs and would also give us the best placement of aircraft for countering or launching an attack against MIGs. The formation for best ECM protection was too tight to handle a MIG attack. Yet, if the wingmen, elements, and flights were spread too far, there was no mutual ECM protection among aircraft. We discovered that a flight of four was the smallest group that could be adequately protected by pods. Breaking a flight down into individual elements of two wouldn't give the needed protection, but with four aircraft appropriately spaced, the protection among aircraft was satisfactory. Based on this knowledge, we developed a strike force of sixteen aircraft. This arrangement provided maximum protection among flights, and if a flight became separated, its members could still count on adequate coverage so long as they maintained a specified spacing between aircraft.

The evidence is clear that ECM pods had a profound effect on our vulnerability to SAMs. Many SAMs "went ballistic" (lost all their guidance) because of the jamming and missed their targets. Invariably, though, if a formation of four aircraft broke up while in the SAM belt, missiles became extremely accurate, and as a rule losses went up.

Once a SAM had been fired, a flight leader had to rely on his visual sighting of the missile to decide whether to retain the ECM protection by keeping his flight together or to break up the flight and take individual evasive action. This was one of the roughest decisions a flight leader had to make. There usually wasn't a second chance, so his decision had to be right. Even if he decided to take evasive action, any delay in executing the maneuver could be fatal. Most flight leaders retained the pod formation until it appeared that there was no alternative to a last-minute break.

The best way to escape a SAM was to turn into it with a hard diving turn, then make an abrupt four-G rolling pull-up keeping the speed up throughout the maneuver. If this maneuver was executed at the proper time, the SAM would not be able to follow. When a pilot could see a SAM coming at him, he could outmaneuver it in either the F–4 or the F–105, but since so much altitude was needed for these maneuvers, weather conditions became a critical consideration in our planning for all missions into SAM-defended areas.

Radar Homing and Warning (RHAW) equipment, like fighter ECM, had a profound effect on both the counter air and the interdiction missions. RHAW provided the pilot an indication that a SAM radar was activated and that a missile launch could be expected shortly. From the indication of an electronic strobe, the pilot could tell the direction from

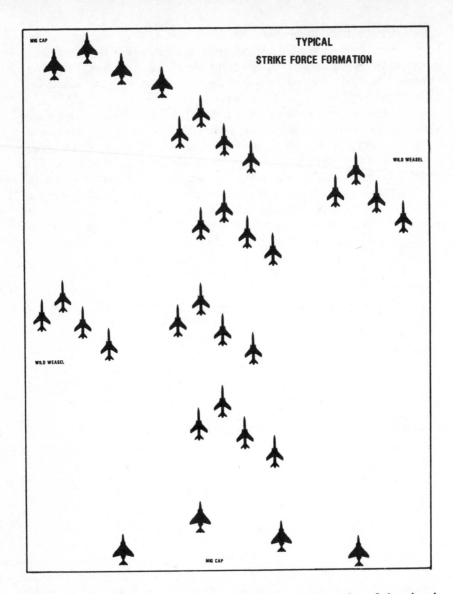

TYPICAL
STRIKE FORCE FORMATION

which the SAM would be launched, and from the intensity of the signal he could determine whether the SAM would be within range of his aircraft. Also, by a series of lights the pilot could tell when the missile was in preparation for launch and when it had been launched.

With the RHAW, our fighters had much greater freedom of action; they could roam over areas that they would have had to avoid if there had been no RHAW. But even with this equipment, a pilot near Hanoi would often have difficulty in determining which SAM site was the immediate threat to him because the SAM sites there were so numerous. Many pilots said they sometimes had so many lights illuminated on the panel that it looked like a Christmas tree. Although it was obviously unsettling at times for the pilots to see so many lights, I felt that the RHAW greatly decreased our pilots' vulnerability to SAMs, and I permitted no aircraft to be sent on a mission without an operational RHAW set.

Another defensive measure used quite effectively toward the end of the war was chaff. Chaff played a major role in the 1972 offensive, but it was used very little in the 1966–1968 campaign because we lacked a suitable dispenser. Some F–4s and RF–4s dispensed chaff from the speed brake well, and chaff cartridges were ejected from the RF–4 in the earlier campaign; however, these techniques failed to significantly affect the enemy's fire control and acquisition radars.

With the 1972 LINEBACKER I and II campaigns, chaff became a major device for reducing the effectiveness of the Fan Song acquisition and tracking radars. Separate F–4 flights were given the primary mission of dispensing the chaff, and they tried a number of different tactics before devising the most effective method for "laying it down." The first plan, used during LINEBACKER I, called for the F–4s to establish a "chaff corridor." Disadvantages of this plan were that the corridor tended to break down rather rapidly, and that the striking aircraft often strayed outside its protection. Either circumstance would allow the planes to be spotlighted by radar, and a firing could be expected shortly thereafter.

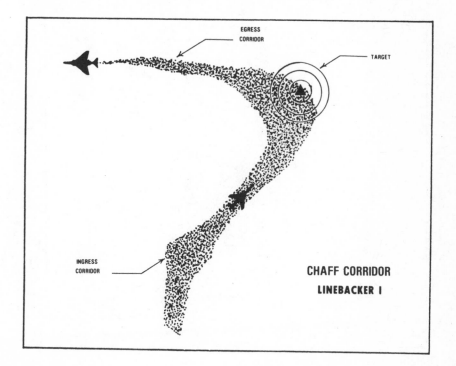

EGRESS CORRIDOR

TARGET

INGRESS CORRIDOR

CHAFF CORRIDOR
LINEBACKER I

Because of the difficulties with the chaff corridor, that plan was abandoned in favor of the more intricate but more effective "single cloud." During the LINEBACKER II operations, chaff clouds were dispensed between 25,000 and 35,000 feet to protect the laser strike forces and the B–52s. Especially important with this technique was the timing of the chaff release for maximum target coverage while the strike forces were in the target area. Wind was another critical factor since it could rapidly dissipate or create holes within the chaff cloud. Neither method was perfect, of course, but the chaff cloud was preferred during LINEBACKER II because it provided the best protection for a striking aircraft in the final seconds of weapons delivery. Effective as the later chaff operations were, though, they didn't reduce the need for organic ECM.

WILD WEASEL—IRON HAND

Another device that was used to counter the SAMs was known affectionately as the "Wild Weasel." The first Wild Weasel was used against a SAM site in North Vietnam in 1965. To make the Weasel, we had modified the back seat of an F–100F to provide an operator with special equipment for determining the location of an active SAM site.[21] With this equipment, the back-seat operator would give directions to the

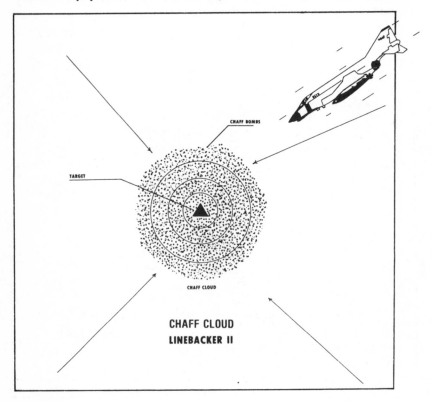

CHAFF BOMBS

TARGET

CHAFF CLOUD

**CHAFF CLOUD
LINEBACKER II**

front-seat pilot to position the aircraft for launching an air-to-ground missile that would home on the beams emitted by the SAM radar.

Unfortunately, the F–100 was too vulnerable for operations in high threat areas, so a replacement was urgently needed. The two-placed version of the F–105 strike fighter was selected and designated the F–105G. These aircraft were employed in the SAM-busting role throughout the war, with the F–4C also making an appearance in this role in 1971.

The mission was complicated and challenging, but most effective. A Wild Weasel formation consisted of four aircraft: two Wild Weasels (F–105Gs or F–4Cs) carrying air-to-ground missiles, and two wingmen (F–105s or F–4s) loaded with conventional bombs. During the early days of this mission, the Wild Weasels carried four SHRIKE missiles; later, these were replaced with nine Standard Arm missiles having a longer range and larger warhead. The Standard Arm permitted a stand-off launch from as far away as 30 miles from the SAM or the early warning radar site.

The Wild Weasel flights of four aircraft were code-named "Iron Hand," and Iron Hand flights required the top pilots in the command. They had the most demanding job and the most hazardous, for these flights were the first into the target area and the last out. It was their task to attack any active SAM site that was a threat to the strike forces. To successfully carry out this assignment, the pilots had to have a detailed understanding of both the tactics employed by SAM crews and the mission and tactics to be employed by the strike forces. With the strike forces and escorting fighters flying at 500 knots and higher, seconds became critical. If a Weasel's timing were off, it could well mean that a member of the strike force would be shot down or that the enemy's defenses would force unacceptable bombing errors.

Usually the Iron Hand flights were ahead of the strike force by about five minutes. If the Weasels arrived in the target area too early, the SAMs would stay off the air to avoid revealing their location. The challenge was to draw out the SAM just before the strike flights were due to reach the target. Then all depended upon a high speed "cat and mouse" game. As soon as the SAM site activated, a Weasel would attack it with a SHRIKE or Standard Arm. The missile attack was followed immediately with an attack by F–4s or F–105s loaded with conventional weapons. This combination of tactics provided the highest probability of keeping the SAMs suppressed throughout the strike mission. Since the strike forces were extremely vulnerable during withdrawal as well as attack, Iron Hand flights also covered the egress routes and caught many of the SAM launches that were directed at the strike fighters.

The Iron Hand missions and tactics remained about the same during the 1965–1968 campaign and the 1972 offensive, though the effectiveness of the flights was often debated. It was particularly difficult to confirm the destruction of a SAM, and some critics claimed that the Weasels were not effective since a relatively small number of destroyed SAM sites could be attributed with certainty to these flights. However, as much as

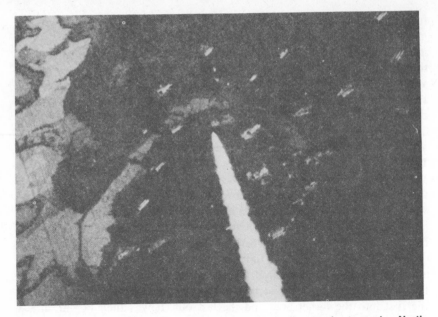

Launched from a USAF F–105, an air-to-ground missile streaks toward a North Vietnamese SAM site.

we wanted to destroy the sites, the effectiveness of the Iron Hand flights must be measured against a criterion of suppression as well as one of one of destruction. If a SAM site could be suppressed so that it couldn't fire against strike aircraft, the mission of the Iron Hand flight was a complete success. With better air-to-ground missiles, the Weasels could probably have destroyed, rather than merely suppressed, more of the SAMs. Because the terminal guidance of our missiles was limited, though, suppression of SAM sites became the realistic objective for Iron Hand flights, and they did this superbly in both the 1966–68 and the 1972 campaigns.

SAM OPERATORS' COUNTERTACTICS

The enemy's SAM tactics changed during the course of the war in response to our countermeasures. Against fighters, there was always a tendency to use more barrage fire because our highly maneuverable fighters could usually avoid a single launch. The enemy constantly sought better ways of dealing with our ECM pods. Frequently, fighter formations would encounter barrage fire until there was a split, at which time aimed fire was more prevalent. When we were out of formation, our vulnerability increased appreciably because the pods then provided ineffective coverage. Even with two pods per fighter, individual protection was inadequate. Another of the enemy's favorite techniques was to launch one missile as a feint, hoping to entice the flight to turn into an area where three or four SAMs could be launched in rapid succession.

SAM tactics were closely coordinated with those of the enemy fighters and the AAA. The MIGs usually operated on the periphery of a thirty- to forty-mile circle around Hanoi, attempting to intercept our fighter escorts some sixty to seventy miles from the target area. When our fighters penetrated the MIG defenses and moved into the target area, they usually encountered heavy anti-aircraft fire. Sometimes this fire was aimed, while at other times it was simply barraged. Irrespective of the type of AAA fire, it was always coordinated with the SAMs for maximum intensity over the target and along our most probable routes of entry and departure. Around such targets as the Doumer bridge, for instance, AAA had first priority for engaging us until one of our aircraft rolled in for attack. As a strike aircraft rolled in, his ECM coverage was temporarily degraded, and the SAMs had their best chance to score a kill.

Pilots in the 1966–1968 campaigns couldn't do much to avoid the SAMs with erratic flight (or "jinking") if they wished to have any hope of getting the bombs on target. But our later bombing systems which compensated automatically for speed, altitude, and a moderate amount of jinking provided pilots much more protection. Also, with the laser weapons used in the 1972 offensive, strike forces had greater freedom of maneuver and could release their weapons from a much higher altitude.

Against the B–52s in LINEBACKER II, SAMs were most effective during the aircraft's final turn off target. On the night of 26 December 1972, SAMs did considerable damage by firing a barrage just as the attacking aircraft made its break to depart the target area. During these turns, the maximum profile of the aircraft was exposed to acquisition radar. However, after the first two nights we changed our tactics to permit the B–52s to operate their ECM equipment at peak effectiveness throughout the mission. After this change, the effectiveness of the SAMs against B–52s was no longer sufficient to influence the conduct of the 11-day offensive.

RULES OF ENGAGEMENT—A CONTINUING SAM PROBLEM

Knowing that U.S. rules of engagement prevented us from striking certain kinds of targets, the North Vietnamese placed their SAM sites within these protected zones whenever possible to give their SAMs immunity from attack. Within 10 miles of Hanoi, a densely populated area that was safe from attack except for specific targets from time to time, numerous SAM sites were located. These protected SAMs, with an effective firing range of 17 nautical miles, could engage targets out to 27 miles from Hanoi. And most of the targets related to the transportation and supply system that supported the North Vietnamese troops fighting in South Vietnam were within 30 miles of Hanoi. Thus the SAMs could hit us whenever we came after one of the more significant targets near Hanoi, but our rules of engagement prevented us, in most cases, from hitting back. Outside the 10-mile zone, but within 30 miles of Hanoi, we could hit SAM sites only if they were preparing to fire on us and if they were not located in a populated area. If they were located in a populated area between 10 and 30 miles from Hanoi, we could hit them only if they were actually firing.

Similar restrictions prevailed near Haiphong. There was a ten-mile restricted area around the city with an inner four-mile circle in which all flight was prohibited except as specifically authorized. Thus SAMs defending Haiphong had even greater freedom from attack than those near Hanoi. By approaching and departing over the sea, however, aircraft hitting targets near Haiphong could avoid prolonged exposure to SAMs.

One of the best known humanitarian policies of the U.S. was that we would not destroy the dikes associated with the North Vietnamese irrigation system. If these dikes had been struck, most of the Hanoi delta with its dense population would have been flooded, killing innocent

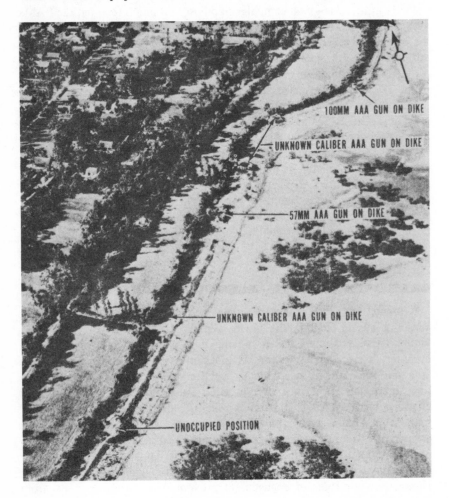

Reconnaissance photo taken 22 July 1972 shows a series of anti-aircraft emplacements along a dike near Thai Binh, some 30 miles south of Haiphong.

civilians and destroying most of the rice crops. The North Vietnamese took advantage of our policy and located SAMs and AAA on a number of the dikes. Actually these sites were authorized for attack if they were firing, but our pilots exercised considerable restraint about hitting them. Whenever a pilot hit one of these sites, the North Vietnamese invariably alleged that we were attacking the dikes. Usually, such allegations were followed by investigations into the legitimacy of the attacks, and with so many higher headquarters involved, pilots much preferred to avoid the AAA and SAM sites on the dikes unless our strike forces were directly threatened.

LOSSES TO SAMS

The argument that aircraft could not operate in an environment with surface-to-air missiles was advanced with increasing frequency after exercise DESERT STRIKE was held in 1964. Conducted in the California desert, this exercise involved more of our army and air forces than any previous exercise. Army Hawk missile battalions played a significant role in the defense of one of the opposing forces. Afterward, many who weren't acquainted with the details of the exercise insisted that tactical aircraft could no longer survive in a SAM environment. The Vietnam War, however, produced no evidence to support such a view. The war did demonstrate the need for special equipment and tactics to counter

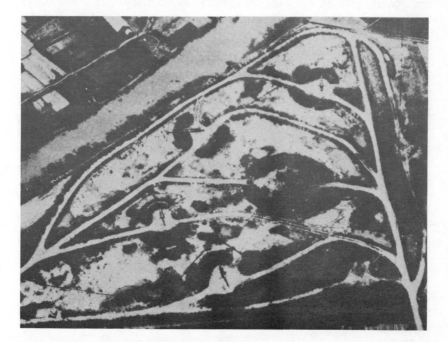

SA-2 missile site along river bank. All launch pads are occupied. Radar and supporting electronic equipment located in revetted area at center of site.

SAMs, but with such measures our fighters breached the SAM defenses as effectively as they had penetrated AAA defenses in World War II and Korea.

In 1967 the loss rate to SAMs was one aircraft lost for fifty-five missiles fired. This eventually went up to one aircraft lost for a hundred missiles fired. During the 1972 offensive when chaff was used extensively, more than 150 missiles were fired for each aircraft shot down.[22]

Although the loss rate to SAMs was not excessive, this fact should not lead us to conclude that the weapon was ineffective. The SAM's effectiveness must be evaluated within the context of its contribution to the overall North Vietnamese air defense system. Because of the SAM threat alone, we had to change the preferred altitudes of operations for our strike forces significantly. To stay out of range of the medium altitude AAA systems, we wanted to operate between 25,000 and 35,000 feet. However, aircraft operating between 20 and 40 thousand feet without extensive countermeasures were too vulnerable to SAMs. From 1966 until the 1968 bombing halt, we compromised by conducting most missions between 12,000 and 15,000 feet. SAM effectiveness was reduced considerably because a SAM is still accelerating at 12,000 feet and doesn't reach its ultimate speed of 2.4 mach until about 25,000 feet. But at the medium altitudes our vulnerability to AAA particularly the 57 mm, was significantly greater. So even though the SAMs didn't get many direct kills, they contributed importantly to the overall defense system by

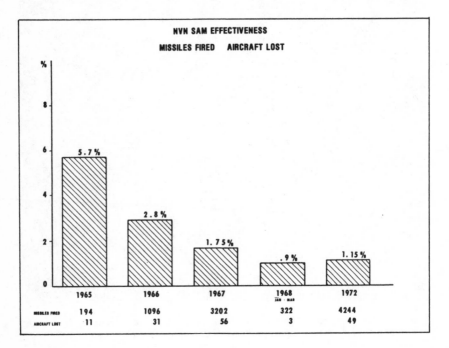

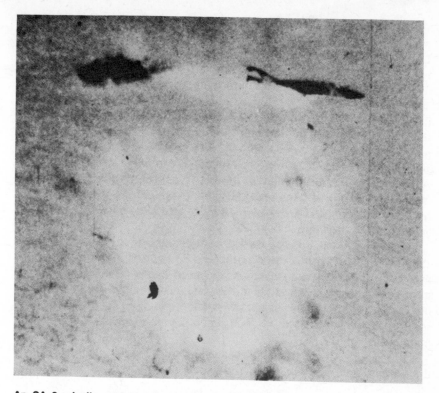

An SA–2 missile explodes just under a USAF RF–4C somewhere over the Red River area of Hanoi.

forcing our operations down to an altitude at which another part of the system was more effective.

MIGs

The North Vietnamese Air Force was small, as air forces go, and patterned after the Soviet Air Defense Force. Many of the features of the force were the same as those we had seen in the North Korean Air Force during the Korean War. Basic pilot training took place in the Soviet Union, and final training for combat was accomplished in China. The quality of pilots, although good, was never up to the standards of our Air Force and Navy. There were, however, some individual pilots who would have been outstanding in any air force.

On 7 August 1964, just a few days after the Tonkin Gulf incident, 30 MIGs flew into North Vietnam from China and landed at Phuc Yen.[23] Soon other MIG units were outfitted in China and moved to one of nine airfields in North Vietnam. The MIG strength in July 1966 numbered about 65 aircraft. There were about 10 to 15 MIG–21s and the remainder were MIG–17s. Losses were rapidly replaced so that the operating inventory was kept fairly high. By mid-1967 the North Vietnamese Air

Force numbered over 100 aircraft. Of this number, about 40 to 50 were MIG–21s and the remainder MIG–15s and 17s. With the onset of LINEBACKER I in May of 1972, the force numbered about 200. Of this total, 93 were MIG–21s, and 33 were MIG–19s. The MIG–19 was the only Chinese fighter introduced into the war by the North Vietnamese. The remainder of the force totaled about 80 MIG–15s and 17s. Surprisingly, although the size of the NVAF doubled between 1967 and 1972, the total number of MIGs up for battle didn't change significantly. During the period from April to June 1967, 42 MIGs were destroyed.[24] During a similar period in 1972, 30 were destroyed. The number of fighters in 7th Air Force was not much different in the two periods either, although two wings of F–105s had been replaced with F–4s and A–7s for the strike role. The number of F–4s for purely fighter activities, however, remained approximately the same.

The MIG–21 was updated during various periods of the war. In late 1965 and 1966, its primary armament was 23mm and 37mm cannon. By mid-1967 the primary armament was a cannon and two Atoll heat-seeking missiles. These missiles had about the same performance as our Sidewinder. By the time of the resumption of the bombing in May of 1972, the standard armament for the MIG–21 was a 23mm cannon and four Atoll missiles.

The MIG–17 was never considered the primary fighter, and changes to this aircraft were never evident. It remained essentially a gun aircraft and

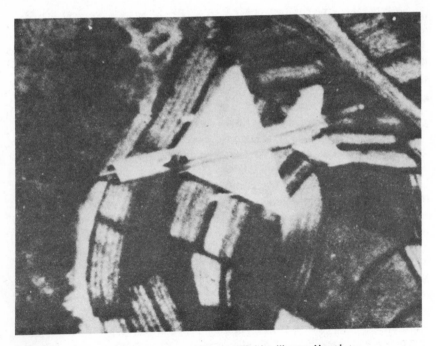

North Vietnamese MIG–21 "Fishbed" near Hanoi.

was employed most of the time at low altitudes where the gun could be used to advantage in a turning fight in which the MIG–17 excelled.

The MIG–19s appeared only after the 1968 bombing halt. These aircraft were supplied by the Chinese, and we assumed that the Chinese trained the North Vietnamese pilots. The MIG–19 carried three 30mm cannon and two Atoll missiles. It didn't have the speed and maneuverability of the MIG–21, and it was easier to defeat in a dog fight. Whereas the MIG–21 had a top speed of about mach 2, the MIG–19 could reach only about mach 1.3 at 25,000 feet.

During the Korean War our best fighter, the F–86 was superior to the enemy's best fighter, the MIG–15, in level flight below 30,000 feet and definitely superior at diving speeds greater than mach .95. The MIG–15 was superior in maneuverability at all altitudes and in acceleration and level flight above 30,000 feet. In Vietnam, the F–4 and the MIG–21 compared in much the same way. The F–4 had a slightly higher top speed, much better zoom quality, better maneuverability at the higher supersonic speeds, and general superiority in level flight below 20,000 feet. The MIG–21 was superior in acceleration and maneuverability at all altitudes at low air speeds and at high supersonic speeds above 25,000 feet.

Because Soviet fighters from the MIG–15 through the MIG–21 were intended for relatively short missions in defense of the homeland, their designers kept them small and highly maneuverable. U.S. strategists, on the other hand, assumed that our fighters would have to go long distances and penetrate the defenses of an enemy. Our designers therefore envisioned larger aircraft capable of great range and speed with some sacrifice in maneuverability. Also, our tactical fighters were either designed or extensively modified to perform all three of the tactical air missions: air superiority, interdiction, and close air support. In the jet age, the F–86, F–100, and F–4 have all been products of this multi-mission concept.

The F–4 was by far the most versatile fighter in the war. In its ability to perform all three classical missions of tactical airpower it excelled all other planes. The versatility of the F–4 provided 7th Air Force commanders a ready capability to meet a sudden increase in the MIG threat or a concentration of North Vietnamese ground forces for an all-out assault.

The F–105 was probably the fastest aircraft in the war below 10,000 feet. It was designed primarily for low altitude nuclear missions in which speed is essential. An ideal situation for the meeting of an F–105 and a MIG occurred when an F–105 had bombed and was coming off the target above mach one. With the F–105's speed advantage, it could make a single pass and then terminate the fight. Its limited maneuverability in comparison to the MIG–21 or MIG–17 made it a very poor plane in a dogfight, however, and tactics were employed to avoid such engagements wherever possible.

MIG AIRFIELDS

As the war progressed, the North Vietnamese expanded the number of jet-capable airfields from nine to thirteen. Most of these fields were in the vicinity of Hanoi. The North Vietnamese were able to expand and develop new airfields without any counteraction on our part until April 1967 when we hit Hoa Loc in the western part of the country and followed with attacks against Kep. The main fighter base, Phuc Yen, was not struck until October of the same year. Gia Lam remained free from attack thoughout the war because U.S. officials decided to permit transport aircraft from China, the Soviet Union, and the International Control Commission to have safe access to North Vietnam. The North Vietnamese, of course, used Gia Lam as an active MIG base.

An argument frequently advanced for not striking airfields was that our aircraft losses would be disproportionate to the damage we could inflict upon the North Vietnamese Air Force.[25] But our strike forces were already penetrating the areas where airfields were located, and there were no major changes in the defenses the enemy could have employed that would have made our losses greater than they already were against other targets in the area.

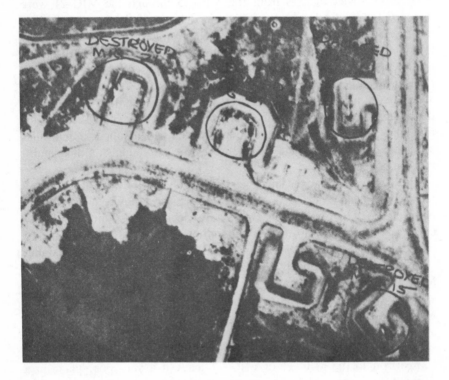

North Vietnamese Air Force's Phuc Yen airfield. Note dispersed and bunkered aircraft parking locations.

140

The southern-most airfields in North Vietnam were developed after the bombing halt in 1968 and were used to a limited extent for the Easter offensive in 1972. During the 1975 final offensive, these southern airfields were in excellent operational condition to support MIGs in a fighter-bomber role in the event such support were needed. For the final offensive, the airfields at Khe Sanh and Dong Ha were fully developed to handle MIGs, although there were no reported MIG flights from either one. These forward airfields could have played an important part in the North Vietnamese offensive if there had been a containment of the assault at Pleiku and Hue.

By not permitting hot pursuit of the enemy into China, our rules of engagement provided a sanctuary for MIGs that were blocked from returning to their home bases. When our fighters established a barrier patrol between Phuc Yen and the border, MIGs often recovered in China rather than confront the F-4 fighter screen. Thus when the North Vietnamese wanted to hold attrition rates down, they withheld fighters from combat by sending them into China until our raid was completed. Even though this tactic saved their fighters, it provided local air superiority for our strike forces by default.

MIG TACTICS

The MIGs were used sparingly at first, and throughout the war reactions to strike missions varied according to the losses suffered. Usually, MIG units would standdown after a couple of days of heavy losses. They did this late in 1966, in the summer of 1967, in December 1967, and during the final offensive in December 1972. During these standdowns, which in some cases lasted two or three months, the enemy developed new tactics.

Regardless of the tactics employed, though, the MIGs were always under GCI control during the entire intercept mission. The major advantage enjoyed by the MIGs was their integrated air defense radar system that completely covered North Vietnam. Because of the system's redundancy, we were unable to deceive or surprise it. The system knew where our fighters were, and because of the limited number of targets and the small area in which the targets were located, it had a very precise idea of which targets would be under attack. North Vietnamese ground controllers must have found it rather easy to position MIGs for an engagement and to feint attacks on our strike force, causing it to jettison its weapons.

Compared to the dogfights of the Korean War, those over North Vietnam were relatively small. In the Korean War, the Communist Chinese Air Force sometimes launched as many as two hundred MIGs against our F-86s and fighter-bombers. MIG formations in that war were loosely controlled, at least until 1952, and were held in high orbits of 30 or 40 aircraft above the F-86s. When the time appeared right, four to six fighters would make a high speed hit-and-run attack coming from out of

141

NORTH VIETNAMESE AIRFIELDS & AOB

NVN AIR ORDER OF BATTLE		
	DEC 1966	OCT 1972
MIG-15/17	50	66
MIG-19	0	40
MIG-21	16	39
TOTAL	66	145

the sun. Formations of MIG–17s and MIG–21s over North Vietnam were always much smaller. The MIG–17 and the MIG–21 employed different tactics, but both types of aircraft were controlled much more closely by ground radar than their MIG–15 counterparts had been over Korea.

The MIG–17s generally defended North Vietnamese airfields and patrolled at low altitude along our approach and departure routes. We believed that North Korean pilots flew most of the MIG–17s, particularly those that covered the airfields during takeoff and landing of the MIG–21s. Most of the MIG–17 formations had only two aircraft, and their favorite tactic was to lure the F–4s into a low altitude dogfight when the F–4s were low on fuel. The MIG–17s were heavily concentrated in the Navy's area of responsibility east of the northeast rail line, and most of the Navy kills were against the MIG–17.

The greatest concern of the 7th Air Force pilots throughout the war was the MIG–21. In early and mid-1966, most of the MIG–21 attacks were made by formations of four to six aircraft. Dogfights were characteristic of the period since MIG–21s were not yet armed with Atoll missile, but the average dogfight lasted less than thirty seconds. It was evident that GCI control had not been perfected and that coordination with AAA and SAMs was poor. Most of these MIGs patrolled near Thud Ridge,* near Thai Nguyen, and southeast of Hanoi. Their tactics appeared designed to pick off aircraft that had received battle damage and were not completely covered by an escort.

By the end of 1966, MIG pilots were beginning to show increasing signs of aggressiveness. From the earlier flights in which losses were running heavily against them, in some cases as high as four to one, they had learned how to improve their tactics. Formations became more flexible and employed the GCI that control had developed in the last

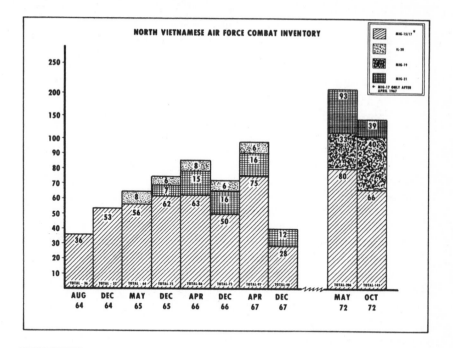

---

*Thud Ridge—A line of hills to the northwest of Phuc Yen airfield that the F–105s (Thuds) used as a shield against North Vietnamese radars as they made a low altitude penetration to targets within the 30-mile Hanoi circle. This low altitude tactic changed with the introduction of pods in 1967.

143

quarter of the year. Furthermore, the new Atoll missile allowed them to attack with less risk using very high speed hit-and-run tactics.*

By the end of February 1967, MIG tactics had become pretty much set. Operating in elements of two, they were vectored by GCI to a position behind the strike force. Then they launched a high speed attack with Atoll missiles. During the attack and breakaway, they attained speeds above mach 1.4; then they used the high speed to zoom above and away from the strike force. Sometimes an element of two MIG–21s would act as a decoy to draw off the F–4 screen, and then another two MIGs would follow with a stern attack. In all cases, the primary weapon was the missile with guns being used only when necessary for close-in fighting.

During the same period, MIG–17s, usually protecting the North Vietnamese airfields, were often vectored away to attack the belly of the strike forces. If this tactic succeeded in disorganizing the strike force, MIG–21s from above would drop down on any element that had split away from its cover. These tactics had limited success, however, because of the disparity in speed between our fighters and the MIG–17s. Because of the MIG–17s' slower speeds, they were seldom able to split up our formations. But the tactic did show a degree of sophistication in the coordination of high and low altitude flights with AAA and SAMs. These tactics complicated our strike missions and thereby represented a decided improvement in North Vietnamese air defense operations.

SEVENTH AIR FORCE DEALS WITH THE MIGS

Since the MIG threat was relatively low in 1965 and early 1966, and since our authorization to hit any particular target was often withdrawn after a short time, our plan was for all strike aircraft to carry bombs. We meant to destroy as much of the target as possible with each strike. Consequently, instead of establishing a separate fighter escort, we simply instructed F–4 pilots to jettison their bombs if they had to engage MIGs before attacking the target. After jettisoning their bombs, the F–4s acted as escort for the attacking F–105s.

Many of the F–4 pilots weren't happy with these tactics—the tactics were defensive in nature and cast them in the unpleasant role of targets waiting to be hit. They wanted to go after the enemy fighters from the outset, but the mission priorities established by CINCPAC and JCS for Southeast Asia provided little opportunity for free-lance operations or fighter sweeps over North Vietnam. Usually the requirement to attack

---

*It is of interest that Soviet fighter tactics as flown by the Koreans and Chinese in the Korean War developed in the same pattern. Initially, there was a tendency to fight in large formations of sixteen or more aircraft. But as more experience was gained, the formations broke down into flights of eight to twelve from which elements of two would break off for a high speed firing pass. The two aircraft would not return to the fight unless the same tactic could be employed again. The tactic provided an opportunity to exploit the better maneuverability and acceleration qualities of the MIG–15.

certain targets within a specified period demanded the full effort of all the available forces. There weren't enough fighters available to conduct random fighter sweeps and also protect the strike forces during their specified times in the target area.

Even if fighters could have been spared for daily sweeps, the shortage of tankers would have precluded such a tactic. From 1966 through 1968, it took about 27 tankers twice a day to refuel the attacking force. With the additional requirements for refueling B–52s and for maintaining an ability to support the first phase of the SIOP,* SAC's tankers simply couldn't support any additional fighter operations in Southeast Asia. We couldn't draw from the strategic tanker fleet any further since it was imperative that our general war forces be able to carry out their assigned missions regardless of the demands of the war in Vietnam.

BOLO—A FIGHTER SWEEP

From our observations of fighter engagements in 1966, we determined that a properly designed fighter sweep might destroy a number of MIGs. Since airfields could not be struck during this period, a large air battle, or series of battles, was the only way to reduce the MIG force appreciably. We began planning in mid-December 1966 for a large fighter sweep which would take place immediately after the expected Christmas and New Year standdown. Previous experience showed that the MIGs would put up a substantial effort after a standdown; their in-commission rate would be high, their tactics refined, and their skills polished by training sorties accomplished with little fear of attack.

The fighter sweep was designed to appear as a normal F–105 strike force with escorting fighters. Call signs of the formation and all other indicators were designed to give the impression that the penetrating force was just another daily strike force. The sweep was to consist of F–4s and F–105 Wild Weasels penetrating simultaneously from Laos and the Tonkin Gulf. Our plan was to create a pincer to close off any MIG flights attempting to recover at Chinese bases. We assigned the F–4s specific areas to sweep, areas where the MIG–21s normally orbited before launching attacks against our entering and departing strike forces. The F–4s coming in from Laos were to handle the MIGs at the higher altitudes while those coming in from the Gulf of Tonkin would take care of the MIGs at the lower altitudes.

The weather was marginal over North Vietnam on the morning of 2 January 1967. It appeared good enough for our mission, however, so we launched 20 flights of F–4s and F–105s. Soon after our planes were launched, the weather became worse than expected; the flights coming from Laos were able to sweep the assigned areas, but those coming from

---

*SIOP—Strategic Integrated Operational Plan: SAC is the executive agent for the JCS in maintaining an integrated plan for the employment of SAC and Navy submarine forces in a strategic air offensive in case of general nuclear war.

Formation of F–105s from the 355th TFW returning from a mission over North Vietnam.

Formation of F–4s (Es & Ds) heading for a tanker somewhere over Southeast Asia.

the Gulf of Tonkin had to turn back after repeated attempts to get into their assigned areas. The MIGs reacted as we had anticipated: in elements of two, they popped up through the overcast to jump what seemed to be a normal strike force. MIG pilots were caught completely by surprise when they encountered not heavily laden F–105s but F–4s with tanks jettisoned, ready for a fight. After a few brief minutes the battle ended with seven MIGs shot down.[26]

BOLO gave us the largest single MIG battle of the war; but as air battles go, it was a small one. In the Korean War, by comparison, our largest air battle took place on 4 September 1952, when 39 F–86s engaged 73 MIG–15s in seventeen separate engagements. In that series of dogfights, thirteen MIGs were shot down and four F–86s were lost.

THE BATTLE CHANGES—MISSILES AND "HIT-AND-RUN"

After January 1967 we had to stop giving F–4s the dual mission of carrying bombs as part of the strike force and providing fighter cover if a MIG attack developed. With the increased MIG activity and the MIGs' high speed hit-and-run tactics, our F–4s couldn't jettison bombs and get in position quickly enough to defeat all of their attacks. If we had had better radar coverage of North Vietnam and if our rules of engagement hadn't required our pilots to visually identify the enemy aircraft before firing, it would have been feasible to employ the F–4 in the dual role on some missions where there was little probability of significant MIG reaction. Given the rules of engagement and the limitations of our radar coverage, though, we had to protect our strike forces by assigning an increasing proportion as escorts throughout the year.

Seeking other ways to provide better protection for our strike forces, we experimented with a number of different tactics and formations in 1967 and 1968. One of our favorite formations consisted of 16 F–105s and two flights of F–4 escorts, one flight positioned ahead of the lead strike flight and the other positioned to the immediate rear of the force. These escort flights weaved back and forth across the path of the F–105s, protecting against attacks from either front or rear. This weaving used considerable fuel, of course, and meant that fuel consumption was always a critical factor. Initially, we instructed these flights to stay close to the strike force and to turn MIGs away rather than engage them. As the MIGs continued to improve their hit-and-run tactics, though, we gave the F–4 escorts freedom to pursue and engage the MIGs on their own initiative. This tactic proved to be more effective in breaking up the MIG attacks.

What we learned during 1967 and 1968, really, was that the best tactics for escorting a strike force had not changed much since 1944. In Europe, Korea, and Vietnam aggressive enemy fighters always had a decided advantage; if they got close enough to press the attack, they were extremely difficult to stop. During the bombing raids in Europe early in 1944 we soon learned that the best way to stop the ME–109 and FW–190 attacks was to let the escorting fighters roam on the flanks and above the bomber formations, seeking out the enemy fighters before they could assemble for an attack. In Korea the best tactics were much the same although the escort problem there was made much more difficult by the disparity between the performances of the new jet fighters and the older B–29s. The relatively high speed and maneuverability of the MIG–15s (not to mention their sheer numbers, which varied from 40 to 80 or more against our strikes near the Yalu) helped them to penetrate our fighter screens, and the brief loiter time of our F–86s (only 25 minutes in the

A MIG-17 jumps an F-105 right after the USAF fighter has struck the Doumer bridge. Another F-105, firing at the MIG-17, records the action with his gunsight camera. Action took place on 19 December 1967.

target area) limited our freedom to seek out the MIGs. Although these disadvantages caused us to lose as many as five B–29s to the MIGs in a single month (October 1951), our experience demonstrated conclusively that a roaming fighter escort was the best defense against interceptors.[27]

Our experience with the F–86s in Korea taught us several lessons about achieving air superiority against enemy interceptors, but in the 11-day offensive of December 1972 the B–52s' need for fighter escort confronted our fighter force with some tough new problems. Prior to these bombing raids, it wasn't necessary to provide extensive fighter cover for the B–52s since they were bombing in South Vietnam, Laos, Cambodia, and the lower part of North Vietnam. Although there was usually at least one flight of F–4s providing area coverage for these missions, the MIGs did not pose a significant threat. It was only when they started bombing around Vinh in North Vietnam that fighter cover became essential.

All of the B–52 strikes in North Vietnam in December 1972 were flown during the hours of darkness. Three flights of F–4s were assigned various areas of patrol to support each of these night strikes. Patrol locations were based on predictions of where the MIGs would most probably launch attacks against the bomber stream. In addition to these screens, four flights of F–4s (16 aircraft) were responsible for close escort of the bomber cells.

The problem of protecting the B–52s in such a narrow air space at night was quite difficult. We had only limited GCI coverage: the radar in the EC–121 was unable to see airborne targets at low altitudes over land. Also, CROWN, a Navy ship providing radar information in the Gulf of Tonkin, had limited capability to see MIG traffic in the Hanoi area. Thus our fighters were forced to depend largely on information from their own radar in making a final decision as to whether an enemy attack was developing.

The difficulty of ascertaining which of the many aircraft on one's radar screen belonged to the enemy prompted considerable debate about the use of IFF.* When our planes "squawked," or turned their IFF transponders on, our pilots knew each other's radar positions clearly, but the enemy GCI controllers also received a brilliant radar display of our positions. If we "strangled" our IFFs, the enemy radars only received a much less distinct "skin paint" of our aircraft, but then we had much more difficulty identifying each other's position. Obviously neither solution was completely satisfactory. However, given the small amount of airspace involved, the redundancy in the enemy's coverage, and the small number of enemy fighters up for any one battle, I would choose to prevent interference between elements of our force by leaving the IFF on during the last phase of the penetration into and the first phase of the withdrawal from the target area. Disagreements continued about the IFF issue, but the important fact is that despite the difficult problem of identification, our escorts did prevent the MIGs from posing any real threat to our strike forces.

Technology affected our escort operations in yet another way in 1972. Because of the effectiveness of newly introduced laser weapons (with a combat CEP** of approximately 30 feet, as compared to the dive bomb CEP of 420 feet in the 1966–1968 period) fewer strike aircraft were needed for point targets such as the Canal Des Rapides bridge or the Hanoi Railcar Repair Shop. However, with a strike force consisting of only three

---

*IFF: "Identification, Friend or Foe." An electronic device first used by the U.S. during the Korean War which would transmit from our aircraft coded electronic signals.
**CEP: Circular Error Probable. "The radius of a circle within which half of the missiles/projectiles are expected to fall." JCS Pub. 1, DOD Dictionary of Military and Associated Terms.

flights of F–4Ds carrying laser weapons, the loss of one aircraft weakened the force much more, proportionately, than had the loss of an F–105 in earlier years. For this reason, if for no other, good tactical judgment required that the fighter cover be greater than it had been. Additionally, MIG caps were given much more freedom of action to screen ahead of the laser flights and to establish barriers along the flanks where an attack could be expected.

CONTROL IN THE TARGET AREA

The avoidance of a border violation with Communist China was always a factor in our planning of fighter and strike missions. Throughout the war, a buffer zone approximately 25 to 30 miles wide was established along this border. During 1966 and 1967, our aircraft were permitted to maneuver in this area only when positioning for attacks against targets outside the buffer zone. Specific operations were scheduled from time to time within the buffer zone, but armed reconnaissance or fighter sweeps were prohibited. Aircraft in hot pursuit of the MIGs were expected to break off their attack upon entering this zone.

To further ensure against border violations, an elaborate system was established to warn aircraft approaching the Chinese border. But when these calls were coupled with numerous MIG and SAM calls, pilots had difficulty determining just who was about to violate the border. When the situation was complicated further by reduced visibility and a pilot's

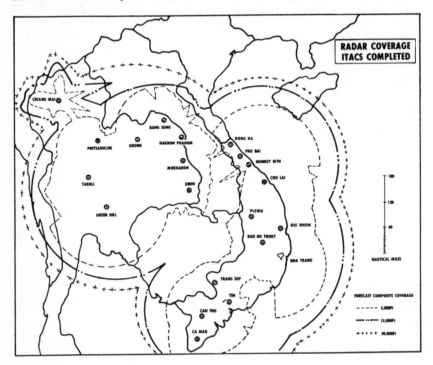

temporary disorientation as a result of a dogfight, it wasn't surprising for a pilot to come close to violating the border. However, because of the navigational equipment in both the strike and fighter aircraft, the excellent warnings given by MOTEL, TEABALL, CROWN,* and the EC-121s, and a great deal of command emphasis, there were only a few violations throughout the war.

The ground based radar system in Thailand and South Vietnam didn't have the range to provide control of air operations much above the nineteenth parallel. It was a fairly complex radar control system and worked well within the performance expectation of the equipment. But each radar site had only partial coverage of North Vietnam, leaving gaps in many target areas. These gaps in coverage, dictated by available geographical locations for radar sites, limited positive control in an engagement.

Our most northern radar was the control and reporting post (CRP) located at Dong Ha, called WATERBOY. Fighters coming up the east side of Vietnam would come under the control of another radar, the control and reporting center (CRC) at Danang. This radar had good coverage to the south and to the east but had blank spots looking to the north. As a result, the CRP at Dong Ha became the primary ground based radar for controlling aircraft enroute to Route Package VI (RP VI) from the Gulf of Tonkin. It was about 350 miles from WATERBOY to the primary target area in RP VI. With its radar limited to about 180 miles, WATERBOY could only handle fighter engagements against MIGs that came as far south as Thanh Hoa, NVN, and rarely during our extended operations in the delta did the MIGs come this far south. WATERBOY also functioned as the primary radar for effecting rendezvous of the fighters with the KC-135 tankers over the Gulf of Tonkin. Additionally, it tied in with the Navy radar system in Task Force 77 (TF-77) operating in the vicinity of the seventeenth parallel in the Gulf of Tonkin. This exchange of tracks provided a consolidated picture of the air situation as seen by the surface based radars.

Once the fighters passed out of the control of WATERBOY, they were picked up by the airborne radar of the EC-121. This aircraft served as an airborne command and control center (ABCCC) and operated under the code name COLLEGE EYE. During all operations above the twentieth parallel, an EC-121 was on station. It had excellent coverage over the water, but when its radar looked down over land, "land clutter" blocked out aircraft movements. The EC-121, therefore, had to depend upon other sources of information for directing MIG intercepts.

*MOTEL—Command headquarters of 7th Air Force that controlled air operations in North Vietnam from 1965 to 1968.
TEABALL—Command headquarters of 7th Air Force that controlled air operations in North Vietnam during 1972 air offensive (LINEBACKER I and II).
CROWN—Navy early warning vessel in Gulf of Tonkin above 18th parallel.

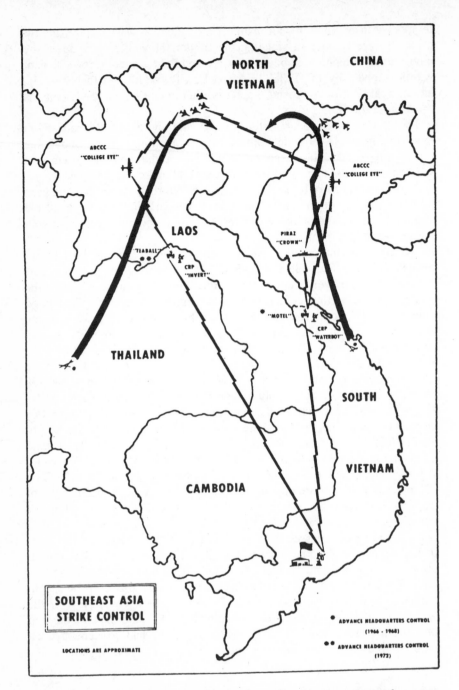

NORTH VIETNAM

CHINA

ABCCC "COLLEGE EYE"

ABCCC "COLLEGE EYE"

LAOS

PIRAZ "CROWN"

"TEABALL"

CRP "INVERT"

"MOTEL"

CRP "WATERBOY"

THAILAND

SOUTH VIETNAM

CAMBODIA

SOUTHEAST ASIA
STRIKE CONTROL

LOCATIONS ARE APPROXIMATE

● ADVANCE HEADQUARTERS CONTROL
   (1966 - 1968)

●● ADVANCE HEADQUARTERS CONTROL
   (1972)

Despite its limitations, the EC–121 was the closest command post to the Air Force's operations in RP VIA. It had the responsibility for controlling intercepts (as well as it could), issuing MIG alerts, warning pilots of potential border violations, and, when other measures were not

"WATERBOY" site at Dong Ha Air Base, South Vietnam, September 1966.

USAF EC–121 "COLLEGE EYE" in flight over South Vietnam.

adequate, issuing SAM warnings. Also, these command posts played a very important role in search and rescue missions for downed pilots. One of the things the EC–121 airborne controller could not do, however, was call off a strike. The final decision about continuing with any given

153

mission rested solely with the mission commander, who normally flew in the lead aircraft of the strike force. He made the decision for the entire force. The ABCCC acted as an information center to keep the strike force commander completely informed of the current friendly and enemy air situation and of any weather or special intelligence reports that he might need to know. In effect, the ABCCC had all the responsibilities of any other operational command post except for the limitation regarding decisions about the conduct of the mission.

For assistance in the detection of MIGs threatening the strike forces, COLLEGE EYE was tied into the Navy's early warning ship, *Piraz*, which was located in the vicinity of the eighteenth parallel. This vessel provided early warning to the carrier task force and MIG information for Navy fighters operating in RP IV and VIB. Normally, the Navy didn't delegate control of intercepts to *Piraz* code named CROWN. Instead, CROWN provided information to the Combat Information Center (CIC) which controlled the intercept from its location on board one of the carriers in the task force. During the 1972 offensive, CROWN was given limited GCI control for both Air Force and Navy fighters. However, CROWN was limited in its capability to control Air Force fighters over Hanoi. Because of its distance from the battle area, it could only see traffic above 10,000 feet, and even then it couldn't see more than about 50 miles inland. As a consequence, it was most valuable in providing information on potential intercepts only in the most southern part of RP VIA.

Located near Danang on Monkey Mountain was the primary CRC for all air operations in the northern part of South Vietnam and RP I in North Vietnam. Adjacent to the CRC was a special advanced operational headquarters of 7th Air Force. This headquarters was known as the TACC, North, code named MOTEL. MOTEL did only limited mission planning; most of the mission planning was done at 7th Air Force in Saigon and then distributed as fragmentary orders to all the units. But MOTEL had the responsibility for controlling the missions entering North Vietnam. Orders were issued from 7th Air Force to MOTEL, which then executed through the ABCCC over the Gulf of Tonkin.

Regardless of whether the forces were coming into the target area from Laos or the Gulf of Tonkin, MOTEL, during the 1965–1968 period, was the controlling agency. All intelligence was collated at MOTEL for on-going missions and then made available to the EC–121 COLLEGE EYE in contact with the strike and fighter forces. Mission results were flashed to MOTEL, which sent them to the 7th Air Force out-of-country command post. In order to have communications with the forces throughout a mission, a C–135 relay platform was maintained over the Gulf of Tonkin prior to and during scheduled missions. MOTEL was in constant contact with the strike forces, relaying messages either through COLLEGE EYE or through the C–135. With this system, 7th Air Force was able to follow a mission from takeoff to landing.

The ground based radar system in Thailand was organized in the traditional manner. There was a CRP at Udorn, BRIGHAM, which provided navigational assistance to the forces going north and was the primary radar facility for effecting a rendezvous of tankers and fighters. Because of terrain masking, it was not capable of control over North Vietnam, but it was kept busy just handling the tanker rendezvous and taking care of aircraft requiring refueling coming back from a mission.

The radar station in Thailand closest to operations in NVN was a CRP located in Nakon Phanom, INVERT. This site had a function similar to that of WATERBOY, the CRP at Dong Ha. INVERT exercised control of the strike forces to the limit of its radar, which was about 150 miles. Thus it was not possible to control the strike forces through this facility once they approached the western border of North Vietnam.

In the 1972 air offensive, EC–121Ts were stationed over Laos and functioned in the same role as COLLEGE EYE C–121Ds had in previous years. This radar was called DISCO; although somewhat improved, it had limitations similar to those of COLLEGE EYE when operating over land. Fortunately, these limitations have been eliminated in the new radars going into the E–3A AWACS (airborne warning and control system). DISCO could control fighter intercepts at the higher altitudes, but was limited in the medium altitudes where most of the fighting was taking place. It was dependent upon other sources of information to fill out its knowledge of the air situation.

With the resumption of air operations over North Vietnam in May of 1972, a new control facility was developed at the CRP at Nakom Phanom. This facility became known as TEABALL, and it performed the same function during this campaign that MOTEL did in the 1965–1968 campaign. However, TEABALL was much better equipped to act as the primary advanced headquarters for control of ongoing operations over North Vietnam. TEABALL was a very effective control headquarters for the 1972 offensive thanks to the integration of all radar and intelligence information into a single facility. This integration resulted in greatly improved control of our fighters in intercepting MIGs. The kill rates do not reflect this improvement, but the 1972 campaign was of such short duration that there was insufficient time for this improved headquarters to have full effect.

CONSIDERATIONS AFFECTING KILL RATIO

Given the advanced fighter weapons available to us in Vietnam, we might have expected to achieve much higher air-to-air kill rates than we had in World War II and Korea. In the earlier wars, the only armament was the 50 caliber machine gun. But in Vietnam, the F–4E, for example, had a standard air-to-air configuration consisting of 630 rounds of 20mm, four AIM–7 radar-guided Sparrow missiles and four AIM–9 heat seeking Sidewinder missiles. With this variety of weapons, our F–4s were capable of bringing an enemy aircraft under fire at distances ranging from five to ten miles down to a thousand feet. The diversity of attack opportunities provided by these weapons should have given our pilots a chance to shoot for a kill under most conditions of air-to-air combat. But the increased opportunities offered by the advanced weapons were largely canceled by other factors.

The necessity for a visual identification of the enemy hindered successful shoot-downs by reducing the frequency of opportunities for employing, for example, the Sparrow. Referred to as an "all-aspect" missile, the Sparrow could be fired from any direction relative to the target aircraft. But the fact that visual identification was required meant that we forfeited our initial advantage of being able to detect a MIG at thirty to thirty-five mile range and launch such a missile "in the blind" with a radar lock-on from three to five miles. Many kills were lost because of this restriction, particularly during periods of reduced visibility, or at times when so few of our fighters were in the area that almost anything on the radar was an enemy aircraft.

History suggests that this rule might have had less impact on the exchange rate if we had had good GCI coverage of the Red River delta area. In the Korean War our kill rate went up sharply when a Tactical Air Direction Center (TADC) was established at Cho-Do Island to provide complete coverage of the MIG base area along the Yalu. With the establishment of this direction center in June 1952, F–86s could be vectored into a position where they had the option of initiating the attack rather than reacting to the MIGs. Having neither precise GCI control

USAF F–4E still carrying air-to-air missiles after bombing mission over North Vietnam.

USAF AIR-TO-AIR WEAPONS KILLS VERSUS FIRING ATTEMPTS
MAY 1972-JANUARY 1973

ORDNANCE	MAY (72) K/A*	JUN K/A	JUL K/A	AUG K/A	SEP K/A	OCT K/A	NOV K/A	DEC K/A	JAN K/A	TOTAL K/A	% KILLS
AIM-4D-8	0/4	0/1	-	-	-	-	-	-	-	0/5	0/0
AIM-9E	2/14	1/9	1/22	0/0	1/15	1/9	-	-	-	6/69	9.0%
AIM-9J	-	-	-	-	3/17	1/5	0/0	0/9	0/0	4/31	12.9%
AIM-7E-2	8/54	0/20	5/28	4/13	1/35	2/35	0/1	2/28	1/2	23/216	10.6%
20 MM CANNON	2/2	1/3	0/1	0/0	3/5	2/3	0/0	0/0	0/0	7/14**	50.0%
TOTAL										40/335	11.9%

* K/A--KILLS/FIRING ATTEMPTS (TRIGGER SQUEEZES).

THERE MAY HAVE BEEN MORE GUN FIRINGS--NOT ALL WERE REPORTED TO THE 1ST TEST SQUADRON.

such as the F–86s enjoyed in Korea, nor permission to fire missiles "in the blind" using our on-board radar, we regularly conceded the initiative, probably the most important factor in air-to-air combat.

Nevertheless, most of our kills were made with missiles, and in fact 57.5 percent were made with Sparrows. Navy fighters, on the other hand, made almost all of their kills with the Sidewinder. The difference between the kill ratios of the two forces can probably be accounted for by differences in the areas they covered, the type of MIGs up for battle in those areas, and the type of engagements encountered. The Navy kills were predominantly MIG–17s, and they were made in close-in engagements. Such engagements required more frequent employment of short range weapons, and since the Navy F–4s had no guns, the Sidewinder missile was their primary weapon. Their F–8s, however, did make a significant number of kills with guns.

The Air Force F–4E with an internal gun didn't make its debut in the war until 1968. Consequently, most of the kills by the Air Force were made with Sparrow and Sidewinder missiles against MIG–17s and MIG–21s. During the 1972 campaign, however, 50 percent of the kills were made with guns.

A surprisingly large number of missiles were fired for the number of enemy aircraft downed. The numbers don't fairly represent the kill rates of the missiles, however, for it was a standard tactic to fire missiles as a deterrent, even though the pilot knew he was out of range. Obviously this tactic tended to corrupt the statistical base on the relative effectiveness of missiles.

The low kill rates for missiles may also be explained in part by the fact that the AIM–7 was designed as an anti-bomber weapon and didn't have the broad envelope for firing or the maneuverability that are needed in a

157

fighter-versus-fighter engagement. A bomber has limited maneuverability, which makes it a fairly easy target for a fighter; extremes of position aren't needed during most firing passes. On the other hand in fighter-versus-fighter engagements, the opportunities for a level shot are very few, with most shots, even those from the 6 o'clock position, involving abrupt maneuvering. Thus our fighter pilots fired the Sparrow out of its envelope very frequently and many times intentionally. The net result was an 11 to 12 percent kill rate. It was an understandably frustrating experience for a pilot to work into a firing position and then have such a low probability for a kill. But the AIM–9 kill rate was somewhat better, about 20 percent, during the latter part of the 1965–1968 campaign.[28]

Another problem that stemmed from the complexity of our weapons had to do with the apparently mundane matter of the F–4E's switches. After some early experience with the difficulty of setting up the gun and missile switches in the F–4E, we improved both the pilot procedures and the switch panel. The switches were rearranged so that the pilot could go from the long range of the Sparrow to the short range of the gun with minimum switch movement. Although these changes helped reduce pilot errors, there was no real change in the rather disappointing kill rates of the AIM–7 and AIM–9 missiles.

During the campaign of 1965–1967, the ratio of enemy fighter losses to USAF fighter losses was between $3^{1}/_{2}$ and 4 to 1. From late 1967 until the bombing halt in October 1968, the rate was about 2 to 1. Although this is a very acceptable rate, it is not as high as either the USAF rate during the Korean War (10 to 1) or the Israeli rate in the 1973 War (approximately 50 to 1). But the different circumstances of the wars in Korea and the Middle East prevent us from making responsible judgments about the relative quality of pilots or equipment. For one important consideration, there were no political constraints in those two wars that required positive visual identification before the pilot could open fire. Although this restriction was lifted from our pilots to some extent during the 1972 campaign, this freedom came too late to have a significant effect on the overall exchange rate. In addition, radar coverage of the battle areas in both Korea and the Middle East was much more complete than our coverage of North Vietnam, and good radar coverage equates directly to favorable shooting positions. Thus both political and technological factors tended to depress our kill ratio in Vietnam, with political constraints being probably the most significant factor.

SUMMARY

The most precious thing an air force can provide to an army or navy is air superiority, since this gives to surface forces the ability to carry out their own plan of action without interference from an enemy air force. Without air superiority, tactical flexibility is lost. Our Army and Navy enjoyed complete immunity from attacks by the North Vietnamese Air Force. Our deployments of troops, locations of supply points, and concentrations of ships in ports were never restrained because of a threat from the North Vietnamese Air Force. Thus the air superiority that was established and maintained in World War II and Korea was even more pronounced in Vietnam.

The necessity for having positive control of the forces operating over enemy territory was demonstrated many times. With jet aircraft operating at such high speeds and with missiles permitting a greater variety of firing opportunities, control of the battle is more critical, complex, and demanding than ever.

Through pilot skill, improvisation, and training, the air battle over the skies of North Vietnam was fought and won. Even though the exploitation of our superiority was limited by political decisions, the end result for the North Vietnamese Air Force was that we could use their air space to perform combat missions, and they couldn't use ours. That is what air superiority means.

CHAPTER IV

FOOTNOTES

[1] Dictionary of Military and Associated Terms, Joint Chiefs of Staff Publication 1 (Washington, D.C.: Government Printing Office, 1974), p. 20.

[2] The Lord Tedder, Air Power in War, The Lees Knowles Lectures for 1947 (London: Hodder and Stroughton, n.d.), p. 32.

[3] Dwight D. Eisenhower, Crusade in Europe (Garden City, N.Y.: Doubleday & Company, 1948), p. 29.

[4] The Lord Tedder, With Prejudice, The War Memoirs of Marshall of the Royal Air Force (Boston: Little, Brown and Company, 1966), p. 546.

[5] The United States Strategic Bombing Survey, Overall Report (European War) (Washington, D.C., 1945), p. 21.

[6] The Lord Tedder, With Prejudice, p. 638.

[7] Trevor J. Constable and Raymond F. Toliver, Horrido and (New York: Ballantine Books, Inc., 1968), p. 335.

[8] Matthew B. Ridgway, The Korean War (Garden City, N.Y.: Doubleday & Company, 1967), pp. 75–76.

[9] United States Air Force Operations in the Korean Conflict, USAF Historical Study No. 72, 1 November 1950—30 June 1952 (Washington, D.C.: Department of the Air Force, 1955), p. 107.

[10] Ibid., p. 129.

[11] Korean Conflict, No. 127, (1 July 1952-27 July 1953) p. 58.

[12] Korean Conflict, No. 72, p. 69.

[13] William C. Westmoreland, A Soldier Reports (Garden City, N.Y.: Doubleday & Company, 1976), p. 85.

[14] Working Paper for CORONA HARVEST Report, Out-Country Air Operations, Southeast Asia, 1 January 1965—31 March 1968, book 1, p. 34.

[15] CORONA HARVEST Report, Out-Country Operations, p. 178.

[16] Wesley F. Craven and James L. Cate, eds., The Army Air Forces in World War II, vol 2: Europe: Torch to Pointblank (Chicago: The University of Chicago Press, 1949), p. 682.

[17] CORONA HARVEST Report, Out-Country Operations, pp. 147, 241.

[18] Ibid., p. 181.

[19] Ibid., p. 227.

[20] Ibid., p. 259.

[21] Ibid., p. 210.

[22] Ibid., p. 285A.

[23] Illustrated History of the United States Air Force in Southeast Asia, Office of Air Force History, Hq U.S. Air Force, July 1974, p. 24. (draft).

[24] CORONA HARVEST Report, Out-Country Operations, p. 168A.

[25] John P. McConnell, General, USAF (Ret), interview held at Pentagon, Washington, D.C., March 1976.

[26] Paul Burbage, et al., Battle for the Skies Over North Vietnam, USAF Southeast Asia Monograph Series, vol. 1, monograph 2 (Washington, D.C.: Government Printing Office, 1976), p. 145.

[27] Korean Conflict, No. 72, p. 115.

[28] CORONA HARVEST Report, USAF Air Operations in Southeast Asia, 1 July 1972—15 August 1973, vol 2 (Hickam AFB HI: Department of the Air Force, Hq Pacific Air Forces, 1973), p. IV–351.

[29] CORONA HARVEST Report, Out-Country Operations, p. 168A.

CHAPTER V

INTERDICTION
(WORLD WAR II, KOREA AND VIETNAM 1964–1968)

From World War II emerged the three basic missions of tactical airpower: counter air, interdiction, and close air support. Although their priority depended upon the battle area and the stage of the war, it was generally in the order listed, because air superiority allowed the other missions to be conducted without interference from the enemy air force.

Once control of the air could be maintained, interdiction proved to be the most effective mission for tactical airpower in World War II and Korea. During World War II, 80 to 85 percent of the tactical air effort was devoted to counter air and interdiction. In Korea, almost 48 percent of Air Force sorties in 1952–53 were allocated to interdiction and armed reconnaissance. [1]

The interdiction campaign begins with attacks against the production sources of war materiel. It continues to bring that materiel under attack as it moves through the air, sea, and land lines of communication to the battle area. Although intense attacks at the source of production have the highest potential for long-term decisive effects, such attacks do not immediately affect the fighting ability of the forces already in the field. Consequently many attacks concentrate on vulnerable supply lines and storage areas to destroy material before it reaches the combat area.

Once forces and supplies arrive in the forward area, they are difficult to destroy except during a major ground action by either enemy or friendly forces. With such action, supplies and forces are concentrated in the battle area, where their vulnerability to air attacks increases sharply.

"OVERLORD" AND INTERDICTION

In World War II the destruction of the synthetic oil plants and related war goods had such a deleterious effect on the fighting ability of the German armed forces that Reichsminister Albert Speer said the oil attacks

163

of 1944 brought about the decision of the war.[2] It had become only a matter of time until the German Army would have ground to a halt for lack of supplies for its armored and mechanized divisions. The Allies, thus, had placed the German Army in serious logistical problems before the invasion.

The object of the initial phase of the interdiction campaign was to reduce the supplies available to the German armed forces so drastically that they would be unable to contain an Allied landing; furthermore, if a counteroffensive developed, insufficient fuel, ammunition, and vehicles would limit that counteroffensive to only a few days.

As Operation OVERLORD (the Normandy invasion) neared, most of the Allied bombers and fighters were diverted from strategic bombing to interdicting the transportation system that was essential for moving *Wehrmacht* reserves to the French beaches where Allied ground forces were to be landed.

The commanders of the American and British bomber forces agreed, some of them reluctantly, to redirect the total air effort for the last three months before the invasion to the destruction of the transportation system in France. Air Marshal Leigh-Mallory, Allied Expeditionary Air Force commander, developed the plan with overall guidance from Air Chief Marshal Tedder, General Eisenhower's deputy. As might be expected, both General Spaatz and Air Chief Marshal Harris, the bomber commanders, vigorously objected to being placed under the operational control of Leigh-Mallory. As a compromise, Tedder acted for Eisenhower as the operational commander for overall planning of the air war. For the invasion, the bomber forces came under the operational control of Eisenhower.

Results of AAF bombing attacks on railroad yard somewhere in France.

Leigh-Mallory's staff developed a detailed interdiction plan based on about 80 targets along the Seine and Loire Rivers and in the Orleans Gap. These targets created an arc of 100 to 150 miles from the beachhead. The bridges across the two rivers and the rail lines through Orleans provided a natural barrier through which the major reserves of Von Rundstedt's army group would have to move. Air Vice Marshal Kingston-McCloughry makes clear the significance of the interdiction mission: "The success of the invasion would clearly depend upon two factors: first, upon elimination of the enemy's power to interfere with the Allied landing, reinforcement and supply; and secondly, upon ability to ensure that the enemy forces attacking the bridgehead did not increase at a more rapid rate than the Allied forces defending and extending it."[3]

In March 1944, attacks began against these targets. All elements of the German logistical system were brought under attack from either the RAF 2nd Tactical Air Force or the American 9th Air Force. The B–17s and B–24s of Spaatz's strategic air force pounded them during the day, and Harris' Lancasters picked up the task during the night. Fighters of the 9th Air Force were scheduled against all roads and railroads throughout the area knocking out trucks, locomotives, and freight cars. The end result, according to Kingston-McCloughry, was that "during the first four weeks after D-Day, the enemy was able to dispatch to the Normandy bridgehead an average of only four troop trains per day; and even of these, the majority never reached their destination."[4]

Because of this campaign, the Germans were unable to move sufficient troops to overwhelm the Allied landing forces. The reserves that did arrive were so weakened that they lacked the fighting stamina demanded

Bomb damage to Folligny Marshalling Yards, France, 8 August 1944.

165

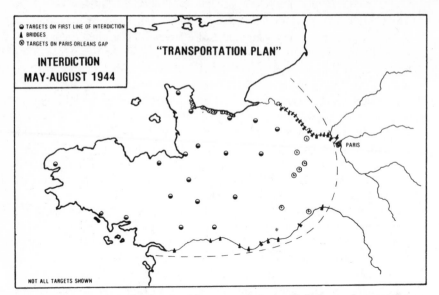

INTERDICTION
MAY-AUGUST 1944

"TRANSPORTATION PLAN"

⊖ TARGETS ON FIRST LINE OF INTERDICTION
▲ BRIDGES
⊗ TARGETS ON PARIS-ORLEANS GAP

PARIS

NOT ALL TARGETS SHOWN

by the intensity of the battle. Nor was there sufficient equipment to re-outfit and prepare them for battle. Von Rundstedt, overall commander of German forces, stated:

> It was all a question of air force, air force and again air force. The main difficulties that arose for us at the time of the invasion were the systematic preparations by your air force; the smashing of the main lines of communications, particularly the railway junctions. We had prepared for various eventualities . . . that all came to nothing or was rendered impossible by the destruction of railway communications, railway stations, etc. The second thing was the attack on the roads, on marching columns, etc., so that it was impossible to move anyone at all by day, whether a column or an individual, that is to say, carry fuel or ammunition. That also meant that the bringing up of the armoured divisions was also out of the question, quite impossible. And the third thing was this carpet bombing. . . . Those were the main things that caused the general collapse.[5]

OVERLORD'S LESSONS

Every major ground campaign through the remainder of World War II was coordinated with an interdiction campaign. After the landings, the primary interdiction campaign fell to the fighters and bombers of the U.S. 9th Air Force and the RAF 2nd Tactical Air Force, but strategic bombers also were used extensively at the beginning of major offensives. From these interdiction campaigns, commanders learned that heavy pressure had to be put on the *Wehrmacht*, forcing it to consume supplies at an accelerated rate. With the interdiction campaign destroying critically needed supplies, the *Wehrmacht* was then forced to fall back, or if units stood and fought, their positions could be overrun because of the logistics

Retreating Nazi forces found AAF tactical fighters a constant menace. An American jeep drives through the twisted wreckage of what used to be the transportation of a German Panzer division.

failure. Regardless of their will to fight, the lack of needed weapons, food, and ammunition made it infeasible for German units to stay in the battle.

From these lessons of World War II, the concepts of interdiction developed: (a) Strike the source of the war material; (b) concentrate the attacks against the weak elements of the logistical system; (c) continuously attack, night and day, the major lines of communication supporting the army in the field; (d) inflict heavy losses on enemy logistics and forces before they approach the battlefield where the difficulty of successful interdiction is greatest; (e) keep continuous ground pressure on the enemy to force him to consume large quantities of logistics.

RETREAT TO PUSAN

With the outbreak of the Korean War, many of these lessons had to be learned again. When the North Koreans crossed the 38th parallel on 25 June 1950, there were insufficient forces to stop the penetration. South Korean forces fell back in disarray; no one knew whether the advance could be stopped. No significant natural barriers stood between retreating South Korean and U.S. forces and the Pusan stronghold. However, as the North Korean logistical lines extended, 5th Air Force fighters and bombers already in control of the air began to impose a heavy toll on the enemy forces and supplies.

At the point in an offensive movement or retreat when supply lines are expanded, airpower can have the most profound effect on an enemy ground force. As the ground force becomes increasingly exposed on open roads, at bottlenecks at bridges, fords, and defiles, fighters can destroy a significant part of the force with repeated bombing and strafing attacks.

As the Allied forces withdrew into the Pusan bridgehead, 5th Air Force inflicted such high losses on North Korean personnel and equipment that the enemy ground forces had neither the strength nor supplies to crack the perimeter. Airpower gave the Allied army time to bring in enough reinforcements and logistics to begin a breakout and pursuit. When the Allies began the pursuit north, and after the Inchon landing, the North Korean Army was near defeat. Airpower was crucial in preparing for the Allied offensive; massive close air support at Pusan, interdiction of all the major LOCs, and isolation of the area around Inchon from enemy ground forces all took place under the complete air superiority established earlier by 5th Air Force.

A major factor in the effectiveness and success of any interdiction effort is careful planning. At the direction of FEAF, 5th Air Force early developed an interdiction plan in four phases. The first phase, designed to slow the enemy advance, covered the Allied retreat and involved the destruction of 60 rail bridges north of Seoul. These attacks were coupled with armed reconnaissance of all major roads used by the advancing enemy army. Later, the second phase prepared the objective area for the 8th Army drive into North Korea. In this phase, 33 bridges north of a line between Pyongyang and Wonsan were targeted. Because the 8th Army advanced rapidly, air commanders considerably modified this phase. When the Inchon landing ended any significant resistance, the interdiction moved further north. The third phase of the plan required B–29s to strike 34 additional bridges on the east and west coasts. The fourth phase, for December 1950, was not implemented because of the intervention of the Chinese Communists. [6]

North Korean supplies burning. Box cars were strafed by USAF F–80s.

CHINESE CROSS THE YALU

While the buildup of Chinese forces took place across the Yalu, the Allies differed as to whether the Chinese would invade North Korea. The answer was not long in coming. The Chinese crossed into North Korea on 1 November 1950, with some 200,000 troops.[7] Almost immediately MacArthur changed the mission priority of FEAF; he directed maximum effort to close air support of ground forces now under heavy attack. MacArthur requested authority not only to destroy the bridges the Chinese used to cross the Yalu but also to attack military targets and airfields in Manchuria. He thought it might be necessary to give up all of Korea if the authority were not given. As the United Nations forces began withdrawing on 1 December 1950, the JCS denied the request and directed that MacArthur establish a series of defense lines. He was to use maximum airpower to interdict and attack the advancing army. But when the 8th Army, "committed to action without a long-range campaign plan, found itself retreating southward in a 'critical' situation, FEAF was directed to employ its aerial predominance in close support of ground units to the exclusion of all else. This absolutely precluded a proper interdiction program. Had FEAF not finally persuaded the CINCFE staff that curtailing the forward flow of communist troops and materiel was essential to the war effort, the 8th Army might never have recovered from its 'critical' position."[8]

The combined efforts of 8th Army and 5th Air Force finally stopped the enemy and stabilized the line of contact on 22 December 1950. Except for some limited adjustments, the line remained essentially the same

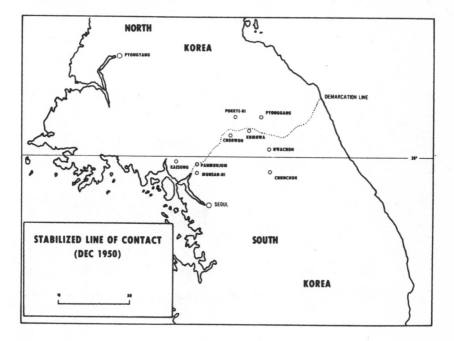

throughout the remainder of the war. Stratemeyer, the commander of FEAF, turned his air force to exert maximum interdiction pressure on the enemy logistical structure plus key targets supporting military forces deployed along the battle line. This shift was an attempt to convince the Chinese that they could not win the campaign. It supported the changed United Nations objective: rather than unification of North and South Korea by political and military action, the UN sought a negotiated settlement along the 38th parallel.

The enemy supply system to be interdicted was extensive. More than 600 miles of North Korean rail lines supported the combined and reconstituted North Korean and Chinese Armies. Planners estimated that, because of the interdiction campaign, the enemy would be able to fight offensively for only two or three weeks before lack of supplies would force him to give up the offensive. If the 8th Army, then, could contain the attack for this period, attrition would compel the enemy to give up the offensive. Weyland, Stratemeyer's successor, observed that "although close air support contributed, the major effect upon the enemy was produced by airpower applied in the rear of his front line combat zone."[9] Once again the lesson that emerged from World War II was relearned; it would later emerge from the Vietnam War.

As the Korean War turned into a stalemate on the ground, the JCS turned to airpower to break the impasse and compel the North Koreans and Chinese to negotiate a settlement. Talks at Kaesong were suspended indefinitely on 25 September. All offensive ground force actions stopped after October 1952. Airpower was the only force sufficient to convince the Chinese that they couldn't win the war and that their only alternative was to reach a settlement as proposed by the United Nations negotiators. In November 1952, General Omar N. Bradley, Chairman, reflected the JCS view when he stated that United Nations Command airpower "constitutes the most potent means, at present available to UNC, of

F–51s from the 18th Fighter-Bomber Wing "Truckbusters" being readied for another day of interdicting the enemy's LOCs.

maintaining the degree of military pressure which might impel the communists to agree, finally, to acceptable armistice terms."[10]

During the spring offensive of 1953, FEAF strategy was to inflict maximum destruction at key points. Pyongyang, not only capital of North Korea, but also psychological symbol of communist strength, was subjected to a two-day maximum attack. The Sui Ho power plant, largest in North Korea, was targeted for attack. These air attacks culminated in the North Koreans and the Chinese agreeing to the essential terms proposed by the United Nations Command.

For what was to come later, it is significant to note that there were few restraints on FEAF's employment of forces or selection of targets. FEAF was limited in attacks only along the Yalu and the North Korean border with Russia. The decision about whether to bomb the Korean dams was left to the United Nations commander. Weyland, who was given the authority for the final recommendation, "felt himself morally compelled to rule that these dams could not be attacked for the purpose of

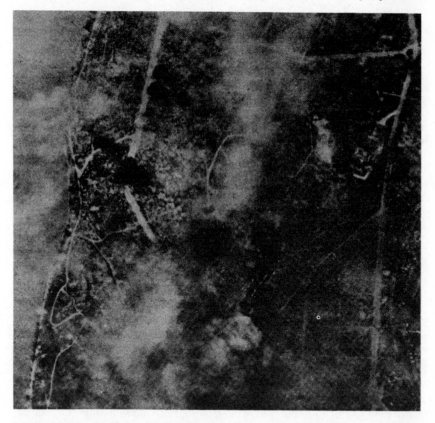

Smoke from bombs delivered by Fifth Air Force aircraft clouds a portion of the city of Pyongyang during heavy all-day attacks the summer of 1952 (above). U.S. Navy and Marine carrier-based aircraft joined USAF fighter-bombers in destroying a machine factory in the heart of Pyongyang the same summer (below).

171

destroying the North Korean rice crops, and he permitted air strikes against only those dams whose released flood waters would wash away rails and military supplies."[11] Except for the dam system, no other targets of significance were restricted from attack. This freedom to target and to use airpower brought the war to an acceptable conclusion. Interdiction was the fundamental mission that pressured a settlement.

INTERDICTION BEGINS—VIETNAM

The aims of airpower in the Vietnam War were much like those of the Korean War; however, the very severe restraints on targets and employment made those aims much more difficult to reach. This similarity of objective appears also in Presidential statements about each war. President Truman stated on 10 April 1951 that the U.S. military objective is "to repel attack . . . to restore peace [and] . . . to avoid the spread of the conflict."[12] About Vietnam, President Johnson, on 23 February 1966, stated, "Some ask if this is a war for unlimited objectives. The answer is plain. It is 'NO.' Our purpose in Vietnam is to prevent the success of aggression. It is not conquest; it is not empire; it is not foreign bases; it is not domination; it is to prevent the forceful conquest of South Vietnam by North Vietnam."[13]

172

From the statements of Truman and Johnson, it is evident that both Presidents saw the two wars in much the same perspective. Both wars' purposes were to halt the spread of communism, stop the aggression, negotiate a settlement, and permit Korea and South Vietnam to determine their own futures without external interference. But, where President Truman permitted latitude to his field commanders in prosecuting the war, President Johnson narrowly limited the latitude of the commanders' decisions about the air war over North Vietnam. As a consequence of this difference between the two Presidents, the air war in North Vietnam was directed and prosecuted much differently from the air war in North Korea. This difference existed even though the capabilities of the forces were similar and, except for the SAMs, the target conditions were remarkably alike.

The objectives of the air campaign in North Vietnam were never changed significantly throughout the war. Though frequently restated in different words, they all added up to the same thing. As stated by Department of Defense representatives and by then Secretary of Defense McNamara in Congressional testimony, the objectives of the bombing campaign were:

(1) To reduce the flow and/or increase the cost of infiltration of men and supplies from North Vietnam to South Vietnam;

(2) To make it clear to the North Vietnamese leadership that as long as they continued their aggression against the South, they would have to pay a price in the North;

(3) To raise the morale of the South Vietnamese people. [14]

Of these three, only the first is a military objective. The other two are psychological; they result from gaining or failing to gain the military objective.

The interdiction campaign, thus, was the heart of the air strategy. Its purpose was to destroy equipment and supplies, and to disrupt, delay, and harass the movement of men, equipment, and supplies to the battlefield in South Vietnam. Strategic planners believed that the level of destruction of all the war-related activities in North Vietnam would be so extensive and debilitating that the North Vietnamese would negotiate rather than continue to pursue the war militarily. Thus, the interdiction campaign in the Vietnam War had objectives much like those of the interdiction campaign in the Korean War: "to interfere with and disrupt the enemy's lines of communication to such an extent that he will be unable to contain a determined offensive by friendly forces or be unable to mount a sustained major offensive himself." [15]

A further likeness between the two wars was the destruction of key psychological and military targets to drive home the hopelessness of continuing the military campaign. Attacks against Pyongyang of 5 July 1952 and periodically thereafter until an armistice was signed paralleled our later all-out bombing of the greater Hanoi area in May and December 1972. These attacks had the same psychological objectives as those in

173

403-892 O - 83 - 13

Korea: to convince the enemy by such overwhelming destruction of the heart of their country that the best way out of war was at the conference table and not on the battlefield.

NORTH VIETNAMESE LOGISTICAL SYSTEM

The complex political factors and the equally complex command organization in Vietnam led to many misconceptions about the interdiction campaign. Some observers viewed the campaign as four separate but somewhat interrelated operations—the bombing campaign in North Vietnam, the interdiction of the Ho Chi Minh Trail in southern Laos, the attacks against the LOCs in northeastern Laos, and the attacks against the roads and trails in South Vietnam. We had, however, only one interdiction campaign and it embraced all these areas. We used different rules of engagement and different tactics in each area, but all of the parts made up the total campaign. Because of the confusion about the totality of the campaign, few observers appreciate how each element of the campaign had an effect on the others.

The campaign, to be effective, had to begin with attacks on the head of the system in North Vietnam. At that point the lines of communications were most vulnerable to an attack, and there the supplies and repair and support facilities for the entire logistics system were located. Sortie for sortie, there the most devastating attrition on supplies could be achieved, and there the most vulnerable bottlenecks were located. Interruptions in the flow at those points would create a greater delay and disruption of materiel moving through the rest of the network. Approximately 30% of the important transportation targets were in Route Packages IV, V, and VI. If we were significantly to reduce the logistics flowing to the North Vietnamese and Viet Cong in the south, we had to deal that part of the system, with its high density of critical items, the most damaging blow.

As the transportation system threaded its way south in North Vietnam, we found fewer vulnerable segments that could be blocked for any length of time. Furthermore, the nature of the terrain allowed the North Vietnamese to relieve backed up traffic with by-pass routes. In the southern part of the system, traffic dispersed and moved in such small segments that we could not achieve a satisfactory destruction rate per attack. As the supplies moved closer to the battlefield in South Vietnam, less sophisticated forms of transportation reduced further their vulnerability to air attack. While freight trains of 40 or more cars transversed the northeast railroad leading from China to Hanoi in Route Package VI, supplies made their way across the DMZ in trucks; during 1966–1968, many supplies were delivered into South Vietnam on bicycles and by porters with "A" frames. The less vulnerable means of transporting supplies required intensive reconnaissance, even to locate the trails. It was yet more difficult to attack such movements at night.

Materiel moving through the system in southern Laos posed a different problem. The Hanoi delta and the LOCs along the coast to the DMZ were in open terrain. We could more readily detect movement and use air attacks to halt the flow in such areas. However, the roads in Laos were concealed in many places by triple canopy trees,* which, along with clever camouflage, helped hide truck movements. When attacked, trucks moved off the road into the surrounding jungle. Because of the jungle cover, successful interdiction in Laos was extremely difficult. Hence, failure to stop the supply flow at the head of the system (in North Vietnam) made it most difficult to pinch off supplies for the enemy's army in South Vietnam. Unlike other parts of the transportation system, the South Vietnamese section used no major road systems. From 1965 to 1972 most of the material was moved by porters and some limited number of vehicles. (However, in the 1972 Easter offensive, the enemy used well-developed roads built during the bombing halt to support forces in Military Regions (MR) I, II, and III.) By the time the logistics had moved into base areas within South Vietnam, interdiction was not a productive effort.

Because of the relationship of the parts of the LOCs, the place to put pressure on the system was in the heart of North Vietnam. Blockading or bombing the ports was essential for a decisive campaign. Without eliminating the ports where the bulk items entered, reducing the flow to South Vietnam was again made more difficult.

Even without eliminating the ports, the interdiction campaign was able to limit the number of forces the North Vietnamese could support in the south. Not until the interdiction campaign ended with the termination of U.S. involvement could the North Vietnamese logistically support and deploy their full strength of 18 to 20 divisions. Before the 1975 offensive, they never deployed more than 11 or 12 divisions, apparently for fear of the destruction they would suffer by exposure to our airpower. A similar thing happened in the Korean War when the Chinese Communist Army was stalemated with more than 1,000,000 reserve troops who could have been thrust into the battle to break the stalemate. The destructive toll of the 5th Air Force interdiction campaign probably led to the decision by Chinese leaders that UNC airpower would make it unfeasible to sustain such a deployment without air superiority. There was no evidence that the Chinese lacked Russian support for deployment of larger ground forces. Nor was there any evidence that the North Vietnamese lacked Russian support if they had elected to deploy another 100,000 troops.

VARIABLES INFLUENCING THE INTERDICTION CAMPAIGN

As much as terrain or political restraints, weather was a key factor in planning and executing the air campaign. The enemy altered logistics movement according to the weather. From May until September, during the southwest monsoon season, heavy rain showers and thunderstorms

---

*The trees in this area grew to basically 3 heights, 50', 100', and 150', creating 3 separate canopies of foliage that barely permitted even sunlight to reach the ground.

reduced visibility as they do in southern Florida in the summer. Though never halting air operations during these times, the weather did require some adjustments. The heavy rains also flooded the roads to Laos, thereby curtailing the enemy's use of these roads to move supplies to the northern two military regions.

Rather unexpectedly, the best weather for air operations in North Vietnam occurred during this southwest monsoon. On relatively few days were air operations restricted. The only real problem was low visibility during the early morning and late afternoon. Since the roads were dry as far south as the DMZ, the enemy moved most of his traffic during these times, thus offering us lucrative targets. The southwest monsoons did offer some problems for the rendezvous of tankers and fighters because it was necessary to maneuver around thunderstorms. Such maneuvers were critical at those times when strike forces were coming back from a fight and were short of fuel; there was little time to delay an air refueling. Likewise, recovery at home base could be a problem if a thunderstorm were passing through when a strike package of 40 or more aircraft were eager to get on the ground, or worse, were having to land as quickly as possible because of low fuel or battle damage.

Perhaps the most direct effects of the weather on the recovery of the force were the pools of water on the runway. They resulted in aircraft planing* with all the accompanying problems of directional control. In some cases, it was necessary for the pilot to make what amounted to a carrier landing by engaging a cable across the approach end of the runway. When this happened, the entire tempo of recovery and staging for subsequent operations slowed down.

The northeast monsoon, on the other hand, caused the most problems for air operations over North Vietnam. When this weather moved in, flight conditions for a visual attack on the upper route areas were very restrictive. Thus, when the weather was bad in North Vietnam, from September to May, it was good over South Vietnam and Laos. The weather would have been less a problem had it not been for restraints imposed upon our forces for positive visual identification of targets. However, even when these restraints were eased, pressure to hold down collateral damage to civilian areas on the fringes of a target demanded that fighters have a positive visual identification of the target before attacking. Such restraints made the task much more difficult for our pilots.

Westmoreland recounts in his book A Soldier Reports, "In the case of Senator Stuart Symington of Missouri, a staunch supporter of the war, what he saw changed his views. Visiting an aircraft carrier, he was shocked by the extreme precautions imposed on attacking aircraft crews

---

*Planing: On a wet runway the water can create a separation between the aircraft's tires and the runway. When this happens, the aircraft "planes" on the water much like a motor boat "gets up on the plane." Directional control becomes extremely difficult.

in an effort to avoid civilian casualties. Planes attacking targets in North Vietnam had to follow perilous, circuitous routes that exposed them much longer than necessary to enemy anti-aircraft fire. While sympathizing with the objective of avoiding civilian losses, Senator Symington resented the additional danger that the policy imposed on American airmen."[16] This policy, however, did not apply for radar bombing attacks. In those cases, the radar return from the target and the capability of the aircraft's radar were the determining factors in executing the mission.

During the northeast monsoon, we could expect that we would have only about four to six days when visual attacks could be made in the greater Hanoi area. During 1965–1968, we sought a 10,000 foot ceiling and no more than 5/10 to 6/10 cloud coverage so that our pilots had both sufficient visibility to see a SAM launch and adequate ceiling for maneuvering to avoid the SAM.

We were able to operate in adverse weather because of LORAN* and radar. When LORAN was introduced in March 1968, it became feasible to strike into most areas of North Vietnam regardless of the weather. However, range limitations determined by the siting of LORAN stations restricted full coverage of all the target areas, particularly those in Route Package VI A. LINEBACKER I operations, therefore, had some of the same all-weather restraints as ROLLING THUNDER, especially for fighter operations. For the B–52s in LINEBACKER II, the "11-day offensive," weather was not a major consideration since that operation was scheduled against targets which provided sufficient radar return. There was no difficulty in identifying and bombing such radar-significant targets under all weather conditions.

While the B–52s attacked radar targets during LINEBACKER II, the tactical forces hit pinpoint targets like the Canal Des Rapides bridge, the Paul Doumer bridge, and the smaller railroad yards and spurs in the buffer zone around Hanoi. These targets were most difficult to strike under weather conditions because of the threat of unwanted damage to adjacent areas and, consequently, were never scheduled for attack under poor weather conditions. Nevertheless, though the tactical forces struck only during good weather, they were able to keep such targets neutralized during ROLLING THUNDER and LINEBACKER.

ALL WEATHER ATTACKS

The most profound difference between the interdiction campaign of 1965–1968 and that of 1972 was our use of B–52s at night and during marginal weather conditions in 1972. This mass concentration of airpower around-the-clock made a major difference in the psychological impact of the bombing. Earlier in 1965–1968, B–52s were withheld from bombing in the Hanoi delta, a void tactical airpower alone could not fill. The F–4s

---

*LORAN—A long range radionavigation system for fixing a position by using the time difference when receiving pulse transmissions from two or more fixed stations.

and F–105s were able to block major movements in the rail and supporting logistical system with day attacks and some limited night attacks. But when the northeast monsoons began, tactical aircraft could operate only four or five days a month against the high value targets such as the Hanoi Railcar Repair Shop. Therefore, during such weather, the main pressure on the logistical system came at the periphery of the high value targets since ground control radar, LORAN, and to some extent, aircraft radar could be used against these targets.

By filling the gap that was apparent in the 1965–1968 campaign, the B–52s allowed the full force of the air campaign to be driven home to the North Vietnamese day and night. Thus the 1972 campaign had all the elements of airpower that had been present in World War II and Korea. Around-the-clock bombing gave the North Vietnamese little opportunity to repair damage to the logistical system, thereby limiting their ability to meet the needs of their forces in South Vietnam. We knew from the bombing campaign in World War II that highly organized labor forces could return a damaged target to partial productivity in a relatively short time if that target were not struck repeatedly.[17] Thus, in the Vietnam campaign, we scheduled the targets for reattack about the time repairs had been completed. With B–52s striking during bad weather and at night, we kept most elements of the important targets neutralized during the 1972 campaign.

EARLY DEVELOPMENT

In 1967, a radar bomb facility was established at Site 85* in Laos. The equipment, the AN–MSQ–77 Radar Bomb Directing Central (MSQ), was like that in South Vietnam used for directing aircraft to preselected targets in bad weather. Site 85, deep in Pathet Lao territory about 25 miles from Sam Neua, perched on a 5,200-foot mountain top 160 miles west of Hanoi; its equipment had been moved in by helicopter. A detachment of skilled Air Force personnel manned the radar which provided guidance to targets within 30 miles of Hanoi. Strike aircraft were equipped with beacons to enhance the radar's tracking ability.

World War II had provided the initial development for this type of radar bombing. An SCR–584 gun-laying radar was modified for bombing during the Italian campaign. This basic technique for bombing in bad weather and at night became standard procedure during the European campaign of World War II and the Korean War. It was used as well in South Vietnam, Laos, and Cambodia. This same system was also used during the 1965–1968 campaign in the area below the 18th parallel (Route Package I). Most of the targets in this area, however, were not suitable for the kind of blind area bombing the system demanded. Vehicles, artillery, and watercraft were the most frequent kind of targets; they

---

*Locations of airstrips, navigational facilities, and other military bases in Laos were identified by number.

required visual attacks to destroy them. In Route Package VI, though, the marshalling yards, military barracks, Thai Nguyen steel mill, depots, and transshipment points were suitable targets for MSQ bombing. With a CEP of approximately 500 feet for the bombing aircraft, these large area targets were good candidates for attack during bad weather.

The MSQ at Site 85 which directed those attacks was unique because of certain political problems. Ambassador William H. Sullivan was reluctant to permit the site in Laos to provide control of aircraft over North Vietnam. His position was that to direct air strikes over North Vietnam from Laos would appear an escalation of the war, in that Laos could be viewed as a base of operations for attacks against North Vietnam. A unique technique was devised to satisfy this political objection; a C–135 relay aircraft, positioned in the Gulf of Tonkin near the 19th parallel, would relay instructions from the MSQ site in Laos to the strike aircraft. The short time delay in the relay operation was accommodated in timing the instructions to release the bombs.

But MSQ bombing had other problems. Strike aircraft were extremely vulnerable to SAMs during this type of attack. The last 60 miles into the target had to be a steady bomb run with speed and altitude held very precisely. Variations in heading, speed, or altitude would produce gross errors in bombing. With such a stabilized run, the force was easily brought under intense barrage fire from SAMs and anti-aircraft artillery. However, to keep the pressure on the enemy during bad weather, the Air Force tried this technique on 18 and 19 November 1967 in the Hanoi area. [18] These missions were unsatisfactory because the MSQ was unreliable at its range limits. Because of these limitations of the MSQ, further attempts to bomb by MSQ in the high threat area were cancelled. However, MSQ bombing continued throughout the war as the main method of all-weather bombing in Laos, Cambodia, South Vietnam, and the lower route packages of North Vietnam.

Many attempts were made during the 1965–1968 campaign to adapt some of the techniques of World War II and Korea to the weather problem. Because of the pinpoint targets such as bridges, airfields, power plants, or repair facilities, most of these measures were not satisfactory. The pressure to hold down damage to civilian targets prevented missions that would have been run in World War II and Korea. These restrictions made it essential to limit the probability that bombs would impact outside the prime target area. Although North Vietnamese propaganda hammered the theme that U.S. bombing was directed at civilian targets, never in the course of the war was a target selected for any reason other than its military significance. Inevitably, in combat some bombs are dropped out of the target area. These cases, however, result from improper release, jettisoning because of enemy fighters, malfunctioning bomb racks, or, in some cases, from releasing ordnance when the aircraft is in an improper attitude, thereby causing gross errors. Many of the North Vietnamese claims of civilian damage came about because their own anti-aircraft rounds and SAMs missed their mark and impacted the ground. As far as we know, the North Vietnamese had no restriction on where SAMs or AAA could fire. They were sited for the best defense of a target, irrespective of the effect such weapons would have on their own civilians, despite the fact that a SAM that didn't detonate in the air was potentially a live bomb when it hit the ground. Unfortunately, we were unable to measure the self-inflicted damage from such firings.

Because of the MSQ deficiencies and the target restrictions, we tried alternative weapons and methods. The F–105, originally designed for the nuclear low level mission, was an excellent fighter-bomber for the day strikes over North Vietnam. Its radar, however, had insufficient discrimination to make it suitable for night and weather bombing missions. Nevertheless, during 1966 and 1967, some all-weather missions were run against Yen Bai, a railroad marshalling yard on the northwest railroad. However, this experiment was abandoned because the accuracy was insufficient to be effective.

Similar missions were tried with the F–4, which had radar designed for air-to-air fighting. It lacked the needed target definition for air-to-ground radar bombing. Furthermore, defenses in the Hanoi delta demanded the same suppression for a night raid as for a day raid. The fighter simply wasn't accurate enough to sustain the campaign at night.

In the Korean War, it had been feasible to use pathfinder techniques developed during World War II by Bomber Command. In those procedures, a highly select crew was scheduled into the target area ahead of the main bombing force. The pathfinder located the target and marked it with varied colored flares for the bombing force. In Korea, hunter-killer teams of B–26s were similarly employed against key parts of the LOCs. A lead B–26 would locate, identify, and mark the target. Another B–26 or a flight of fighter-bombers would attack the target under the flares laid down by the B–26. This tactic, suitable for a lightly defended area, could not be a standard procedure for missions north of Pyongyang and along the Yalu where flak was extremely heavy. For those missions, SHORAN* bombing was used with B–26s or, later, with B–29s.

In Vietnam, missions to bomb the northeast rail line in 1966 and 1967 used the same hunter-killer concept. F–4s flew as both hunters and killers against prime targets like the Kep Marshalling Yard. A lead F–4 illuminated the target; then a flight of F–4s carrying five 750-pound bombs each followed up with an attack. Although these techniques had been refined in a test series at Eglin AFB, the problem of running such missions against a SAM environment hadn't been thoroughly explored. Once the flare aircraft ejected its first flare, the enemy had a good prediction of the flight path over the target and was then able to put up a

---

*SHORAN—A precise short-range electronic navigation system which uses the time of travel of pulse transmissions from two or more fixed stations to measure slant range distance from the stations.

very high volume of fire. Because these missions were run at 6,000 to 7,000 feet, aircraft were highly vulnerable. By mid-1967, recognizing our equipment limitations, we altered such missions to ones of harassment rather than destruction. This change, however, was not a substitute for a more concentrated night effort.

F–111 OPERATIONS

The next solution to the all-weather problem was the F–111. Because of the need to get an airplane into operation in the Hanoi area that could fly when the weather was bad and still deliver munitions with a low CEP, the F–111 was moved into combat on a compressed schedule. When the F–111s deployed to Thailand in March 1968,[19] we planned to use them against targets within the 30-mile circle almost exclusively. Its excellent radar and high speed at low altitude made this weapon system a prime candidate for these missions, the most demanding of all. At the time the F–111 was introduced into the theater, it had been in tactical units only a limited time, and operational procedures had not yet been tested in combat. Also, there were the usual equipment problems which occur with all new aircraft entering the active inventory. Nevertheless, the F–111 low-level bombing system (below 200 feet if needed) was a revolutionary breakthrough in an all-weather delivery system.

To determine the operational performance of the system under less stringent target conditions, the initial flights were flown into Route Package I. Although these targets were not as heavily defended as those in the Hanoi area, the route to the target required flight over rough terrain in Laos and gave the terrain-avoidance radar a good test. The flight path out of Takhli went first to Nakon Phanom and then to Mu Gia Pass and on to Dong Hoi, the target, using terrain following at altitudes under 500 feet for the last leg. Climb to altitude was made over the Gulf of Tonkin before returning to home base south of the DMZ. This flight path had all the attributes of an attack into the Red River delta except for the SAMs. The light to moderate anti-aircraft fire, however, provided a basis for examining the tactics that were finally adopted.

In the course of the missions, three F–111s were lost, two in March and one in April.[20] Technicians determined that the severity of the terrain and the high concentration of moisture during the monsoon period caused false indications in the radar presentation to the pilots. As a result of this very limited combat test (only 55 missions), the F–111 returned to the U.S. for further work on the radar system. Some modified F–105s were then brought in to determine their suitability for the task.

These aircraft, THUNDERSTICK II, received a very limited evaluation. They required extensive radar modification and, once modified, required a high level of maintenance skill to remain operational. The system had to be at peak performance to achieve the sought-after CEPs of less than 500 feet. The modified F–105s, nevertheless, showed promise against marshalling yards and did make a number of attacks against the

Smoke billows from secondary explosions caused by the bombing attack of USAF F–105s on the Kep Marshalling/Railroad Yard, 39 miles northeast of Hanoi. This strike took place 20 April 1967.

northwest railroad. However, they were never used against the northeast railroad before the President decided to halt the bombing north of the 20th parallel on 31 March 1968.

The B–52s were not used in Route Package VI during the 1965–1968 campaign because of the administration's concern that the North Vietnamese would view employment of this strategic weapon as an escalation in the conflict. An additional argument for not using the B–52 on these raids was the concern about the effect losing even a single aircraft would have on the image of our strategic deterrent. Airmen have long known that a penalty must be paid to penetrate any defense system, and several of us commanders felt that the loss of some B–52s to a relatively small defense system (as compared to that of the USSR) would not bring into question the ability of our strategic force to perform its mission against the Soviet Union in the event of nuclear war. Nevertheless, officials within the Department of Defense and the State Department maintained that such losses could have a negative impact on the image of our strategic forces. For these reasons, the B–52s were not to go into the Red River delta until 15 April 1972.

RAILROAD SYSTEM

The North Vietnamese railroad system consisted of nine segments, the most important parts of which were north of the 20th parallel. Almost 80% of the major targets were in this area laced together by the rail system. The most important contribution of the system was to move the main fighting weapons from China to redistribution points at Kep, Hanoi, Haiphong, Nam Dinh, and Thanh Hoa. They were then further distributed by truck and watercraft to the intransit base areas whence porters carried the weapons, food, and ammunition on the final leg into combat.

The northeast rail line, termed RR #2, ran from the southeast border of China to Hanoi. Because it was the most important segment of the system, the North Vietnamese exerted the most effort to keep it open. Eighty-two nautical miles long with a capacity of about 27,000 short tons daily, this line was near almost all the major targets in North Vietnam. If the interdiction were to succeed, it had to embrace systematic attacks on this line from the Chinese border to the heart of downtown Hanoi. Besides the repair effort, the North Vienamese demonstrated the importance of this segment of the system by concentrating SAMs and AAA along the first 30 miles of the line and then gradually thinning them out nearer the buffer zone. The 25-mile buffer zone was a self-imposed restriction to minimize possible U.S. violation of Chinese territory. The North Vietnamese took advantage of this sanctuary to stage and marshal trains during the day for night runs into Hanoi. They realized that we would strike in the buffer zone only for a particular target; we didn't permit armed reconnaissance in the buffer zone except for specific limited periods.

The two bottlenecks on RR #2 were the Canal des Rapides and Paul Doumer bridges. For supplies to move through the southern part of the system, they had to pass over these bridges. Thus, destroying the bridges and keeping them from repair was basic to slowing down the movement of war goods to troops in Laos and South Vietnam. When pilots struck these targets, they expected and received the maximum firepower the enemy could put up. The bridges were defended with the highest SAM and AAA concentration of any targets in the war. While most of the defense was trusted to SAMs and AAA, the MIGs loitered on the periphery of the SAM ring. These conditions held true during both the 1965–1968 and 1972 campaigns.

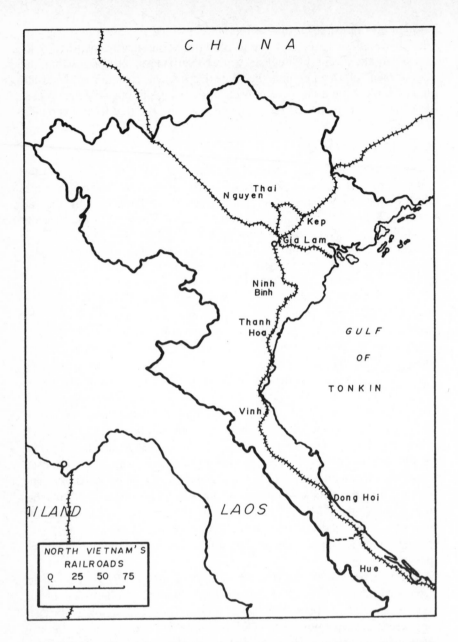

C H I N A

Thai
Nguyen

Kep

Gia Lam

Ninh
Binh

Thanh
Hoa

GULF

OF

TONKIN

Vinh

Dong Hoi

THAILAND

LAOS

Hue

NORTH VIETNAM'S
RAILROADS
0   25   50   75

Though the Canal Des Rapides bridge was released for attack in April 1967, the Paul Doumer bridge was not released until August 1967.* Thus, for the first two years of the bombing campaign, these key bottlenecks were not brought under attack. As a result, until mid-summer 1967 the

---

*See USAF SEA Monograph Series, Vol I, for complete story of the Paul Doumer and Thanh Hoa bridges.

184

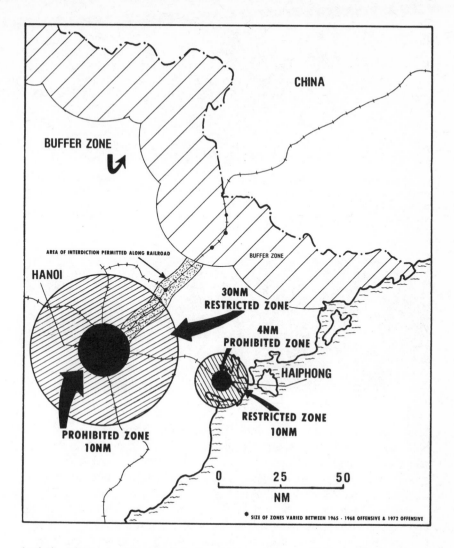

CHINA

BUFFER ZONE

AREA OF INTERDICTION PERMITTED ALONG RAILROAD

BUFFER ZONE

HANOI

30NM
RESTRICTED ZONE

4NM
PROHIBITED ZONE

HAIPHONG

RESTRICTED ZONE
10NM

PROHIBITED ZONE
10NM

0   25   50

NM

* SIZE OF ZONES VARIED BETWEEN 1965 - 1968 OFFENSIVE & 1972 OFFENSIVE

logistics flowed with relative freedom on the northeast rail line, even though most of the other smaller bridges had been struck. Because these small bridges could be repaired in relatively short time, we gained no significant reduction in logistics capability, with North Vietnamese repair teams working around the clock. But with the Canal des Rapides and Paul Doumer bridges, the Vietnamese faced a different problem. The Paul Doumer bridge, for example, was 5,532 feet long and 38 feet wide. When such bridges were dropped, to work around them with a by-pass was a major effort. Even though the North Vietnamese tried the same ruses the Koreans had earlier, photographic evidence showed what was happening. They commonly attempted to give the appearance that a bridge was not being repaired; however, during the night a floating span would be moved to fill the place of the downed span. Traffic could then flow over the

Reconnaissance photo shows five of the original spans of the Paul Doumer bridge destroyed. Over 1,900 feet of the eastern portion of the bridge were destroyed by crews from the 355th and 388th TFWs. The lower photo shows partial rebuilding of the western portion of the bridge which was damaged by attacks in October of the same year (1967).

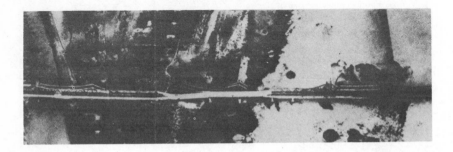

bridge throughout the night until the span was floated to an adjoining bank some yards up or down stream before sunup. The photo interpreter faced a significant challenge to locate these well-hidden and camouflaged spans. Once we confirmed this technique, though, our aircraft attacked the floating spans and the heavy cranes used to move them.

Another segment of the rail system from Kep to Thai Nguyen was important for the movement of finished steel from the Thai Nguyen steel mill into the main transportation system. When the mill was struck on 17 January 1967, however, this rail line had less significance to the North Vietnamese war effort.

The Kep-Thai Nguyen line remained blocked during both the 1965–1968 and 1972 campaigns from repeated attacks against marshalling yards at various points. These yards were the largest the Vietnamese had except for the Yen Vien and Hanoi classification yards. Immediately after each attack, the North Vietnamese always made a major effort, particularly at Kep, to get a through-line open. This work was little different from the German and Chinese attempts to restore traffic on main arteries in Europe and Korea. The fact that the line was blocked one day didn't mean we could write it off for even a week.

Because of the frequency of repair, reconnaissance missions and reattacks were all-important considerations. Reconnaissance was essential

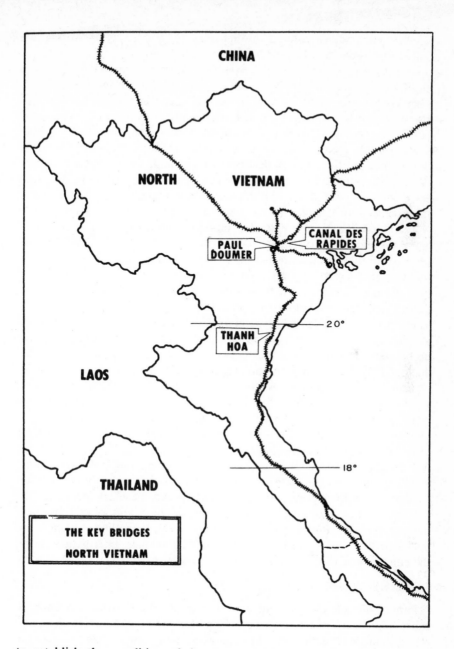

CHINA

NORTH   VIETNAM

CANAL DES
RAPIDES

PAUL
DOUMER

20°

THANH
HOA

LAOS

18°

THAILAND

THE KEY BRIDGES
NORTH VIETNAM

to establish the condition of the target, for it wasn't prudent to run a strike into a high threat area, even when cleared by the JCS, until we knew the strike would have a worthwhile target.

Another rail line segment, the Haiphong-Hanoi line, was frequently attacked along its forty miles. Task Force 77 was responsible to keep this line interdicted, especially at Hai Duong, the key choke point on this line. Since most of the bulk products such as food, clothing, fuel, and essential civilian commodities entered through Haiphong, interdiction of this line put stress on the distribution system centered in Hanoi.

The longest segment of the overall rail system ran from Hanoi to Vinh; from Vinh to the DMZ the line was unusable. From Hanoi to Vinh the line was 165 miles long, parallel to the coast, and in open terrain most of the way. It was, therefore, vulnerable to attack by air and from naval gunfire. Generally, we kept the line interdicted so extensively below Thanh Hoa that the enemy resorted to a shuttle operation between the cuts. Trucks bridged the gaps of destroyed track, and, as everywhere except during the bombing halt, movement took place during darkness. Because the mountains were about 25 miles from the rail line, moving supplies into the excellent cover along the karst was a relatively safe operation.

The North Vietnamese unloaded supplies and armament in all the cities along the entire railroad and road network. Since most of these cities, particularly those in the northern area, were prohibited from attack (even though we had reconnaissance photographs showing large concentrations of supplies and vehicles in these cities, towns, and villages), we had to destroy the supplies before they could be stored and dispersed in these sanctuaries. All residential areas of Hanoi and Haiphong were filled with supplies stacked on each side of the street. Photos showed vehicles lined up bumper to bumper. My feeling was that since such material could be deployed by the enemy to inflict casualities on our forces, it should constitute a legitimate target. Such targets in World War II and Korea would have been cleared for attack without having to query higher headquarters. In Korea restraints were placed on attacks against Pyongyang on separate occasions, but these restraints did not apply to targets of military value within Pyongyang proper.

The concern for civilian casualties in bombing raids allowed many legitimate military targets to go free. By stockpiling supplies in the sanctuaries of towns and cities, the North Vietnamese compensated for the time lost in by-passing a downed bridge or a temporarily blocked marshalling yard. Furthermore, the enemy shuttled supplies at night from one sanctuary to another, a strategem which made interdiction more difficult because of the lower effectiveness of night attacks.

ROAD NETWORK—KEY ELEMENTS

Part of the interdiction campaign was to attack key road junctions to cause bottlenecks in the truck flow. The blocking of a key point on Route #1 or Route #15 leading into Mu Gia Pass, for example, was to force the trucks into the open so that a series of strikes could take a large toll of the trucks and supplies. The intent of the interdiction campaigns from 1965–1972 was not to "strangle" the flow of traffic. This misconception led some to believe that the interdiction campaign was not succeeding because the flow of traffic wasn't stopped. Traffic wasn't stopped in the European or Korean campaigns, either, but it was reduced to such an extent that the enemy couldn't get enough supplies for sustained

USAF F–105s make direct hits on blast furnaces and boiler plant facilities during an attack on the Thai Nguyen Steel Mill on 18 April 1967.

operations. This, too, was the objective for the campaign in Vietnam; by slowing the traffic with a series of calculated choke points in the rail and road system, we could destroy trucks and supplies piled up by the blockage.

By striking the most vulnerable part of the logistical system in Route Package VI, we could begin the attrition of supplies at a relatively high rate. As remaining supplies moved through the southern route package and into Laos, further attrition made movement very costly. The North Vietnamese couldn't put enough supplies into the system to get enough supplies to wage a war of much greater sophistication and size than they had at the Tet offensive of 1968.

The existing road network had been developed by the French over a number of years. The North Vietnamese continued to expand this net to accommodate traffic moving materiel to South Vietnam. Estimates put the system at 5,800 miles of motorable roads in 1964, about 1,070 of which were all-weather. This figure increased, however, throughout the

403–892 O – 83 – 14

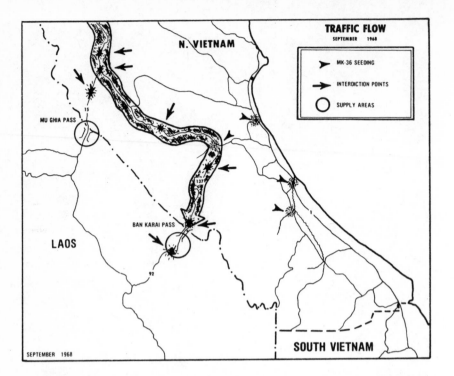

N. VIETNAM

MK-36 SEEDING

INTERDICTION POINTS

SUPPLY AREAS

MU GHIA PASS

BAN KARAI PASS

LAOS

SOUTH VIETNAM

SEPTEMBER 1968

war until by the 1972 cease fire, all main roads were in the all-weather category.

Roads, per se, have never been a good target because they can be repaired in a short time, or alternate routes can be developed. These conditions were particularly true in North Vietnam and Laos, for roads were continually repaired and even improved. With an immense labor force available 24 hours a day, most of the major roads were reopened within a day and in some cases, only a matter of hours after attack. In many places alternate routes circumrouted a blockage so effectively that the blockage delayed traffic for only a short time. Nevertheless, the enemy continued to use the same basic road structure throughout the war.

Interdiction had an added benefit: It reduced the available enemy manpower. The labor force devoted to the maintenance of both rail and road systems included an estimated 500,000* troops and civilian militia plus another 175,000 committed to the country's air defense system.[21] These were troops who could have been in combat units if not diverted to this task. The significance of diverting troops for air defense was recently expressed by Albert Speer, the former Reichminister of Armaments and War Production in Hitler's Germany. In his book, The Secret Diaries, he says,

---

*Other estimates put this figure around 300,000.

Two photographs of the Ban Laboy interdiction point in the Ban Karai Pass. Close-up is of the bend in the river at the center-left of the overall shot (arrow).

The real importance of the air war consisted in the fact that it opened a second front long before the invasion of Europe. That front was the skies over Germany. The fleets of bombers might appear at any time over any large German city or important factory. The unpredictability of the attacks made this front gigantic; every square meter of the territory we controlled was a kind of front line. Defense against air attacks required the production of thousands of anti-aircraft guns, the stockpiling of tremendous quantities of ammunition all over the country and holding in readiness hundreds of thousands of soldiers, who in addition had to stay in position by their guns, often totally inactive, for months at a time.

As far as I can judge from the accounts I have read, no one has yet seen that this was the greatest lost battle on the German side. The losses from the retreats in Russia or from the surrender of Stalingrad were considerably less. Moreover, the nearly 20,000 anti-aircraft guns stationed in the homeland could almost have doubled the anti-tank defenses on the Eastern front.[22]

The number of civilians and troops maintaining LOCs in Laos rose from 15,000 to 20,000 in 1966 to more than 35,000 by the time of the bombing halt above the 20th parallel in 1968. During one of his visits to Vietnam, President Johnson recognized the effect of the interdiction campaign in reducing the enemy's combat capability:

> Through the use of air power, a mere handful of you men—as military forces are really reckoned—are pinning down several hundred thousand—more than half a million—North Vietnamese. You are increasing the cost of infiltration. You are imposing a very high rate of attrition when the enemy is engaged, and you are giving him no rest when he withdraws.[23]

During the southwest monsoon (May–October), most of the traffic moved down Route 1 to Vinh. From Vinh, supplies destined for northern Laos (BARREL ROLL) were routed over Route #7 and through Barthelemy Pass. The logistics moving over this road supported the Pathet Lao as well as more than a division of North Vietnamese troops fighting in and around the Plain of Jars. However, this route was not heavily used during the rainy season, since a large portion of the North Vietnamese troops withdrew into North Vietnam. With the dry season (October–May), however, this route became the prime one for supplying the troops. Towards the end of 1971, requirements in the south increased, and the North Vietnamese kept their troops in Laos in combat even during the rainy season. At that time, Route #7 became a prime target during the rainy season as well.

Vinh was the hub of the road and supply system supporting troops in Laos and South Vietnam. Route #15 was the prime supply route leading from Vinh into Mu Gia Pass, at the head of which were a number of natural caves where supplies could be stockpiled safe from air attack. This route was an all-weather road, and, as the war progressed, was used heavily during both the dry and rainy season.

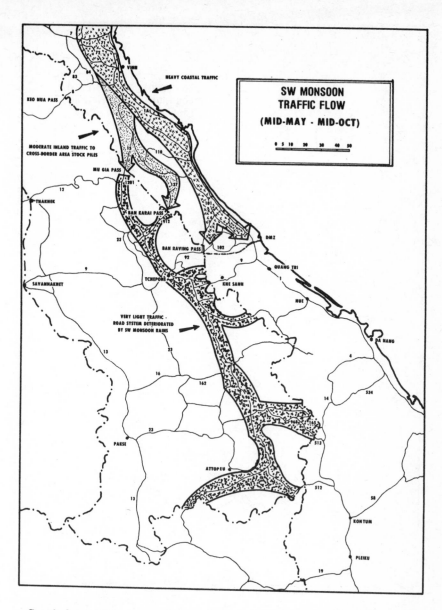

SW MONSOON
TRAFFIC FLOW
(MID-MAY - MID-OCT)

South from Vinh, the main supply route was #1A during both seasons. During the southwest monsoon when the Laotian roads were under water, the flow of traffic was very heavy on this coastal route. In addition, coastal shipping discharged supplies at Ron and Quang Khe where they were loaded on trucks for further movement into South Vietnam.

Until the spring of 1968, most of the traffic going into Laos during the southwest monsoon moved over Route #137, through the Ban Karai Pass and then into the Laotian network. Despite the continued bombing by 7th

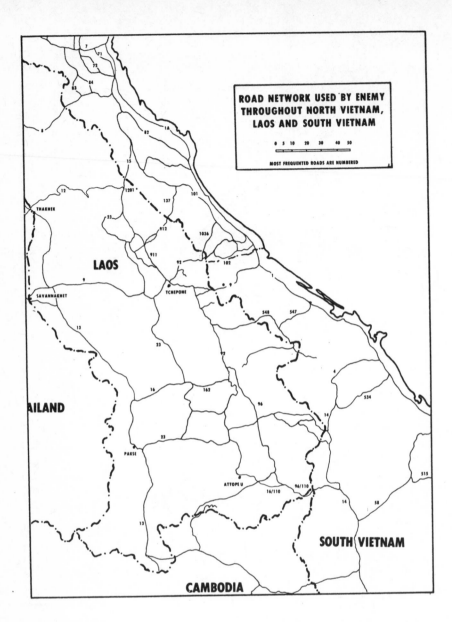

Air Force for almost a year, a new road into Laos south of Dong Hoi was constructed. This tortuous route through the mountains, designated #1036, joined with the Laotian roads at Tchepone. Thus, during the southwest monsoons this new route avoided much of the Laotian road network that was subject to flooding. Although the major roads were obviously of primary importance, during the 1965–1968 campaign some

Truck convoy moving along Route #137 in the Ban Karai Pass.

supplies filtered through the DMZ on a series of feeder roads such as #103.

However, by the spring offensive of 1972 a large portion of the supplies moved over secure routes all the way from Hanoi to the DMZ. This security was possible only because of the bombing halt of 31 October 1968. It provided the North Vietnamese with relatively secure lines of communication for an offensive buildup. The heavy traffic on these secure routes resulted in a series of limited strikes, termed "protective reaction,"* designed to deter continued buildup for a future large scale offensive. These strikes, however, were of such limited strength and duration that they had only a very temporary effect. Thus, by 1972 the North Vietnamese had so improved the road network that it could support the main offensive—the invasion of South Vietnam by 40,000 troops on 30 March 1972. Unfortunately, at the time of the invasion we could not limit the amount of supplies flowing into Haiphong and then down the railroad to Vinh.

Before President Johnson's partial bombing halt, 31 March 1968,[24] most of the support for troops in South Vietnam during the northeast monsoon was routed through the Laotian panhandle. It came through the Mu Gia, Ban Karai, and Ban Raving passes, which fed the major north-

---

*"Protective reaction strikes" were originally authorized to counter the increasing SAM and MIG attacks against air reconnaissance flights over North Vietnam. These strikes, however, were expanded to include attacks on the buildup of supplies above the DMZ.

south artery in Laos. The so-called Ho Chi Minh Trail was not simply a trail but a well-developed series of roads that supported major base camps in Military Regions I and II, and joined with river and other road nets in Cambodia to support Military Regions III and IV. The trail also consisted of a number of paths with well-conceived rest stops along the way for porters and infiltration groups.

The main road through Laos changed designations in relation to the three passes. It was designated Route #23 in the north, #911 in the center, and #92–96 in the south. Three other main roads connected to this route. The junction of each of these roads was a key interdiction point. One of our more effective tactics was to put in heavy strikes at day's last light using delayed action bombs, and to follow up with attacks throughout the night. During the interdiction campaign these and other basic tactics remained essentially the same with slight modification to take advantage of new equipment or weapons.

INTERDICTION ZONES

On 3 April 1965, Laos was divided into a series of zones north of Mu Gia Pass to clarify primary interdiction responsibility. The U.S. embassy in Vientiane was the civilian agency to establish policies for the conduct of air operations in Laos, while 7th Air Force was the air headquarters to conduct and provide most of the aircraft for missions in the area. The embassy, however, had certain air assets that it controlled for the direct support of Vang Pao's operations; many of these resources such as helicopters and forward air controllers came from 7th Air Force. However, when 7th Air Force conducted operations in BARREL ROLL, for example, it did so at the request of the air attaché acting for the Ambassador.

Not until the latter part of 1971, when the ground situation around the Plain of Jars had deteriorated so badly, were large quantities of strikes devoted to interdiction on other than an individual request basis. Until that time, most of the support 7th Air Force provided was to defend an outpost, suppress flak, escort a helicopter assault, destroy supply caches, neutralize artillery positions, or fly armed reconnaissance of selected areas.

The area south of Mu Gia Pass to Route #9 or the 17th parallel was designated STEEL TIGER. Although 7th Air Force had responsibility for determining the targets along the LOCs in this zone, the embassy in Vientiane established the rules of engagement for any targets more than 200 yards off the road. This split responsibility caused many problems, because when truck parks, for example, were discovered some distance from the road, the target couldn't be struck unless the U.S. embassy in Vientiane approved. All strikes within Laos, as in South Vietnam, were under the control of a Forward Air Controller (FAC). This control system accelerated target approval, especially when a Laotian representative was

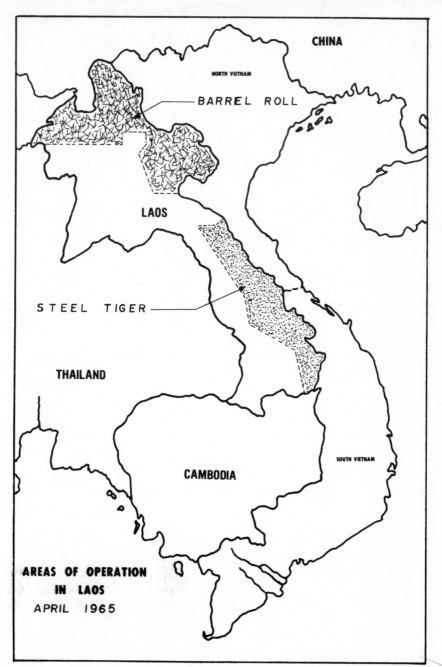

CHINA

NORTH VIETNAM

BARREL ROLL

LAOS

STEEL TIGER

THAILAND

SOUTH VIETNAM

CAMBODIA

AREAS OF OPERATION
IN LAOS
APRIL 1965

on the ABCCC controlling the area. The Laotian could validate a target for attack immediately after it had been discovered by a FAC.

The area from the 17th parallel or Tchepone south to the Cambodian border was designated TIGER HOUND. The Commander, U.S. Military

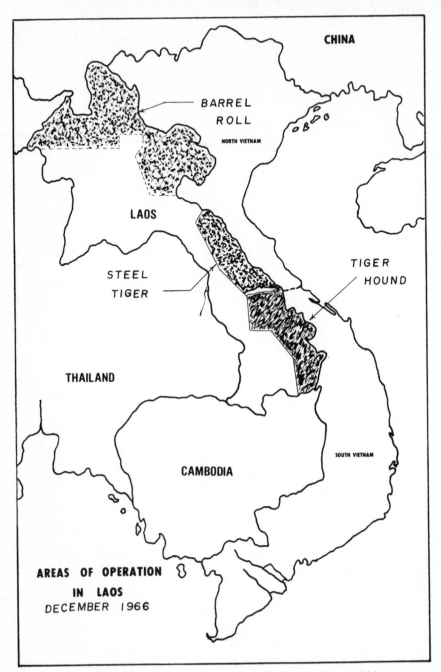

CHINA

BARREL
ROLL

NORTH VIETNAM

LAOS

STEEL
TIGER

TIGER
HOUND

THAILAND

SOUTH VIETNAM

CAMBODIA

AREAS OF OPERATION
IN LAOS
DECEMBER 1966

Assistance Command, Vietnam (COMUSMACV) considered this area an
extension of the battlefield in South Vietnam and requested that it come
under his jurisdiction. CINCPAC approved this arrangement, accepting
COMUSMACV's argument that the area was of more immediate interest

to the war in South Vietnam than was the air campaign to the north being directed by CINCPAC.[25] Actually, the entire interdiction campaign was of immediate interest to COMUSMACV for the simple reason that the effectiveness of the air attacks against the enemy's logistical system largely determined the quality and size of enemy forces facing him, as well as the length of time the enemy could sustain an offensive. Inescapably, TIGER 'HOUND, STEEL TIGER, and BARREL ROLL were as much a fundamental part of the interdiction campaign as were ROLLING THUNDER and LINEBACKER.

In December 1965, a TIGER HOUND task force from 7th Air Force was established to plan and control strikes against the southern routes into MR I and MR II. During the short life of the TIGER HOUND task force the trucks destroyed or damaged during daylight movement totalled 1,430.[26] By the beginning of the dry season (October 1966), however, 7th Air Force absorbed the TIGER HOUND task force into its regular staff. At the same time the quality of the enemy weapons defending the LOCs ended our use of low speed FACs (0–1s), and most of the movement shifted to Route Package I where the roads were dry because of the peculiar qualities of the monsoons. When the 1966 dry season returned to Laos, TIGER HOUND and STEEL TIGER were treated as simply another element of the interdiction campaign and not as separate entities. Even though COMUSMACV retained operational command authority, the commander of 7th Air Force decided on employment of airpower in the area.

In mid-1966, COMUSMACV became concerned about the increase in artillery rounds and rockets fired against U.S. Marine and ARVN forces in I Corps (eventually redesignated Military Region I). As with TIGER HOUND, he viewed the area above the DMZ as an extension of the South Vietnam battlefield and wanted to focus air strikes there. Although the area was already receiving about as much attention as the availability of forces would permit, CINCPAC agreed to assign Route Package I to the operational command of COMUSMACV. In reality, only the southern half of the package, about 30 miles, was of immediate concern to MACV.

A special task force concentrated on this 30-mile zone, TALLY HO, above the DMZ during the 1966–67 dry season in North Vietnam. Because of heavy defenses, the tactics used in Laos were not suitable for North Vietnam. For a short period 0–1 and later 0–2 FACs could be used to locate targets, but as the enemy increased his use of the LOCs (because of the flooded LOCs in Laos) he drove the FACs out of the area with AAA fire and, in a few isolated cases, SAM firings. TALLY HO ceased to exist as a separate entity by the winter of 1967; strikes were handled in that area as in other parts of the interdiction campaign.

After the 1965–1966 dry season, the North Vietnamese, except occasionally, stopped moving logistics along the LOCs in daylight. The ever-increasing effectiveness of our air strikes made such movements prohibitive. The ability of the 7th Air Force and Navy fighters to patrol the roads in Laos was reminiscent of similar operations in World War II and Korea. In the Korean War, the enemy stopped moving in daylight because of the high kill rate against trucks by our F–51s, F–80s, F–84s, and F–86s. Knowing the weather limited our kill rate, the North Vietnamese would watch the weather very closely; if flying weather were bad enough, a large convoy of trucks and other vehicles would venture forth during daylight. One of their favorite tactics was to try to run a large convoy through Mu Gia or Ban Karai passes in daylight if they thought the weather would force 7th Air Force to bomb by MSQ. MSQ coverage of the pass areas was good, but for pinpoint destruction of a convoy, individual attacks were the only way. On numerous occasions despite the weather, fighters caught such convoys in the open and destroyed most of the vehicles.

SORTIE FREQUENCY

In Laos, about 200 sorties daily flew against the route structure, most of them during darkness. During the day we ran a number of armed reconnaissance missions to locate truck parks and other targets suitable for attack, and to develop specific information about the conditions of the choke points. Many of these missions were flown at night with some aircraft held on air alert to verify truck kills and to provide current information on activities around the choke points.

The comparatively low level of anti-aircraft defenses and the absence of SAMs until 1967 along the major passes permitted much greater freedom of action than in the upper route packages of North Vietnam. Pilots could fly at lower altitudes with a very limited threat from the anti-aircraft defenses, particularly during 1965–1967. As a result, the large support forces required to penetrate a target in North Vietnam were not needed for most operations in Laos, although some support forces such as Wild Weasels, EB–66s for ECM missions, F–4s for flak suppression, and a few F–4s for fighter patrols were scheduled on all missions. The high level of defenses in North Vietnam dictated an entirely different method of operation; these defenses restrained night operations in North Vietnam as compared to the extensive night operations in Laos.

During the dry season in Laos, the scheduled day effort often was augmented by strike forces diverted out of Route Package VI because of adverse target weather there. Through diversions from their primary to secondary or tertiary target, additional strikes could be put into BARREL ROLL, STEEL TIGER, TIGER HOUND, and Route Package I. Since there was very little vehicle traffic, most of the augmenting strikes went against selected choke points and supply points. This situation was most frustrating for the pilots bombing these selected areas, but it was the only way we could keep hammering away at the logistical system. The restraints of weather and policy in the northern area gave us no choice.

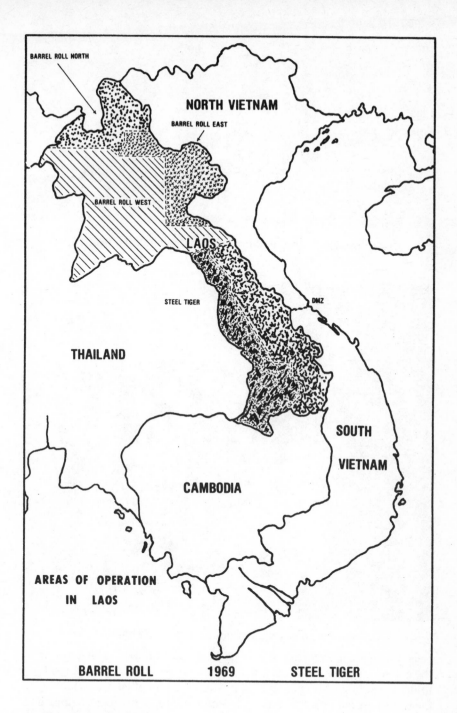

BARREL ROLL NORTH

NORTH VIETNAM

BARREL ROLL EAST

BARREL ROLL WEST

LAOS

STEEL TIGER

DMZ

THAILAND

SOUTH

VIETNAM

CAMBODIA

AREAS OF OPERATION
IN LAOS

BARREL ROLL   1969   STEEL TIGER

COMMAND & CONTROL

All flights except armed reconnaissance were under the control of
FACs. Because the FACs worked the network day in and day out, they

C–130 "Hercules" Airborne Command and Control Center (ABCCC)

Some of the operating stations inside the C–130 ABCCC.

became expert in detecting a change in the condition of a by-pass or in the enemy's success in restoring an underwater ford, a damaged road, or bridge. The FAC was a constant source of intelligence which was funneled into the ABCCC (specially configured C–130s that commanded all aircraft operating against the lines of communication in Laos). Because this ABCCC was acting for the 7th Air Force commander, it was authorized to decide what targets would be struck.

The strike forces under the ABCCC throughout the war followed the same basic procedures whether they came from Thailand, South Vietnam, or Task Force 77. Aircraft taking off from their home base would report to the appropriate CRC or CRP; these radar facilities would vector the strike aircraft into the vicinity of the FAC with whom the pilot would work the mission. Upon entering the area, the strike pilot would call the ABCCC (for central and southern Laos called HILLSBORO).* Since this airborne command post had no radar, it depended upon the fighter to report his position. HILLSBORO would then assign the strike aircraft to a FAC who was working a given sector of the road network. The FAC would control the specific strike and report the results to the ABCCC. Seventh Air Force always maintained contact with the ABCCC by direct communication or through the tactical air control system.

If 7th Air Force intelligence developed a target of opportunity or if reconnaissance produced a "perishable" target, that information was immediately passed to the ABCCC for execution. Sometimes, however, depending upon the other missions in progress, the ABCCC had insufficient forces for the target. Since 7th Air Force maintained a running status of strike aircraft allocated to the ABCCC, 7th Air Force might then divert additional aircraft directly to the ABCCC.

With Navy aircraft, the procedure was much the same. As a rule Navy aircraft were used in the interdiction campaign in Laos only as a result of a divert** because of weather in North Vietnam. These aircraft would report to the CRC (Panama) at Danang which then vectored them to the area in Laos where they would work. As they reached a designated point before entering Laos, they notified HILLSBORO who would assign a FAC to the strike leader. The same methods were applied to Marine aircraft performing the interdiction mission or to VNAF aircraft in Laos. The remainder of the procedure was identical to that of an Air Force strike. However, the VNAF was so hard-pressed during the 1965–1968 period to meet close air support demands in South Vietnam that they flew only in the southernmost part of TIGER HOUND, Route #110, for a short time. Daily planners did not include VNAF resources in planning interdiction missions.

SOME TACTICAL INNOVATIONS

The 1968 halt to the bombing of North Vietnam did not change our tactics in Laos; it actually resulted in a more intense effort of more than four hundred sorties a day against the road and logistical network. Even with standdowns for Christmas and Tet, the interdiction campaign in Laos continued. This campaign continued even during the periodic standdowns from attacks on North Vietnam when the various diplomatic

---

*HILLSBORO—Call sign of C–130 ABCCC in Southern Laos which operated in the daytime.

**divert: Controlling authority would divert airborne aircraft.

overtures were being made to get negotiations started. This apparent anomaly was another peculiarity of the war, but it derived from the refusal by both the U.S. and North Vietnam to admit to engaging in active combat operations in Laos. The typical response to the question of whether the U.S. was carrying out air operations along the Ho Chi Minh Trail in Laos was that armed reconnaissance missions were being conducted at the request of the Laotian government. The North Vietnamese never admitted that they had troops in Laos, although estimates at the time of Lam Son 719 in February 1971 put more than 35,000 troops maintaining the logistical routes and another 30,000 fighting General Vang Pao in the Plain of Jars and the Long Tieng area.

NIGHT TACTICS

During the interdiction campaign approximately 90% of the trucks that we destroyed or damaged were hit at night when most truck movements took place. This achievement is all the more remarkable in view of the terrain. Even in the more open terrain of Europe, the night mission was very difficult. There, the pathfinder area bombing techniques of RAF Bomber Command against a target such as Cologne, or the illumination of the Port of Bizerte in North Africa, were effective against a large, well-defined and immovable target. Of course, striking a small moving target at night in mountains or jungle was a much different challenge.

The tactics used in the Vietnam War had their origins in the Korean War. When that war broke out, the Air Force reactivated a large project at Eglin AFB to develop new techniques for attacking vehicles and trains at night. This project developed the hunter-killer team we used in the Korean War. Basically, the team consisted of two A–26s, one illuminating the target while the other attacked. As long as the AAA was relatively light, this technique was fairly effective, although the number of trucks destroyed per pass was low. Because both hunter and killer were vulnerable, we had to choose carefully the areas where we could use these teams. After the Korean War this tactic remained the basic technique for tactical forces except that fighters working in pairs alternated illuminating and attacking, the shift occurring when the first fighter had expended its ammunition.

With this technique we could use low speed aircraft in 1965 to 1966, when the flak was relatively light on the road network. The most important innovation of this period was to install a starlight scope in a C–123. The starlight scope, an adaptation of the infantrymen's sniper scope for night vision, was mounted in the C–123's belly; an operator stretched out on a mattress in a very uncomfortable position scanned the road through the scope. At about 3,500 feet the C–123 cruised along at about 140 knots. To keep the aircraft over the road was difficult since the road would disappear from time to time as it passed under the jungle, only to emerge in a different direction. Working with the C–123 was a killer aircraft—the old B–26 of World War II and Korean fame; these two aircraft worked very efficiently during this time. But as the enemy moved more AAA weapons into Laos, this low speed combination could no longer overcome the defenses. Though we assigned this team to the more lightly defended target areas, we had to pull the B–26s out of the theater altogether.

As the AAA continued to increase in late 1966, the Air Force devised new techniques for the interdiction problem in Laos. The ABCCC continued as the command element for strikes and reconnaissance along the road network. The FACs, because they were more maneuverable, had greater flexibility for operating against increased enemy fire than the C–123. They were aided by hand-held starlight scopes the Air Force Systems Command had developed. FACs operating in Laos with these instruments would patrol assigned segments of the roads, and when they located a target, they would report it to MOONBEAM.* MOONBEAM would then alert a pair of F–4s or a B–57. As these strike aircraft approached the FAC's position, the FAC would mark the head and tail of the convoy with a ground-burning flare. With this as a reference, the FAC then would bring the strike aircraft into the target. Because visibility was bad a large part of the time, all of these missions demanded the highest degree of professional skill and coordination. During a night attack, regardless of visibility, the recovery was made on instruments, a very difficult maneuver because of the steep attitude of the aircraft pulling up. In F–4s, the back seat pilot stayed on instruments throughout the attack and recovery, permitting the front seat pilot to concentrate on attacking and destroying the target. Because of the weather and recovery procedures, pilots often compared flying conditions in Laos to flying in a milk bottle—they couldn't tell up or down without reference to instruments.

By mid-1967, it became necessary to cover strike forces in Laos with EB–66 ECM aircraft and Wild Weasel F–105s. The picture was changing dramatically from what it had been in early 1966. Then a pilot could fly a mission in Laos with a slightly higher vulnerability than a mission in South Vietnam. But as the enemy moved more forces and more sophisticated weapons into South Vietnam, the threat in Laos increased markedly. The balancing of demands for support aircraft became a major factor in composing the forces that would strike in Laos, Route Package I, V, and VIA. Priority went to those forces going into Route Package VIA because of the number of SAMs and AAA. At the same time, we had to provide some ECM support to the other strike forces. Because we couldn't get enough ECM pods to equip all the force, many of the fighters going into Laos depended upon the EB–66 and Wild Weasels to provide SAM warning and countermeasures. Fortunately SAMs were not the threat in mid-1967 that they were during the 1972 Easter offensive.

The Mark 35 incendiary bomb carried by the B–57 was the best truck-

---

*MOONBEAM—Call sign of C–130 ABCCC in southern Laos operated at night.

403-892 O - 83 - 15

USAF B–57 "Canberra" attacking enemy supply lines.

killer in 1966 and early 1967, when the B–57 replaced the B–26. This was a surprise, since the tacticians had thought a combination of general purpose bombs with a high angle strafing attack using incendiary ammunition would produce the best results. The incendiary bombs, however, set the cargo as well as the trucks on fire. Before pipelines were laid through Laos, a significant part of the essential fuel was carried by truck to distribution points, and in some cases even floated in 50-gallon drums to various pickup points along the rivers. Fuel trucks were, therefore, a very productive target during this period. However, as time passed the North Vietnamese supply system became more sophisticated. As more enemy forces deployed to South Vietnam, after the pull-out of U.S. troops, most of the fuel went by pipeline from Haiphong to a point near Tchepone; it was then moved by truck to MR II, by water or truck into Cambodia, and then trucked the final miles into MR III and MR IV.

We became more and more adept at hitting the fuel supply as our experience and imagination increased. Sometimes, particularly at night when the F–4s were the primary strike aircraft, either a FAC or another pilot would locate the target and the ABCCC would call in a special C–130 (LAMPLIGHTER) to illuminate the target with flares. As more sophisticated techniques were developed, the combination of LAMP-LIGHTER with laser designator aircraft became a routine procedure.

SUMMARY

Techniques for interdicting the flow of enemy material and supplies in Vietnam changed somewhat from those of World War II and Korea. The introduction of new and different weapons permitted these changed tactics. The fundamental objective, however, of the interdiction campaign remained essentially the same throughout all three wars. Restraints imposed on attacks against the heartland of North Vietnam and the long delay in sealing off the ports undercut the full effectiveness of the interdiction campaign during this time. Even though the campaign limited the number of troops the North Vietnamese could support in South Vietnam, and therefore restricted the size of the war, the will of the North Vietnamese to continue the war was not broken. Not until the 1972 campaign were we given the targets that changed the enemy's will to fight.

The beginning of 1968 saw the introduction of a new family of precision weapons. But just when these weapons could significantly increase the ratio of targets destroyed per weapon expended, the President halted the bombing of North Vietnam. As a result of this decision, the interdiction campaign underwent significant changes in targets and tactics. The following chapter discusses the changing campaign and the events leading up to the final offensive in December 1972.

CHAPTER V

FOOTNOTES

[1] United States Air Force Operations in the Korean Conflict, USAF Historical Study No. 71, 25 June—1 November 1950; USAF Historical Study No. 72, 1 November 1950–30 June 1952; USAF Historical Study No. 127, 1 July 1952–27 July 1953; 3 studies (Washington, D.C., Department of the Air Force, 1952–1956), No. 127, p. 192.

[2] Wesley F. Craven and James L. Cate, eds., The Army Air Forces in World War II, vol. III: Argument to V-E Day (Chicago The University of Chicago Press, 1949), p. 179. See also The United States Strategic Bombing Survey, Overall Report (European War) Washington, D.C., 1945), p. 41.

[3] E. J. Kingston McCloughry, The Direction of War (New York: Frederick A. Praeger, 1955), p. 85.

[4] Ibid., p. 86.

[5] The Lord Tedder, Air Power in War, the Lees Knowles Lectures for 1947 (London: Hodder and Stroughton, n.d.), p. 113.

[6] Korean Conflict, No. 71, p. 140.

[7] Robert F. Futrell, The United States Air Force in Korea 1950–1953, (New York: Deull, Sloan and Pearce, 1961), p. 224.

[8] Korean Conflict, No. 72, p. 115.

[9] Ibid., p. 69.

[10] Korean Conflict, No. 127, p. 9

[11] Ibid., p. 37.

[12] Ibid., p. 10,

[13] Weekly Compilation of Presidential Documents, vol. 2, part 1, no. 8, the President's remarks upon receiving the National Freedom Award

from Freedom House, 23 February 1966 (Washington, D.C.: Government Printing Office, 28 February 66), p. 253.

[14] Working Paper for CORONA HARVEST Report, Out-Country Air Operations, Southeast Asia, 1 Jan 1965–31 Mar 1968, book 1, (Maxwell AFB, AL: Department of the Air Force, 1971), p. 17.

[15] Korean Conflict, No. 72, p. 139.

[16] William C. Westmoreland, A Soldier Reports (Garden City, N.Y.: Doubleday & Company, 1976), p. 220.

[17] Strategic Bombing Survey, Overall Report (European War) p. 108.

[18] CORONA HARVEST Report, Out-Country Operations, book 1, p. 137.

[19] The Air Force in Southeast Asia; Toward a Bombing Halt (Office of Air Force History, Sep 1970), p. 58.

[20] Working Paper for CORONA HARVEST Report, USAF Activities in Southeast Asia, 1 April 1968–31 December 1969 (Maxwell AFB, AL: Department of the Air Force, 1972), p. 3–15.

[21] Townsend Hoopes, The Limits of Intervention (New York: David McKay Company, 1969), p. 101.

[22] Albert Speer, Spandau, The Secret Diaries (New York: Macmillan and Company, 1976), pp. 339–340.

[23] Hoopes, Limits of Intervention, p. 301.

[24] Westmoreland, A Soldier Reports, p. 301.

[25] CORONA HARVEST Report, USAF Activities in Southeast Asia, 1954–1964, vol 2, book 1, p. 4–78.

[26] Illustrated History of the United States Air Force in Southeast Asia (Department of the Air Force, Office of Air Force History, July 1974), p. 57. (draft).

CHAPTER VI

INTERDICTION
(VIETNAM 1968-1972)

Each succeeding interdiction campaign from 1968 on was called COMMANDO HUNT. They followed the pattern of earlier years, though more forces were available in 1968 and 1969 because of the bombing halt. But in 1970 and 1971, forces started to phase down as U.S. ground forces withdrew. There was a temporary surge of forces from the United States in 1972 to counter the buildup and subsequent offensive of North Vietnamese forces into South Vietnam when an estimated 200,000 enemy troops attempted to take over large parts of South Vietnam. [1]

LASER & AC-130s—NEW SYSTEMS

Among the new developments in our interdiction efforts during this time was the AC-130 gunship which made its debut in Vietnam in 1967. At first, I was quite skeptical about the advertised capability of the aircraft to kill trucks. Not long after these aircraft were in combat, however, the results more than confirmed the advertised potential. The initial AC-130 gunships, equipped with two 7.62mm and four 20mm guns, were used in the lower part of TIGER HOUND where the flak was light. Later models were equipped with 40mm guns, and one version (Pave Aegis) carried a 105mm gun. The on-board sensors made the aircraft a potent weapon, for it was equipped with low light television (LLTV), infra-red (IR), and radar. Later models also had a laser designator, a device to pinpoint a target and direct a bomb to it.

With these capabilities, the AC-130 became the best truck-killing weapon in the war. The variety of sensors allowed the operator to select the best presentation and direct his fire by that picture. Normally radar was used first because of its longer acquisition range; then, as the target came into closer range, the LLTV or IR was employed. By COM-

211

Side view of AC–130 parked at Ubon AB, Thailand.

Miniguns and 20mm Vulcan cannons protrude from the side of an AC–130 gunship. This weapon system became the king of the truck killers along the Ho Chi Minh Trail.

MANDO HUNT V, which started in October 1970, the truck kill rate per sortie was 9.72 for the AC–130 as compared to 2.30 for the B–57G.[2]

This increase reflects the fact that tactics changed considerably in the 1968–1972 period. Although laser weapons had been introduced before the 1968 bombing halt, they were not employed in North Vietnam until bombing resumed in Route Package VIA in May 1972. However, in 1968 we began to use laser bombs against truck traffic on the routes in Laos.

The system worked this way: The AC–130 acquired the target with its sensor and then designated the target with its laser; two F–4s in a predetermined position with respect to the AC–130 would toss a laser bomb into the laser beam or "basket." This technique was very difficult in that coordination between the AC–130 and the F–4 to identify the target was very exacting. The AC–130 had to describe the target to the fighter in order that the bomb, when tossed, would follow the proper path to intersect the laser beam. To bring off a successful mission using these techniques required extensive conversation between the F–4s and the AC–130.

Although the AC–130 was a tremendous boost to the truck-killing effort, it still wasn't enough. Later on, some F–4s were equipped to designate targets with laser beams and to launch laser bombs from the same aircraft. These F–4s could also designate for other fighters carrying laser bombs. With these advanced techniques, F–4s became much more successful in the destruction of trucks. The experience gained in Laos with delivery techniques and use of laser bombs stood the strike forces in good stead when operations resumed in the Hanoi delta. The CEP of these weapons with laser designation truly brought a new dimension to the employment of airpower.

Because of the success of laser designation, the equipment was installed in other aircraft. The OV–10, next to be modified with the laser gun, became the primary vehicle to designate targets for the F–4s, thereby allowing the AC–130s to act independently. However, as the flak continued to increase, a larger portion of the F–4 force worked to suppress flak so that the AC–130s and AC–119s could devote their full attention to destroying trucks. Because the truck routes were heavily defended, these converted transports found working alone difficult. When anti-aircraft fire came from a given area, either the AC–130 or OV–10 would acquire it with the infra-red sensor. After the guns had been precisely located, a laser beam was laid on the site and an F–4 would deliver a laser bomb against the position. This was the most effective technique developed for the suppression of 37mm and 57mm anti-aircraft fire.

While the enemy moved increased quantities of supplies to support an offensive in early 1972, the lines of communication in Laos filled with the heaviest traffic of the war. In the 1966–1967 campaign some 49,371 truck sightings were reported with 7,194 trucks destroyed and another 3,278 damaged.* The truck kill rate through this early period averaged from 15–

---

*Trucks destroyed/damaged—the statistics were and are subject to differences of interpretation. The difficulty in confirming a truck destroyed plagued the commanders of 7th AF throughout the war. The North Vietnamese were able to move trucks struck during the night into the heavy jungle. Reconnaissance was not able, therefore, to provide the desired evidence on many missions as to the number of trucks destroyed/damaged.

18%.[3] The rate increased to about 20% for COMMANDO HUNT VII which began in October 1971 and concluded with the signing of the cease-fire agreement in January 1973. The rate would have probably gone above 20% if the campaign had continued for the remaining three months of the dry season in Laos. The truck sighting rate then was 1,000 per night which reflected the extensive North Vietnamese effort to build up stocks for the coming Easter offensive. The pattern of buildup for that offensive followed very closely the pattern of buildup for the 1952 spring offensive by the North Korean and Chinese communists.

Seventh Air Force estimated that the North Vietnamese put approximately 60,000 tons in the logistical system in Laos during the 1970–1971 dry season; about 10,000 tons actually found their way to the troops in South Vietnam.[4] This represented about 16% arrival rate. Nevertheless, these supplies were ample to support an offensive for a short time, especially when coupled with logistics that moved through the transportation network in North Vietnam free from sustained attacks because of the 1968 bombing halt. At this arrival rate, the North Vietnamese took at least six months to accumulate supplies for the Easter offensive. Although the main offensive in Quang Tri province ground to a halt within 30 days, the balance of the offensive lost its momentum after less than 30 days, even though the seige of An Loc continued until late summer. A combination of shortage of supplies, stiffened resistance by the ARVN after an initial setback, and the overwhelming air support by U.S. air units and the VNAF forced the North Vietnamese to abort the offensive.

Though the AC–130s and AC–119s had the best truck-killing rate during the early phases of COMMANDO HUNT VII, October 1971, they encountered increasing difficulty with the AAA and the SAMs. As early as 1 January 1971, a B–52 strike in the vicinity of Ban Karai Pass was subjected to a SAM launch. The SAMs became a continuous threat in Laos by the time of the 1971–1972 dry season. More than 160 SAMs were launched in the course of the campaign causing the loss of ten aircraft. The first AC–130 shot down by a SAM in Laos occurred ten miles southwest of Tchepone on 29 March 1972.[5]

In the face of such a threat, we moved the AC–130s to the less defended areas and increased their altitude of operation. The result was a sharp drop in the AC–130's effectiveness as a truck killer. Though it had earlier achieved an average rate of five trucks per sortie, high performing fighters had to take its place because of the fighters' ability to survive in a SAM environment. The AC–130 had been an exceptional weapon system in a semi-permissive defense environment, but it had to give way or become extinct when the enemy brought the full weight of his best defensive weapons against it. The AC–130 operated much longer than many of us expected; we had foreseen that SAMs and AAA would make it prohibitive to try to operate such a low performing aircraft, regardless of the protective and suppressive measures devoted to its security on interdiction missions.

1968 BOMBING HALT—IMPACT ON INTERDICTION

In March 1968, I concluded that we could not stop the North Vietnamese from building up supplies and forces just above the DMZ for their eventual offensive into South Vietnam if the bombing were stopped in North Vietnam. With a partial halt of the bombing, the North Vietnamese repaired all the main bridges and marshalling yards above the 20th parallel. This repair work allowed rapid movement of supplies and replacements by train as far south as Thanh Hoa with no interference from airpower. These repairs to the transportation system increased the capacity beyond pre-war levels.

When the President stopped all bombing in North Vietnam on 31 March 1968, the North Vietnamese accelerated the movement of logistics to forward depots near each of the major passes and the Bat Lake area only 20 miles above the DMZ. This forward placement of logistics provided them the capability to launch the offensive in April 1972, an offensive to test the will of the South Vietnamese to fight without the aid of U.S. troops, then in the final phases of withdrawal. Although U.S. ground forces were scarce, the North Vietnamese again underestimated the role of airpower as they had done earlier at Khe Sanh. One would have thought after their failure at Khe Sanh because of airpower they would have pursued a different strategy, one which would have neutralized the force of U.S. airpower. This airpower the Vietnamese ignored broke the back of the 1972 offensive.

With the 1968 bombing halt, the President had stipulated that reconnaissance aircraft would continue to fly over North Vietnam. These reconnaissance aircraft were authorized to have fighter protection to defend them against the MIGs. But despite the surveillance, the North Vietnamese violated the agreement with an unprecedented movement of men, equipment, and supplies into South Vietnam. They protected these movements by expanding the air defense system to the outer limits of Danang. New MIG airfields were developed at Dong Suong and Quang Lang; the one at Vinh was restored to full operational condition. Dong Hoi was brought up to a limited operational condition. The SAM coverage gradually extended to all the passes and into the northern half of Quang Tri province in South Vietnam.

Expanded SAMs and airfields, however, weren't the only threat. For the first time, MIG–21s became a real threat to both the airfield and port of Danang. Early in the war, MIGs had been a potential threat to MR I by staging out of Vinh, but such operations had been limited. Besides, 7th Air Force and TF–77 had subsequently kept all the potential staging bases under close scrutiny and had attacked whenever it appeared the bases could support MIG operations. The real threat during that period had been from the IL–28 Beagle light bomber based at Phuc Yen. Though they numbered but half a dozen, these aircraft had sufficient range to bring Cam Ranh Bay under attack. Thus, they were a number one target

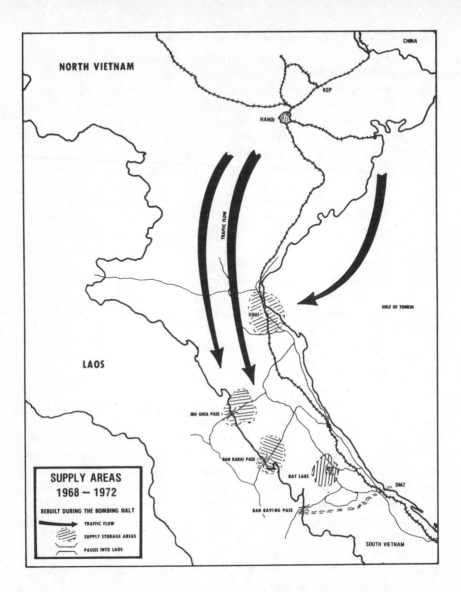

NORTH VIETNAM

CHINA

KEP

HANOI

TRAFFIC FLOW

GULF OF TONKIN

VINH

LAOS

MU GHIA PASS

BAN KARAI PASS

BAT LAKE

DMZ

SUPPLY AREAS
1968 – 1972

REBUILT DURING THE BOMBING HALT

TRAFFIC FLOW

SUPPLY STORAGE AREAS

PASSES INTO LAOS

BAN RAVING PASS

SOUTH VIETNAM

when Phuc Yen was finally cleared for attack in October 1967. To escape attack the North Vietnamese then shuttled the Beagles back and forth to Chinese bases when the raids began.

Even earlier, in the last part of 1966, the MIG–21s had periodically trailed the last element of our homeward-bound flights into Laos. However, they never penetrated very deeply and were not much of a threat. The MIGs would occasionally venture into northern Laos along Route 7 in an attempt to shoot down some of the armed reconnaissance aircraft. These forays were rarely successful and had no real impact on operations.

216

With the 1968 bombing halt, however, the MIGs increased their excursions into Laos against aircraft working the Ho Chi Minh Trail. These flights reached as far south as the Ban Raving Pass, both in daylight and at night. Although most of the attempted intercepts were made at night by a single aircraft, the lack of proficiency of MIG pilots in night interception and the limited operating range of their GCI prevented them from effectively interfering with our air operations.

Less than a month after the bombing halt, the increased belligerency against our armed reconnaissance flights resulted in an RF–4 being shot down by anti-aircraft fire on 23 November 1968 after we had said that reconnaissance aircraft would continue to fly over North Vietnam.[6] After the loss of the RF–4, 7th Air Force was authorized to suppress SAMs and AAA that fired against our reconnaissance flights. Thus, protective reaction strikes took on a much broader perspective than merely protection from the MIGs. From then on, protective reaction strikes were periodically directed against all elements of the air defense system that threatened air operations in Laos and reconnaissance missions over all of North Vietnam.

As the North Vietnamese continued to pour forces into the area above the DMZ, special strikes, which began in a very limited way, expanded until 1,000 sorties were flown against 41 targets above the DMZ on 30 December 1971. From January 1971 until March 1972, we made more than 300 strikes south of the 20th parallel. Besides striking SAMs, AAA, and MIG airfields, many missions went against supply concentrations, particularly those above the DMZ and in the central part of Route Packages I and II. By the size of these concentrations of supplies, we knew that we had to reduce them or the South Vietnamese could be overrun by a superiority of forces in the northern provinces. But these special strikes became principally a political tactic rather than a military maneuver. As a military action, these strikes were insufficient to substantially reduce the buildup. They were used instead to threaten the North Vietnamese with resumed bombing above the 20th parallel if they did not desist from their obvious preparations to mount a major offensive.

STRIKES BELOW THE 20th PARALLEL

Our techniques for operating below the 20th parallel were very much the same during both the 1965–1968 and the 1972 campaigns. In TALLY HO, the southern part of Route Package I, 0–1 aircraft were the primary source of information about the disposition and strength of enemy forces above the DMZ. As more anti-aircraft weapons were moved into the area in the summer of 1966 and fall of 1967, the 0–1 FACs were forced off the major lines of communications into the western part of the route package. Even in this area, the altitude at which the 0–1s were forced to fly was so high they could no longer provide detailed information about routes the enemy was using. With the loss of the low speed FAC coverage we had to devise different techniques to get the information we needed.

For some time before, 7th Air Force had been considering the idea of putting a FAC in the back seat of an F–100F or F–4. The shortage of F–4s led to the first combat test of a high speed FAC taking place in an F–100F in Route Package I. After a number of trial missions, the high speed FAC became standard where there were SAMs, AAA, and the threat of MIGs. They patrolled specified areas below the 18th parallel to uncover targets either for scheduled strike missions or for diverted missions from Route Package VI. By flying over the same area day in and day out, these MISTY* FACs came to know their area in great detail. Because they could spot targets that the strike fighters would normally not see, they uncovered many targets of opportunity. If the FAC suspected a target area, he would request a low altitude reconnaissance mission.

After a photo interpreter had analyzed the reconnaissance photos and found a good probability that there was a SAM, we would schedule a flight of fighters armed with 750-pound bombs. The MISTY FAC, on station before the fighters arrived, would lead the first element into the target area. He would then circle above the target to see the results. Sometimes the 750-pound bombs would blast off the camouflage and reveal parts of a mobile site. When that happened the FAC would call HILLSBORO, the ABCCC, and request a maximum effort against the discovered target. (HILLSBORO had authority to divert scheduled missions for a priority target such as a SAM site, tank staging area, artillery position, or logistical dump.) If HILLSBORO had no aircraft under control working in Laos, or on a diversion from preplanned targets in Route Package VI, the TACC at 7th Air Force headquarters in Saigon would scramble aircraft for HILLSBORO to control.

These techniques became increasingly sophisticated with each period of the war. To maintain an FAC in position longer, it was not uncommon to air refuel him two or even three times. Doing so meant he could spend as much as four hours over enemy territory, of course, always subject to SAMs and AAA. Theirs was a tough mission; these high speed FACs, like their counterparts in Laos, were among the most courageous pilots of the war.

Targets for all-weather strikes during the northeast monsoon were selected from reconnaissance during a break in the weather. We laid the groundwork for bad-weather strikes with a series of missions run in these target areas during good weather. The MSQ radar site at Hue Phu Bai would plot the flight path, and the pilot would verify his position. When the weather turned bad, flights under MSQ control could then strike these areas with a high degree of confidence that their bombs would be within 400 feet of the target. Good radar coverage of Route Package I and parts of Laos as far north as Mu Gia Pass was available during the 1965–1968 campaign. However, during the 1968–1972 campaign a new system,

---

*MISTY FACS—FACs were given different call signs by areas in which they operated. High speed FACs were known as MISTY FACs.

LORAN, became the preferred technique for all-weather bombing in Laos and RP I.

LORAN techniques offered two advantages: The accuracy was better, and a formation could bomb at the same time as the lead aircraft with better bomb spacing than with MSQ. Usually striking below the 20th parallel, a formation of four to eight aircraft could bomb with the lead F–4, with only one or two aircraft having LORAN equipment. The deputy lead, who would take over in the event of an abort by the flight lead, was also equipped with LORAN. Thus in bad weather on 21 September 1971, 196 F–4s in two waves struck five targets 35 miles above the DMZ using LORAN to guide them to the targets. More than 2,000 five hundred-pound bombs and 3,000 CBUs* dropped on petroleum and logistical storage areas and military barracks. [8] This strike was a significant advance in the use of a large number of fighters against targets that would normally not be struck because of the lack of visual flight conditions. Seventh Air Force reported excellent target coverage with major damage to the target area. PACAF, however, did not share the enthusiasm of 7th Air Force about the degree of target damage. Despite these differences in interpretation, this mission developed and confirmed the all-weather technique that was to be used against targets in Route Package VI in May 1972.

BOMBING IN ROUTE PACKAGE VI A (HANOI AREA)

The tactics for operations in the Hanoi area evolved over a considerable period of time. Optimum tactics had to be adjusted to accommodate the political restraints prevalent throughout all strike missions. Many of the tactics developed in World War II and Korea were tried, modified, and adopted. Probably the most significant factor affecting the type of tactics devised was the SAM, coupled with widespread use of electronic countermeasures by both sides.

In many respects, the strike missions in RP VI had all of the fundamentals of some bombing missions of World War II. Although the size of the force and the quality of the targets did not compare with those of World War II raids on Schweinfurt or Regensburg, the fundamentals did. Routing the force into the target, designating the aiming points within the target, reconnoitering to determine the condition of the target before and after the strike, using weather scouts to report in advance the prevailing weather conditions, employing fighter forces as direct cover to the formation, mounting corollary strikes against surrounding airfields to reduce the number of enemy fighters that could get airborne, and protecting the strike unit with ECM through on-board jammers and chaff—all of these facets of the mission were common. At the same time,

---

*CBU—cluster bomb unit: A CBU consists of a number of bomblets that are carried in a dispenser that is burst open at a predetermined height above the ground dispersing the bomblets in a desired pattern. Bomblets are designed for anti-tank, material, personnel, and mines.

however, tactics were constantly changing within these fundamentals to meet the changing political and military situation.

The mission on 29 June 1966 against petroleum, oil, and lubricant (POL) tank farms within four miles of Hanoi marked the beginning of highly complex air operations that continued to increase in sophistication with the introduction of new weapons.[9] The large scale attack against the POL complexes was executed without each aircraft having its own ECM; some units also lacked RHAW equipment for all aircraft. These early missions were planned with great detail and less freedom of action for the strike forces because of the initial uncertainty about how best to operate in a SAM environment with acceptable losses.

During the 1965–1968 period, the force was flown at about a .8 rate—that is, a sortie rate of .8 of the aircraft assigned. This rate of operation was derived from planning factors based on fighter-bomber experience in World War II and Korea. Two primary F–105 fighter-bomber wings of about 55 aircraft each, the 355th located at Takhli and the 388th at Korat, constituted the major strike element for operations in the Hanoi area. From many years of experience, the Air Force estimated that a combat unit would average about 75–80% of its assigned aircraft ready for operations on any given day except when either heavy or light losses put the in-commission rate above or below this figure. Thus, for any given day during the campaign, there were approximately 80–85 F–105s available for combat operations in North Vietnam and Laos.[10]

STRIKE FORCE COMPOSITION

Because the nature of the targets demanded individual bombing, the number of aircraft that we could use effectively on a given target was limited—smoke and debris would quickly obscure the target. As a consequence of these target limitations, the size of the force available, and the number of tankers available at a given time, a strike force of 16 F–105s composing four flights of four aircraft each provided the best balance that a single strike force commander could handle. From early combat experiments a basic force was built around this strike package of 16 aircraft with additional strike packages being added as the size of the target, number of targets, and availability of aircraft dictated.

To give this force the best opportunity to get into the target, we used a fighter cover that usually consisted of two flights of four aircraft. The cover aircraft were F–4s which flew to the front and rear of the strike force. Additionally, two flights of Wild Weasels with four aircraft in each flight provided most of the SAM suppression. One flight would precede the force by about five minutes, and the other flight would cover the withdrawal.

For jamming Fan Song SAM acquisition radar and the early warning radars of the GCI fighter intercept net, EB–66s were positioned on the outer limit of the 30-mile restricted area to cover the approach to the target and the withdrawal route. By being so close to the major target

USAF EB–66E in flight over North Vietnam.

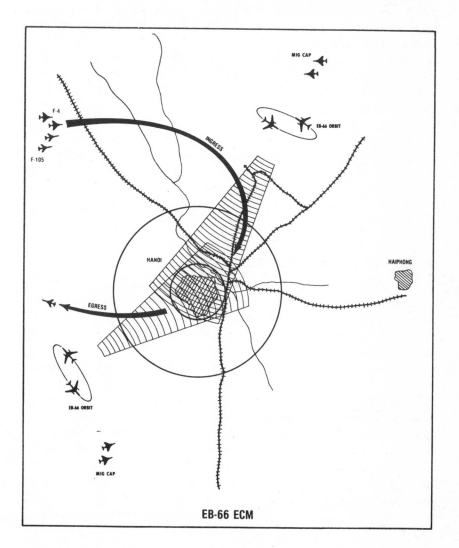

EB-66 ECM

area, these EB–66s could provide effective blocking of acquisition radars. Two EB–66s in the northwest quadrant and two in the southwest quadrant provided, in effect, a jamming beam into and out of Hanoi for the F–105s. During late 1966 and until mid-1967, it was feasible to position the EB–66 at 25,000 feet for EB–66s had their own F–4 fighter cover with others available to reinforce if MIGs became a threat.

As the SAM and MIG threat expanded, the closeness of coverage became too risky. One of the EB–66s was shot down by a flight of two MIGs on 14 January 1968. [11] The difficulty of fending off a jet attack was fully understood from the extensive air battles in the Korean War along the Yalu. With the warning that was available, the risk of having the EB–66s as close as possible to the target area was acceptable as long as the threat came only from MIGs. With the loss of the EB–66, however, and with the increase in SAMs, we moved the EB–66s much further from the target area, either into Laos or over the Gulf of Tonkin. As a consequence, the effectiveness of the EB–66 declined, since the effectiveness of jamming depends upon the power that the jamming aircraft can emit compared to the radiating source. Hence, the farther the jammer is from the target, the less the effective power.

From this experience with standoff jammers, we realized that the EB–66 had to be used in more permissive areas and strike forces had to rely more upon their own jammers to block out the enemy acquisition radars. An aircraft with higher speed was needed to penetrate with the strike forces and have the ability to function where firing of SAMs was frequent. During this period, however, the EB–66 continued to be effective for penetrations from the Gulf of Tonkin and for all operations in the other route packages and Laos.

PENETRATING THE DEFENSES

Until the first of 1967, the strike forces penetrated at 4,500 feet, a compromise between the threat of anti-aircraft fire and the SAMs. Because ECM pods were not available until January 1967, we thought that 4,500 feet would avoid much of the light automatic weapons while giving some protection against the SAMs. At this low altitude the SA–2, the only radar-directed, surface-to-air missile employed in North Vietnam during the war, was still gaining speed, and consequently pilots had a better chance of evading it. However, the low altitude penetration had some undesirable features that made it only an interim tactic. At these lower altitudes, target acquisition was particularly difficult because of the reduced visibility, rough terrain, and the speed of penetration. Further, the aircraft had to pull up rapidly to a predetermined altitude to begin a dive bomb run so that the 750-pound general purpose bombs would be released at the right altitude. These early missions in the high threat areas were very demanding, and the losses were accordingly higher than they were later.

Usually, two packages of 16 strike aircraft each were scheduled in the morning and the same in the afternoon. This type of operation for tactical aircraft differed from that of the 9th Air Force in World War II and 5th Air Force in Korea. In both those wars, the aircraft were scheduled continuously throughout the day with lesser coverage at night. Fighters normally operated in sections of eight or squadrons of twelve, but most of the time there was considerable freedom for flights of four where no enemy fighters were a threat. These extensive operations in World War II and Korea did not plan on air refueling, ECM protection, and target restrictions as we did in Vietnam. As a consequence, in the earlier wars we could keep the enemy under pressure from some elements in the force at all times.

In North Vietnam we would have liked to do the same thing, but with the large requirement for support forces, we weren't able to mount such an operation. The maximum effort we could stage was four strike packages* a day. With the strict controls on targets, we wanted to have the most flexibility within the strike force once it reached the target area. Since most of the targets were within a few miles of one another, we used a central penetration route and then broke a flight off against a particular target. Sometimes the target would require the entire force; at other times, particularly along the northeast rail line, each flight of four aircraft would be assigned a specific target. Yet, all of the aircraft were in such close proximity to one another that they shared support from the fighter cover and the ECM aircraft. An exception to the normal strike level was the Thai Nguyen steel mill. Because of the mill's area and the need for a high number of bombs to achieve the desired level of damage, two packages were scheduled in the morning and two in the afternoon. Each aircraft, however, had an individual aiming point within the target.

After January 1967, when ECM pods were finally available for all strike aircraft, penetration altitude was moved up to 15,000 feet. [12] This altitude permitted better target acquisition and allowed the strike leader to position the force better. At the higher altitudes, the loss rate to AAA dropped considerably, even though AAA was responsible for about 65% of all the aircraft lost during the war in North Vietnam. The fighter cover at the higher altitude also had more freedom to handle the low altitude threat from the MIG–17, forcing it to climb up to the strike force as it was leaving the target.

Throughout the war the MIGs used GCI holding points northwest and southwest of Hanoi. These points were best for intercepting both inbound and outbound flights. Their most strenuous effort came when the strike force was inbound, forcing the F–105s to jettison their bombs to evade the MIG attack. To give the F–105s a higher degree of probability of

---

*A strike package contained 16 strike aircraft carrying air-to-ground weapons with supporting fighter escort, Iron Hand, ECM, refueling tankers, command and control aircraft and intelligence platforms.

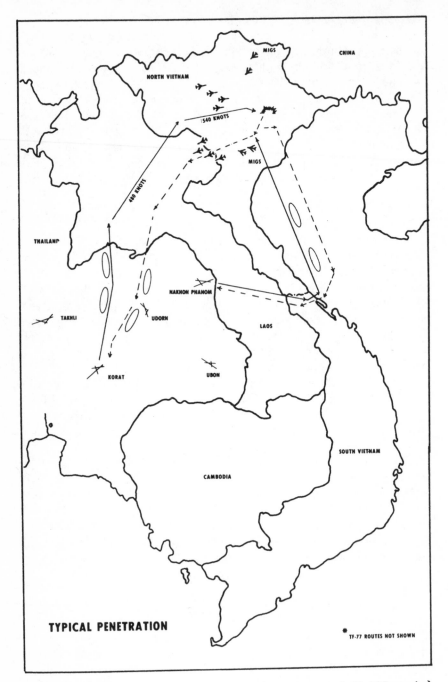

TYPICAL PENETRATION

*TF-77 ROUTES NOT SHOWN

Labels within figure: MIGS, CHINA, NORTH VIETNAM, 540 KNOTS, MIGS, 480 KNOTS, THAILAND, NAKHON PHANOM, LAOS, TAKHLI, UDORN, UBON, KORAT, SOUTH VIETNAM, CAMBODIA

evading a MIG or SAM and still get to the target, each F–105 carried seven 750-pound bombs, far less than the capability of the aircraft. But that reduction allowed the aircraft to retain a high degree of maneuverability.

USAF F–105 Thunderchiefs deliver their bombs under radar control.

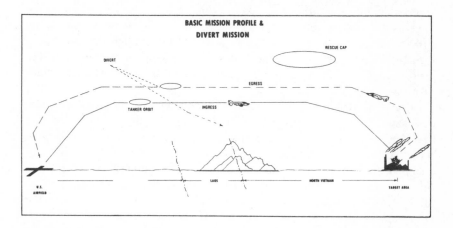

As the strike force entered the high threat area, approximately 40 miles from Hanoi, airspeed increased from 480 knots to about 540 knots. These speeds required the highest degree of professionalism in handling a force of 32 strike aircraft under intensive AAA and SAM fire with MIGs lurking above and to the sides. The force commander had a minimum of time to decide whether to abort the mission because of poor weather or to shift to the secondary target. Many times during the northeast monsoon the weather would appear satisfactory when the force was less than 30 miles from the target, yet in the immediate vicinity of the target, a broken condition with 7/8 cloud cover was present. We had no way of predicting these rapid changes in condition, so I counted on my strike force commanders to make the right decision when they saw the actual weather in the target area.

With the speeds approaching 540 knots, the flights began to position toward the outside of the turn that they would use in popping up to

225

bombing altitude. Subsequently, they made a half roll for a good look at the target as they maneuvered into the bombing run. The bomb release altitude was about 7,000 feet with a pull-out around 4,500 feet. At the time of pull-out, the F–105 speed was often above mach one as they headed for the shortest route out of the target area.

The pilots would not attempt to recover the formation until the flight had escaped the defenses. Then the flight lead would slow down and the other members of his flight would close on him. Unless a wingman had received battle damage, the crews would not attempt a rendezvous until 30 miles from the target. If a member received battle damage, the flight leader automatically circled back to cover the crippled aircraft, and the F–4 fighter screen would drop a flight back to provide close protection to the cripple.

Usually the fighter cover did not fly directly over the target with the strike aircraft. They would cover the strike force until it was within the area usually assigned to SAMs and AAA; they would then screen approaches into the target and take up a position to the rear of the strike force as it withdrew. However, if it appeared that MIGs would attack the strike force in the target area, the F–4s would fly through the target area with the F–105s. As a rule, the MIGs patrolled on the perimeter of the major threat areas and broke off pursuit as the strike forces came under SAM and AAA fire.

During all strike missions, specific flights within the package were designated for flak suppression. We found this technique was better than that used in World War II and Korea when a flak suppression force often preceded the main bombing force. In Vietnam, the relatively small target area and the density of defenses within that area required us to compress the time our forces were exposed. Furthermore, the lack of continuity in the target assignments from higher authority made it mandatory that the maximum number of penetrating aircraft deliver firepower on target. Thus, the number two and four aircraft in the second flight might be the flak suppression element. They carried a maximum load of CBU–24s, an excellent weapon for covering an AAA battery site. The location of the flak suppression element varied, as each wing commander determined the need for such protection. Seventh Air Force, however, specified the number of bombing aircraft and the load these aircraft would carry. This specification was necessary to achieve the precalculated probability that the target would be destroyed. All such calculations were made by 7th Air Force based on detailed study of target characteristics.

Whenever targets on the northwest and northeast rail lines were cleared for attack on the same day, we often penetrated from Laos and the Tonkin Gulf simultaneously. This tactic was an attempt to split the defenses and make coordination of the SAM and AAA fire more difficult. Our concern was not for the MIGs, since they were manageable with the fighter screen, but for the SAM and AAA, constant sources of difficulty. Such missions usually required more tankers; and then weather became a significant factor, since tanker tracks had to be clear over both Laos and the Gulf of Tonkin. Nevertheless, to get as much variation in the attack pattern as possible, these penetrations were made as often as the assigned targets permitted.

The average mission length, three to three and a half hours, had a major impact on the number of missions we could run. This factor alone dictated the limit of two strike missions per day. Ground crews needed about three hours to rearm, reservice, and prepare the force for the second mission. But it wasn't only the time to load and rearm the strike aircraft; we needed time to service and launch the tankers, reposition the intelligence platforms, move fresh bar caps* into position for protection of forces in the Gulf of Tonkin, rebrief for target changes made while the first mission was airborne, and reposition rescue forces if they had been used during the morning mission. These time factors, of course, were in addition to other considerations such as target, rescue, and weather.

Compounding the difficulties of the time constraints was the fact of the weather; usually the best weather occurred between 1000 and 1500 during the northeast monsoon. Because commanders emphasized continually the need to avoid collateral damage, it was imperative that pilots have the best possible conditions for acquiring positive identification of the target. In many cases, directives specified that a target would not be struck without positive visual identification. For example, it was not sufficient that the Yen Vien marshalling yard be identified; the pilot had to see the particular segment of the yard that was to be struck. This demand made the problem for strike crews very difficult when weather was at best marginal, for pressures were strong at all command levels to hit a target once it was released for attack.

Afternoon strikes were particularly difficult if the run into the target area was from east to west. The haze and sun made the target difficult to see; however, making the bomb run from west to east placed the aircraft over the most heavily defended areas longer and during a vulnerable part of the recovery when the aircraft was within optimum range of 37mm and 57mm guns. Thus, bombing targets within the ten-mile circle was best in the morning when better visibility allowed more latitude in approach and withdrawal. Many times, though, this choice was not within my prerogative as 7th Air Force commander; the release of targets so close to Hanoi was controlled from Washington, and the mssion was laid on for a strike as soon after release as possible. Normally, we had already planned for such missions, and we needed only a release date. Release of targets was closely controlled from Washington throughout the war and was not relaxed until the 1972 campaign, when some target restrictions were lessened.

---

*A term used by the Navy for USAF fighter patrols stationed over water between the shore and the fleet for fleet cover.

IMPORTANCE OF TANKER SUPPORT

Because all missions going into Route Package VI were air refueled, the KC–135 refueling tankers were positioned over Laos or the Gulf of Tonkin, depending upon the penetration routes for the day. Four fighters, refueled by a single tanker, took on 10,000 to 12,000 pounds of fuel each. This amount required careful scheduling to get all the forces on and off the tankers in the limited time available. Since the entire misssion was based on a precise time over target (TOT), each element of the force had to meet that time. As a conseqence, aircraft were concentrated in the refueling area where more than 60 aircraft would take on fuel within a few minutes of one another. Refueling altitude, like time, was limited; most refuelings took place at about 15,000 feet because that was as high as the F–105 could refuel with the bomb load it was carrying.

We used two tanker tracks over the Gulf of Tonkin and nine over Laos. In Laos we were limited on the optimum location of the tracks because the U.S. Ambassador was concerned that refueling combat aircraft in the vicinity of Vientiane would show too high a level of U.S. air operations in Laotian airspace for strikes against North Vietnam.[13] We therefore routed the refueling tracks around Vientiane to avoid this political issue. On the other hand we needed to press the tanker tracks as far north as possible so the fighters would have more fuel for extended fights with MIGs. The fighters needed afterburner, for it gave them increased performance for bombing, evading SAMs, and pursuing or evading MIGs; yet its use was limited because of high fuel consumption. At best, most of the strike and fighter aircraft had sufficient fuel for only about 15 to 20 minutes *without* afterburner in the target area before having to head home.

Coming back from a mission, most of the fighters required another air refueling. If they had not engaged the MIGs or deviated from the strike plan, they usually took on 5,000 pounds of fuel from the KC–135. This amount provided sufficient reserve for a traffic delay brought about by a barrier engagement or a battle damaged aircraft. But if not for the initiative of many tanker pilots, who actually went into the western part of North Vietnam to refuel fighters that were nearly empty, many more aircraft and pilots would have been lost. Tankers in the Gulf of Tonkin often went as far north as 19°30' and sometimes even further to respond to the frantic call of a fighter in distress, regardless of instructions as to how far north they could go.

WEATHER INFORMATION

Always on the day of scheduled strikes in the north, a weather scout went into the fringes of Route Package VI at least two to three hours before the first strike mission of the day. Normally they would fly around the periphery of the high threat area and report the existing weather to the ABCCC, or to one of our controlling radar facilities. Their reports, coupled with the normal forecasts and satellite pictures of the cloud

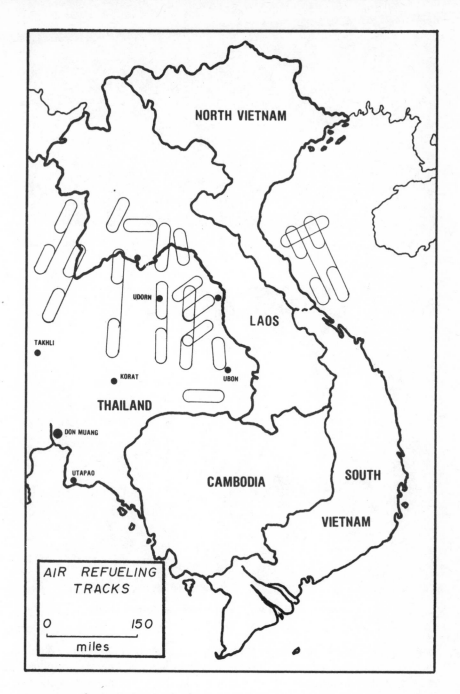

NORTH VIETNAM

UDORN

LAOS

TAKHLI

KORAT

UBON

THAILAND

DON MUANG

UTAPAO

CAMBODIA

SOUTH
VIETNAM

AIR REFUELING
TRACKS

0   150
miles

coverage, formed the basis for a commander's decision to launch the force.

The satellite picture was particularly useful in deciding the launch of the afternoon mission. The morning launch was usually made without

KC–135 tanker refuels F–105s over Southeast Asia. One "Thud" is on the boom while others in the flight await their turn.

F–4s refueling on their way to hit targets in North Vietnam.

satellite information for it wasn't available until about 1100 hours. When satellite pictures were available, however, they became the primary source of determining the cloud condition in the target area.[14] Without them and with only the traditional forecast, many missions would not have been launched. The satellite picture allowed us to launch a mission with a reasonable probability that favorable cloud conditions would prevail at the time the strike forces arrived. Before satellite information was available, sometimes we launched forces based on the assumption that the hole in the clouds over the target area would hold. Many times, however, when the strike force reached the area the hole had closed. Thus, the satellite picture was a major advance in providing the commander with real-time information about the weather his forces would probably encounter.

LACK OF SURPRISE

We had little opportunity to surprise or deceive the North Vietnamese about strike force targets and times. They fully understood the creeping release of targets, and therefore could predict from day to day what was next on the list. Furthermore, the targets were so close together it made little difference tactically whether the target was Phuc Yen airfield or the Hanoi Railcar Repair Shop. The 42 or so SAM battalions were so distributed that they needed little adjustment to give one or the other target a better coverage. Thus, the North Vietnamese defenses had about the same alert conditions, regardless of the target, once they determined our forces were headed for Route Package VI.

The excellent radar coverage of their air defense system permitted the North Vietnamese to gain positive information on force size while it was refueling. Early warning radars, code named BAR LOCK, covered all of the western and southwestern part of Laos; those located at Thanh Hoa and Vinh covered all air activities over the Gulf of Tonkin. From this information, the enemy could compute the time before our forces would be on target. As a result, their entire system was alerted, and usually their fighters were airborne minutes before our forces started their penetration. Because the target area was limited, we could not create deception to the extent we had in World War II and to a lesser extent in the Korean War. Feints could be made to a different part of North Vietnam, but the short distances between targets gave lie to the deception.

We assumed that the North Vietnamese intelligence network was active around all launch bases and that takeoff times and force sizes were relayed back to the operational commanders at Bac Mai or Phuc Yen. Even if one assumes that there was no such network, active DF* sites in North Vietnam could provide much the same information. All air forces depend upon such measures to fill out their knowledge about the activities of the opposing air force. For example, in North Africa on 18 April 1943 information from this type of intelligence activity permitted the Northwest African Tactical Air Force's P–40s and Spitfires to shoot down 50–70 JU–52 transports out of a flight of 100.[15] The *Luftwaffe* was desperately trying to get von Arnim's defeated Afrika Korps out of the clutches of the advancing U.S. and British Army in Tunisia and back to safety in Italy.

---

*Direction Finding—A procedure for obtaining bearings of radio frequency emitters with the use of a highly directional antenna and a display unit.

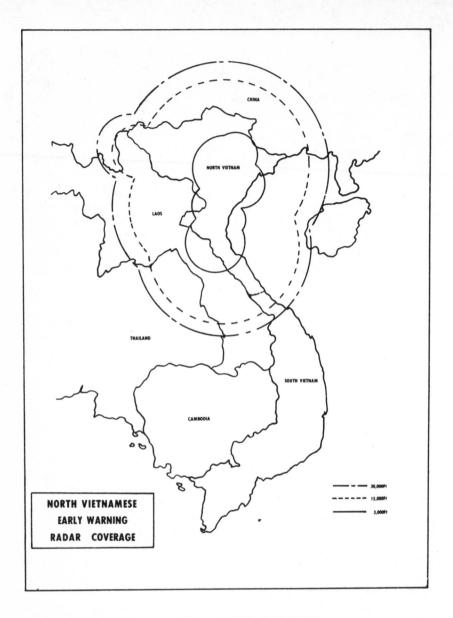

RECONNAISSANCE—COVERING THE TARGETS

Tactical air reconnaissance provided the detailed target information for scheduling strikes and evaluating their results. Although SR–71s and other reconnaissance platforms provided considerable information, their information usually was not timely or pertinent to the targets planned for a particular day. Most of the information produced by these platforms was used by national intelligence agencies for detailed evaluation of the effects of air attacks on the military, political, and economic life of the country.

Further, this intelligence analysis led to the nomination of targets to the JCS that the President approved and then sent to CINCPAC for strike.

For day-to-day operations, I depended upon the tactical reconnaissance force. Although details on specific targets often came from national intelligence agencies, this information was slow in reaching the field and had little influence on the hourly decisions of how best to strike the targets. For the weekly projection of strike operations, however, national intelligence information was used extensively.

For tactical reconnaissance during the early part of the 1965–1968 campaign we relied on RF–101s for photographic coverage of targets in North Vietnam. In 1967, RF–4s took over most of the job, and in the 1972 campaign the task was exclusively the RF–4s'. Throughout the war there was unusual interest in having photographs of the day's strikes in Washington. High level interest in each bombing mission resulted in photographs being flown back to Washington on scheduled courier flights before field agencies had fully interpreted strike results. This procedure led to considerable difference of opinion about strike results, differences which had to be ironed out before the next list of targets was released.

A basic source of information was the Q–34 drone which produced outstanding photographs. These drones flew at both high and low altitude. The low altitude flights, particularly valuable during periods of marginal weather when the RF–4s couldn't get in, produced details not provided from the medium altitude coverage of the RF–101s and RF–4s. But a combination of the manned and unmanned sources provided us the best view of the effects of the bombing campaign.

Again, each weapon system had its strengths and weaknesses. For this reason a combination of systems usually gets the desired results. The drones had to be programmed, and they therefore provided limited flexibility once the mission was launched. If we developed a higher priority target which required immediate updating, we could not change the flight profile of the drone. Further, the drone's low altitude photography was more suitable for point rather than area coverage because of the swath of the camera. On the other hand, the RF–4 was the most flexible tool for reconnaissance that I had. We could change missions while airborne and attain transitory target information more readily. Medium altitude reconnaissance provided both point and adjacent area coverage of immediate value in making decisions for restrike.

RECONNAISSANCE TACTICS

Neither in World War II nor Korea was the debate settled about employing tactical reconnaissance in elements of two aircraft or as a "lone wolf." The same arguments cropped up during both phases of the air campaign in Vietnam. As in World War II and Korea, proponents tried each technique. The argument for two aircraft was that the wingman was needed as a lookout for MIGs and SAMs so that the leader could concentrate on the target. Also, in the event the leader's camera malfunctioned, the wingman could take over the mission without exposing additional aircraft to enemy defenses. Finally, the two-aircraft proponents argued, reconnaissance missions into the high threat area involved such risks that we needed additional backup to be sure we got the job done on the first try; the wingman's cameras would cover the target just as the leader's did, even though the wingman's primary task was to be a lookout.

The lone wolf advocates maintained that a single aircraft had a better probability of getting into the target area undetected and, therefore, was less vulnerable to SAMs, AAA, and MIGs than was an element. One aircraft, they argued, had more flexibility in maneuvering, evading, and escaping a MIG than two aircraft when the leader had to worry about his wingman's position. Those arguing for the lone wolf system were hard-core traditionalists of the school that maintained the reconnaissance pilot lived by his wits in out-foxing the enemy and he did so easier with a single aircraft. The argument was no more settled in the Vietnam War than it was in World War II and Korea. We tried both techniques, and each time the losses changed we altered the technique as we had done in Korea.

In covering heavily defended targets along the Yalu in the Korean War, RF–86s went separately or as a part of the fighter formation covering them. Because of the MIG threat, as many as 18 fighters sometimes covered two RF–86s. [16] A variation of this tactic was to make the leader of a flight of fighters an RF–86. In effect, the reconnaissance aircraft then became a part of the fighter formation, thus hiding the identity of the reconnaissance aircraft and providing close fighter protection.

In Vietnam, the MIG threat to the reconnaissance mission was not as great as in Korea, but SAMs and AAA were significant threats. Because of the pressure to get results against targets these systems defended, especially in the restricted and prohibited areas, reconnaissance pilots endured a high degree of exposure. The best technique was to schedule the reconnaissance aircraft into the target as close as possible to the strike aircraft to take advantage of the shock of the attack and of its ECM and fighter protection. Although this seemed the best method of gaining increased security for the reconnaissance aircraft, it didn't necessarily provide the best photo coverage because of smoke in the target area. If the reconnaissance pilot delayed until smoke cleared the target, he faced loss of fighter and ECM protection. Therefore the timing of reconnaissance missions was a compromise. In general we wanted to cover as many targets as possible in the ten- or four-mile circle on a single mission. For strikes in these areas, the reconnaissance aircraft was often broken out separately to make a series of runs independent of the strike force, thus relying on the general protection rather than being buried in the pack. If we had high level interest in the results of a specific strike, the reconnaissance flight followed the strike flight within five to seven minutes. This timing provided fairly good protection for the reconnaissance aircraft and reasonable target coverage if the wind was right.

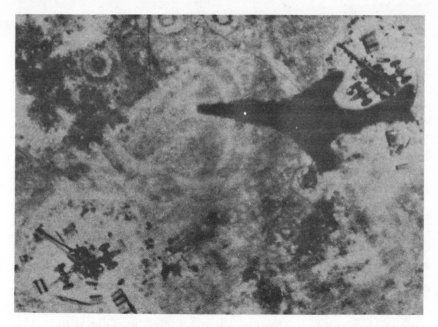

USAF RF–101 flies over 57mm anti-aircraft gun emplacements.

Maintenance men scramble down from an RF–4C reconnaissance aircraft about to take off on a photo mission over North Vietnam.

By mid-1967, fighters began to be equipped with strike cameras to record their own results. Although the photographs produced by these cameras were not as good as those in the reconnaissance aircraft, we

235

could determine within a few hours from landing whether the target would need to be restruck. Strike cameras shot to the rear of the aircraft so that as the pilot pulled off, the cameras could photograph the bombs impacting the target area. A quick examination of the developed print determined the general level of damage or lack of damage to the target without our having to wait for results of the reconnaissance mission. From these strike-camera pictures, each unit not only had a good tool to analyze individual bombing techniques and determine where additional training might be required, but it also had an invaluable asset when priority targets were subject to withdrawal if we did not destroy them within a particular time limit.

The operating techniques of the 1965–1968 campaign formed the basis for employing the force when we resumed bombing in the Hanoi delta in May 1972 although some systems used in the later campaign, the laser weapons and extensive use of chaff, either were not available or were used on a very limited scale in the earlier campaign.

FINAL DAYS—1972 BOMBING OFFENSIVE

With the resumption of bombing, 7th Air Force initially used a single strike force once a day rather than the two we had used during the 1965–1968 campaign. The reason for this change was the increase in the ratio of support forces to strike forces—approximately four to one—and the small number of laser weapons needed to take out the bridges. Of a total of about 80 aircraft in a single mission, only 12 to 16 were strike aircraft delivering laser weapons. For point targets and in good weather conditions, these weapons had nearly a single shot kill probability. If the target could be seen and the target was vulnerable to the explosive power of the weapon, the probability of damage with a single weapon was 80–90%. On 22 May 1972, eight F–4s carrying 16 laser bombs destroyed five bridges and damaged a sixth.[17] During the 1965–1968 campaign, such destruction would have required a much larger number of sorties.

As additional targets were assigned, the single force package was augmented with 32 strike aircraft carrying conventional bombs to hit area targets such as marshalling yards, railroad repair shops, logistical depots, and troop training centers. For a large target area, the pattern of a large number of conventional bombs provides the best coverage. For this reason, LINEBACKER I relied upon the same general techniques that we had used in ROLLING THUNDER when striking such targets.

An event we had long sought occurred on 9 May 1972—the mining of Haiphong harbor.[18] We had argued from the outset of the bombing campaign that a blockade of Haiphong was essential, if maximum stress were to be put on the North Vietnamese logistical system through interdiction. Not only was such a blockade essential to reduce the North's ability to support their fighting forces, but also to strain the economic, social, and political structures of the nation. The blockade of Haiphong was an intrinsic and fundamental part of the air campaign; its absence during 1965–1968 made our task much more difficult and our effects on the enemy's will to fight less conclusive.

Once the missing part of the interdiction campaign was corrected, additional targets, earlier off limits, were cleared for attack. Further, the 30-mile restricted zone around Hanoi was reduced to 10 miles and the 10-mile zone around Haiphong to five. With these new guidelines, the weight of airpower came closer to the heart of the country, and its effects were felt throughout all of North Vietnam.

In brief, LINEBACKER I demonstrated that the U.S. was ready to employ its airpower decisively. The consequence of this employment would be the paralysis of North Vietnam's ability to feed and protect its citizens. Evidence of the strain the North Vietnamese nation was under came from a number of independent sources. Basically, the same situation had existed in the earlier campaign.

Wheeler, Sharp, McConnell, Ryan, and I had argued in the summer of 1967 that the air campaign was on the verge of forcing the North Vietnamese to negotiate a settlement; if Haiphong and the other targets in the 30-mile circle were cleared for sustained attacks, the settlement would come quickly. We had the capability and the bombing attacks had reached the same level of effectiveness against the LOCs as LINE-BACKER I achieved later on. The interdiction argument centered on the issue of where to concentrate the effort. We argued that the campaign should be focused in Route Package VI, the top of the "funnel," whereas those who favored halting the bombing above the 20th parallel felt it should be directed at the lower route package. Secretary of Defense McNamara and his staff contended that the 20th parallel restriction would be inconsequential and that concentrating the interdiction effort in the lower route package would have just as much effect on the enemy.

During the visit of President Nixon to the Soviet Union from 21 May to 5 June, strikes were not permitted inside the 10-mile zone around Hanoi.[19] Such a restriction was imposed any time there were high level political contacts or visits by Soviet officials to Hanoi, but this time the visit of the President to the Soviet Union coincided with a new airpower posture and a new willingness to use the power to force negotiations.

As negotiations became stalled, rules of engagement were altered to permit strikes against more targets near Hanoi. Flights of four aircraft carrying laser bombs struck power plants, command and control centers at Bac Mai, and all the key bridges hit in ROLLING THUNDER. By 13 July, the North Vietnamese were prepared to resume negotiations. The constant pounding of targets in the Hanoi delta and the halting of the North Vietnamese ground offensive in South Vietnam were key factors in this change in attitude.

B–52s were then being used on a continuing basis against targets in North Vietnam. Tactics for the B–52 were altered from those in South Vietnam where no threat of SAMs, MIGs, or AAA existed. There it had been feasible to take more time in the target area, and the spacing

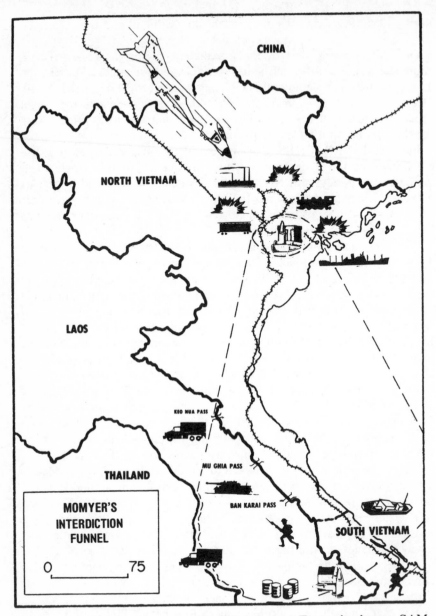

MOMYER'S
INTERDICTION
FUNNEL

0 _____ 75

between bomber cells was not a critical factor. From the heavy SAM firing against the B–52 force striking Thanh Hoa in April 1972, however, we knew that the time in the target area needed to be compressed to minimize exposure. Tactical forces had learned the same lesson when first cleared into Route Package VI in 1965. Regardless of the effectiveness of on-board ECM equipment, the shorter the exposure in the target area the lower the probability of being shot down. Obviously, a B–52 doesn't have the speed of an F–4; therefore, it must depend upon its on-

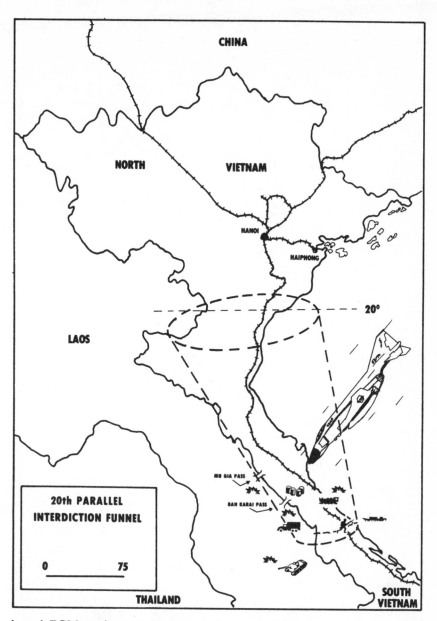

CHINA

NORTH VIETNAM

HANOI

HAIPHONG

20°

LAOS

MU GIA PASS

BAH KARAI PASS

20th PARALLEL
INTERDICTION FUNNEL

0   75

THAILAND

SOUTH
VIETNAM

board ECM equipment plus supporting tactical forces to create the most favorable conditions for penetrating the defenses. It was particularly difficult for commanders to devise entirely different procedures for supporting B–52s rather than a fighter strike force.

The F–111s returned to combat on 28 September 1972. They then played a major role in the night attacks against airfields, SAM sites, marshalling yards, and power plants. Before they returned, the terrain-following radar had all the bugs out; it proved to be a significant new factor in the ever-changing scene of air combat. The low altitude attacks (below the effective altitudes of the SAMs), and the high speed (above Mach .9) made the aircraft almost immune to the anti-aircraft fire that had caused the heaviest loss in the 1965–1968 campaign. As a rule, the F–111s preceded the B–52s to the target to suppress the fighters and reduce the ability of the command system to employ SAMs and AAA in a cohesive manner. Because the F–111s with the F–4 fighter screen and the Wild Weasels, did their job well, the MIGs were never a very important factor in the night attacks. Significantly, the ratio of SAMs launched per aircraft destroyed was approximately 70–1; the ratio in the 1965–1968 campaign was 55 to 1.[20]

On 23 October, because of progress in the peace talks, the President directed the suspension of bombing above the 20th parallel.[21] However, as the peace negotiations dragged on, we accelerated our planning for a three-day maximum effort in Route Package VI using B–52s, F–111s, F–4s, EB–66s, and carrier-based aircraft. Weather soon became a major consideration in the planning, for the northeast monsoon was in full force. This fact alone meant ceilings would run about 1,000 to 3,000 feet with visibility approximately one mile. Based on experience from the 1965–1968 campaign, we could make visual attacks only four to six days a month. Thus, our planning was based on the assumption that all-weather bombing would necessarily be the primary means of attack. That meant using B–52s.

Prior to this time, B–52s had not been employed above the 20th parallel, and had been used sparingly north of the 17th parallel until the resumption of the bombing in April 1972. The seasonal conditions that required all-weather attacks and the need to confront the North Vietnamese with total application of our airpower for the first time led to the decision to employ B–52s.

The plan was based on a proposed three-day effort, with B–52s in three waves attacking throughout the night and tactical forces striking targets in the day under visual flight conditions, if the weather permitted; if not, they would use LORAN. In planning the campaign, now known as LINEBACKER II, instructions were issued to be prepared to follow the three-day effort with a phase of indefinite duration. The planners, therefore, had to pace the attacks on the assumption that the high level of effort could extend over a longer time and would require a sustained in-commission rate of better than 70%. Even though it might have been tactically and strategically sound to bunch the force and go for an all-out effort, this possibility could be disastrous if the campaign were to stretch out. For this reason, we decided to put up initially about 30 B–52s a night and spread the attack over seven to nine hours.

To support these attacks required a major effort from the tactical forces of 7th Air Force. The demands for chaff, ECM, and fighter cover exceeded the assigned resources; 7th Air Force could not sustain night support and full day operations. The day strikes actually required even

240

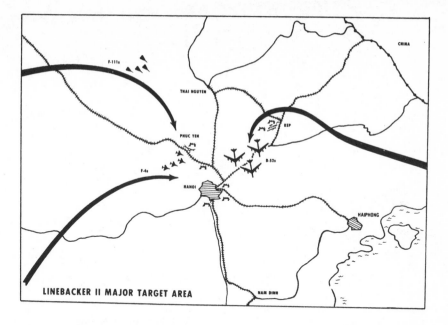

LINEBACKER II MAJOR TARGET AREA

heavier support than did the B–52s striking at night. Night operations limited the MIG threat, and to some extent, curtailed the anti-aircraft threat. The smaller anti-aircraft guns (37mm and 57mm) were of little concern since they were mostly visually directed, particularly during periods of heavy jamming. However, the 85mm and 100mm guns were radar directed and posed a definite threat even at night.

The weather for the operation was about as expected—1,000 to 3,000 feet—but the five-mile visibility below the clouds was somewhat better than predicted. About 60% of the targets were associated with the transportation and logistical system while the remaining 40% covered power plants, airfields, SAM sites, communications installations, and command and control facilities. The B–52s used radar exclusively to bomb their targets—the marshalling yards near Hanoi, airfields, and storage areas. The F–4s carrying laser bombs were employed against the Hanoi power plant, railroad classification yard, and radio station. The A–7s led by LORAN-equipped F–4s bombed the Yen Bai airfield using a combination of visual and LORAN attacks. The F–111s bombed airfields, SAM sites, and marshalling yards.

During the 11-day campaign, tactical forces flew 2,123 sorties of which 1,082 were at night. B–52s flew 729, all at night.[22] Tactical support forces such as the chaff flights, fighter cover, Wild Weasel, and ECM comprised nearly 70% of the total sorties flown by tactical forces. Many of the support forces, however, delivered weapons against the enemy, and thus destroyed targets besides suppressing enemy defenses for the B–52 and fighter strike forces. If the campaign had continued, more tactical forces would have returned to active strikes, for the enemy defenses continued

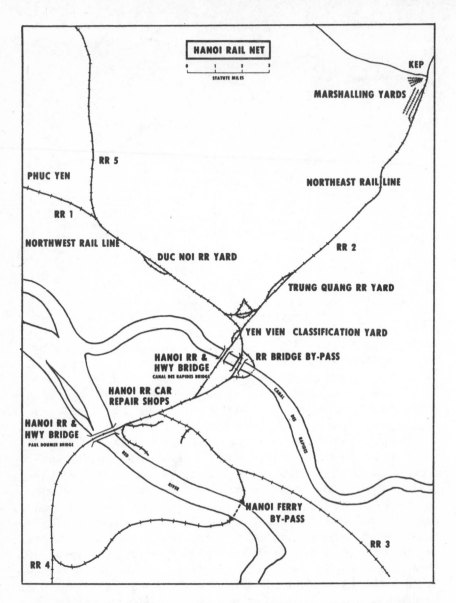

HANOI RAIL NET

STATUTE MILES

KEP

MARSHALLING YARDS

RR 5

PHUC YEN

NORTHEAST RAIL LINE

RR 1

NORTHWEST RAIL LINE

DUC NOI RR YARD

RR 2

TRUNG QUANG RR YARD

YEN VIEN CLASSIFICATION YARD

HANOI RR & HWY BRIDGE
CANAL DES RAPIDES BRIDGE

RR BRIDGE BY-PASS

HANOI RR CAR REPAIR SHOPS

HANOI RR & HWY BRIDGE
PAUL DOUMER BRIDGE

CANAL DES RAPIDES

RED RIVER

HANOI FERRY BY-PASS

RR 3

RR 4

to deteriorate from the unrestricted attacks. The proportion of support to strike forces, as seen here, is a sensitive balance and varies as the type of forces, the quality of enemy defenses, and the priority of the targets change.

SUMMARY

The 11-day campaign came to a close on the 29th of December 1972 when the North Vietnamese responded to the potential threat of continued air attacks to the economic, political, social, and military life of their

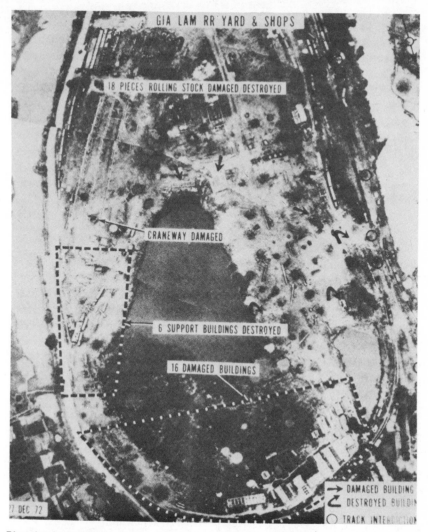

GIA LAM RR YARD & SHOPS

18 PIECES ROLLING STOCK DAMAGED DESTROYED

CRANEWAY DAMAGED

6 SUPPORT BUILDINGS DESTROYED

16 DAMAGED BUILDINGS

7 DEC 72

→ DAMAGED BUILDING
↗ DESTROYED BUILDI⸱
○ TRACK INTE⸱⸱⸱⸱⸱⸱

Pinpoint bombing by B–52s destroys much of the Gia Lam Railroad Yard and Repair Shops.

country. It was apparent that airpower was the decisive factor leading to the peace agreement of 15 January 1973.* The concentrated application of airpower produced the disruption, shock, and disorganization that can be realized only by compressing the attack and striking at the heart with virtually no restraints on military targets which influence the enemy's will to fight.

---

*Although the North Vietnamese would not admit that it was, in fact, the bombing, comments by Henry Kissinger at a January news conference strongly implied that it had been the *coup de grace*. See State Department Bulletin, LXVIII, No. 750, Jan. 8, 1973, pp. 33–41.

CHAPTER VI

FOOTNOTES

[1] William W. Momyer, The Vietnamese Air Force, 1951-1975: An Analysis of Its Role in Combat, USAF Southeast Asia Monograph Series, vol. 3, monograph 4 (Washington, D. C.: Government Printing Office, 1977), p. 44.

[2] CORONA HARVEST Final Report, USAF Operations in Laos, 1 Jan 1970-30 Jun 1971 (Hickam AFB, HI: Department of the Air Force, Hq Pacific Air Forces, 1972), p. 68.

[3] Trucks destroyed/damaged figures vary greatly depending on the source. Reporting criteria for aircrews were changed several times throughout the war to make reporting as accurate as possible. For a thorough discussion, see CORONA HARVEST Reports, Out-Country Air Operations, 1 Jan 1965-3 May 1968, book 1, chapter 2, and USAF Operations in Laos, 1 Jan 1970-30 June 1971, pp. 74-84.

[4] CORONA HARVEST Final Report, USAF Operations in Laos, 1 Jan 1970-30 Jun 1971, p. 74.

[5] Momyer, Vietnamese Air Force, p. 39.

[6] USAF Management Summary, Southeast Asia, Directorate of Management Analysis, 29 Nov 1968, p. 40.

[7] CORONA HARVEST Final Report, USAF Air Operations Against North Vietnam, 1 Jul 1971-30 Jun 1972 (Hickman AFB, HI: Department of the Air Force, Hq Pacific Air Forces, 1973), pp. 42-43.

[8] Working Paper for CORONA HARVEST Report, USAF Air Operations Against North Vietnam, 1 Jul 1971-30 Jun 1972, pp. 28-29.

[9] Working Paper for CORONA HARVEST Report, Out-Country Air Operations, Southeast Asia, 1 Jan 1965-31 Mar 1968, pp. 91-92.

[10] Working Paper for CORONA HARVEST Report, USAF Logistics Activities In Support of Operations In Southeast Asia, 1 Jan 1965–31 Dec 1969, part 2, p. II-2-81-94.

[11] CORONA HARVEST, Out-Country Operations, Book 1, p. 242.

[12] Ibid., p. 116.

[13] Ibid., book 2, pp. 27, 29.

[14] Ibid., p. 158.

[15] Wesley F. Craven and James L. Cate, eds., The Army Air Forces in World War II, vol. 2: Europe: Torch to Pointblank (Chicago: The University of Chicago Press, 1949), p. 191.

[16] United States Air Force Operations in the Korean Conflict, USAF Historical Study No. 72, 1 November 1950–30 June 1952 (Washington, D.C.: Department of the Air Force, 1955), pp. 113–14.

[17] CORONA HARVEST, Air Operations North Vietnam, p. 157.

[18] Weekly Compilation of Presidential Documents, vol. 8, no. 1, President's Address to the Nation, 8 May 1972 (Washington, D.C., Government Printing Office, 15 May 1972), p. 840.

[19] CORONA HARVEST, Air Operations North Vietnam, p. 101.

[20] CORONA HARVEST Final Report, USAF Air Operations in Southeast Asia, 1 July 1972–15 August 1973, vol. 2, book 4, p. 298.

[21] Ibid., p. 44.

[22] Ibid., pp. 249, 252.

CHAPTER VII

AIRPOWER AND THE GROUND BATTLE

INCREASED U.S. INVOLVEMENT IN VIETNAM

With the Ely-Collins agreement* and the fall of Dien Bien Phu the U.S. began to assume more responsibility for equipping and training the South Vietnamese forces.[1] By early 1958, the U.S. had assumed complete responsibility, and the French gradually withdrew from all parts of the advisory program.

As a part of the peace agreement of 1954, there was to be held, within a period of two years, a country-wide election which would provide for the peaceful joining together of North and South Vietnam. President Diem, in 1956, abrogated this provision of the agreement because of the infiltration of North Vietnamese agents into South Vietnam and the fact that North Vietnam was a communist society in which there was no freedom of choice. He therefore maintained that to hold a country-wide election at this time would be tantamount to turning over all of Vietnam to the communists.

From 1956 to 1959 guerrilla activity in South Vietnam remained at a fairly low level. There were indications that the insurgency was under control and that with time South Vietnam would be able to stand on its own feet without the detailed and extensive support it was receiving from the U.S. The U.S. took heart in these developements and likened them to the development of a strong and independent South Korea after the conclusion of that war.

The North Vietnamese were not long in establishing their intent to take over South Vietnam both politically and militarily. In September 1959 this intent was reflected in the ambush by two companies of Viet Cong of a

---

*General Paul Ely, Commanding General of French forces in Indochina, and General J. Lawton Collins, President Eisenhower's special envoy to Saigon, signed a formal agreement on 13 December 1954, which marked the beginning of a new role for the U.S. in Indochina.

South Vietnamese force searching in an area southwest of Saigon.[2] This was the first large scale engagement by the Viet Cong. Until this time most of the military action had been by small bands of guerrillas attacking isolated villages, harassing vehicle traffic in the countryside, assassinating local officials, and disrupting tax collectors.

As the guerrilla action expanded, requests for U.S. training missions increased. The advisory group had doubled by the end of 1960. Most of the increase was in U.S. Army Special Forces which provided training for Army of Vietnam (ARVN) rangers who patrolled the border areas and infiltration routes into South Vietnam.

As early as 1958, it was suspected that the Ho Chi Minh Trail in Laos had become the main infiltration route for forces into South Vietnam. Furthermore, the Russians were providing the Pathet Lao, the communist insurgent force in Laos, with supplies and armament in 1960. These deliveries were made by IL–14 transports, an aircraft similar to our C–47. By 1961 there was evidence that the Russians were delivering armament as far south as Tchepone, Laos, and it is probable that some unidentified flights into the II Corps area of South Vietnam along the Laotian border were being made by Russian AN–2 Colts. These are the same aircraft the North Vietnamese modified and used on 12 January 1968 to bomb Site 85 where the U.S. had a radar and a tactical aid to navigation (TACAN) facility.

With the expanding war, it was apparent that the Vietnamese Air Force (VNAF) would have to gain a capability for supporting its ground forces without having the time to develop and train its pilots in a normal manner. Unfortunately, the French had not given the VNAF the opportunity to develop as an independent air force. Only a limited number of proven combat leaders were available to handle the needed expansion, and very few of its people were trained in air-ground operations. The French Air Force had performed these tasks through its own system, and the size of the VNAF—one fighter squadron, two liaison squadrons, two transport squadrons and one helicopter squadron—was no more than a reinforced wing, much less an air force.

Krushchev's speech of January 1961, indicating that wars of liberation were just and the way of the future, was instrumental in prompting the incoming President, John F. Kennedy, to issue a directive to the Secretary of Defense to develop a capability to deter and, if deterrence failed, to defeat such aggression.[3] President Kennedy's chief military advisor, the former Chief of Staff of the Army, General Maxwell D. Taylor, had for several years disagreed with current military strategy and its emphasis on general war. In his book The Uncertain Trumpet, he stated his case for limited war, particularly low scale, non-nuclear, limited war.[4]

During the 1950's, the U.S Air Force had espoused the strategy that forces equipped for general war would deter most forms of aggression; if deterrence failed, those forces could fight a limited war. The Air Force felt that nuclear weapons were the paramount means of fighting war; and limited war, even if begun as non-nuclear war, would rapidly reach the point where nuclear weapons would be required. Based on this position, USAF tactical forces were trained primarily for nuclear war. Since the close of the Korean War, the U.S. capability for non-nuclear conflict had steadily deteriorated. Little money had been devoted to the development of non-nuclear weapons, and existing stocks were those remaining from the Korean War.

In early 1961, as the emphasis shifted to developing non-nuclear forces for limited war, the lower part of the spectrum of conflict began to receive the most attention. Initially the focus was on "counterinsurgency." This focus was too far down the spectrum for many elements of our tactical air forces. Tactical airmen had serious reservations about putting so much attention on counterinsurgency when there was a need to restore the non-nuclear capability of our tactical forces.

As aggression increased in Vietnam, the Air Force argued that additional resources should be devoted to the tactical air forces to regain the size and capability that existed at the end of the Korean War. The Air Force maintained that these forces would be more suitable as a hedge against a counterinsurgency war since such a war would probably escalate into a non-nuclear, limited war if it continued for any length of time. The fact that the political situation in such conflicts was so unstable as to lead to breakdown in control, with the resultant use of organized armed forces, indicated that the war was already above the counterinsurgency level and demanded a more comprehensive use of military power.

Within President Kennedy's administration, however, there was deep concern over the lack of imagination shown by the services in developing innovative ideas, organizations, and equipment to cope with counterinsurgency. Therefore, the Advanced Research Project Agency (ARPA) within the Office of the Secretary of Defense (OSD) responded to the President's directive for developing new equipment for low scale conflicts.[5] By June 1961, ARPA had established a joint test organization (Joint Evaluation Group Vietnam) in Vietnam to test the combat suitability of new equipment. This testing created considerable tension between the Army and Air Force because the Army had stated that counterinsurgency was primarily ground combat, and the Army was responsible for land warfare.[6] The emphasis on counterinsurgency was further exemplified by the expansion of the Special Forces (Green Berets) and the attention this organization had received including President Kennedy's visit to Fort Bragg, North Carolina, the home of these forces.

Concurrent with the emphasis on counterinsurgency was the Army's desire to use the helicopter to increase battlefield mobility. Studies done by the Army showed that in all future wars, mobility would be even more important than in the past, and that the helicopter would give the soldier greater mobility.[7] The Air Force considered helicopter forces to be extremely vulnerable to enemy ground fire, and, therefore, such forces could only be employed under very restricted conditions. Thus, only a relatively small helicopter force should be developed, one designed for counterinsurgency.

These differing views on the forces necessary for counterinsurgency and the differing perceptions of such wars were being overtaken by events in Vietnam. As the U.S. support for the South Vietnamese and determination to prevent the spread of communism continued to grow, most requests for expansion of the Vietnamese armed forces received favorable consideration. However, the lack of equipment necessary to modernize the VNAF was a major problem. By 1961, there were no propeller-driven combat aircraft in the USAF inventory. All fighter units had converted to jets following World War II, and the only combat propeller aircraft were in the Navy and in storage.

DEVELOPING THE FORCE

The Air Force submitted proposals to the Secretary of Defense to modify the T–37 jet trainer into a fighter-bomber for U.S. and VNAF units. However, there was considerable opposition to this course of action since the introduction of jets into South Vietnam would appear to violate the Geneva agreement. Furthermore, the OSD staff argued that it would be easier for the Vietnamese to maintain propeller aircraft and that these aircraft were better suited for close air support in a jungle environment than jets.

With President Kennedy somewhat impatient at the progress being made with counterinsurgency forces, the Secretary of Defense went all out to gain a capability without further delay.[8] The Army Special Forces, then being deployed to train the South Vietnamese rangers, knew the value of air support and realized that the ARVN was in danger of being annihilated unless the VNAF could provide reliable and constant air support. The VNAF, confronted with these demands, was not able to fight, train, and expand at the same time. By mid-1961, it was apparent that further help was necessary to train and expand its forces.

A few months earlier, the VNAF had been provided funds to increase its fighter force to two squadrons. Additionally, a detachment of the 507th Tactical Control Group (TCG) was moved from Shaw AFB, South Carolina, to Tan Son Nhut AB in January 1962.[9] The VNAF had little capability to operate such a facility; the Air Advisory Group planned from the outset to use people from Tactical Air Command to operate the system. The deployment of the 507th was prompted by reports of unidentified aircraft in the Saigon area. (We were to be plagued with periodic reports of unidentified aircraft in South Vietnam throughout the war.) With the potential for Soviet aircraft to airdrop supplies to Viet Cong forces in the delta, it was prudent to get radar into Tan Son Nhut to monitor such traffic as well as deploy the fighters necessary to deal with intruders. With a Control and Reporting Post (CRP)* established, a command and control structure for the employment of the forces to follow was begun.

The Air Staff had conducted a number of technical studies to determine what propeller aircraft could be put in a combat ready condition as quickly as possible. These studies showed that the T–28, a trainer, could be modified as a fighter-bomber. Even though these aircraft were 12 years old, they still had a few useful years left. A modification line was established and the aircraft structure was modified for delivery of bombs.

Another vintage aircraft was the B–26. It had had a useful role in the Korean War and there were a number of these aircraft still in storage. Considerable modification, however, was needed to restore them to an operational condition. It was originally designed as an attack aircraft, and delivery of weapons was done straight and level or in a slight dive. It was not stressed for any form of dive bombing where three or more "Gs" are common. Nevertheless, it was decided to modify a limited number of these World War II aircraft for counterinsurgency missions.

The final elements of the counterinsurgency force were the tried and proven C–47, of which a number were still around and in good condition, and the U–10 Heliocourier. These aircraft were modified for psychological warfare roles which called for the delivery of leaflets and the broadcasting of taped messages through a loudspeaker system. In addition, a flare dispensing rack was installed in the C–47 to provide illumination of areas at night as a deterrent or for strike aircraft.

U–10 returning home after leaflet mission near Lac Thien Special Forces camp. Note leaflets caught on horizontal stabilizer.

*CRP—control and reporting post: An element of the United States Air Force tactical air control system, subordinate to the control and reporting center, which provides radar control and surveillance within its area of responsibility.

FARM GATE—A FORCE FOR COUNTERINSURGENCY

In April 1961, the 4400th Combat Crew Training Squadron was activated at Eglin AFB. This unit had as one of its missions the development of tactics, techniques, and equipment for counterinsurgency. It was also to be prepared to deploy detachments to such conflicts. The initial strength of the unit was 16 C–47s, 8 B–26s, and 8 T–28s. With the pressures building to develop a force for such low scale conflicts, the unit had a minimum of time to prepare itself. Fortunately, the Air Force had experience with unconventional warfare from World War II and Korea. During World War II, there was an extensive effort in dropping forces into Yugoslavia and the subsequent support of forces by air drops. These missions were conducted by highly trained crews who had developed air drop techniques for supporting irregular or guerrilla forces. After the war they were gradually phased down until there was only a minimum capability for this type of conflict by the time of the Korean War.

During the Burma campaign in World War II, a unit was activated to operate behind Japanese lines. The force was composed of British soldiers supported by an American air task force. The unit was led by British Brigadier Orde Wingate, and was called the "Wingate Force."[10] The air units consisted of P–51s, C–47s, and gliders. The air units became known as air commandos, the name adopted by the initial units that deployed to South Vietnam. Thus, the 4400th CCTS came to be known as JUNGLE JIM.

As the Army continued to press for the expansion of special forces for counterinsurgency, the joint testing in Vietnam took on added significance. The Air Force was eager to establish its role in this type of conflict and pushed for the deployment of a detachment of the 4400th to South Vietnam to train the South Vietnamese in the techniques of air-ground operations and to devise new techniques for incorporation into our own air doctrine.

USAF A–26 parked on the flightline at Korat AB, Thailand. The aircraft, used primarily along the Ho Chi Minh Trail, is "loaded for bear" and awaiting the day's mission.

The versatile C–47 "Gooney Bird."

T–28 taking off from Nakon Phanom AB, Thailand.

On 11 October 1961, Detachment 2A of the 4400th was deployed to Bien Hoa Air Base. It was code named FARM GATE and became the first USAF unit to conduct combat missions in the Vietnam War. This detachment consisted of 8 T–28s, 4 SC–47s, and 4 B26s. The 151 men of the detachment rotated periodically to the parent organization.

The mission of this unit from the outset was ambiguous. The aircraft had VNAF markings, and the unit was not authorized to conduct combat missions without a Vietnamese crew member.[11] Even then, the missions were training missions although combat weapons were delivered. The missions were designed to train Vietnamese pilots to bomb and shoot, and since there were real targets, the situation provided maximum training.

TACTICAL AIR CONTROL SYSTEM BUILDS

With the limited capability of the VNAF, FARM GATE began to fly more of the missions in which close coordination with ground units was required. As the requirement for close air support exceeded the capability of the VNAF, requests were made to let FARM GATE function in the same manner as were U.S. Army and Marine helicopter units which

253

carried U.S. markings. These repeated requests were denied, and the training mission of the unit was maintained.

As additional units were deployed, primarily C123s, the air section of the Military Advisory Group was not adequate to direct FARM GATE activities and the expanded air control system. Detachment 7 of 13th Air Force was therefore activated in November of 1961 and became the operational headquarters for the control of air activities. Nevertheless, the rapid growth of forces created other problems.

The CRP at Tan Son Nhut was not able to regulate and control the increased air activity. To alleviate this deficiency, in December 1962 all the main elements of a tactical control system were moved to Vietnam. The CRP at Tan Son Nhut was expanded to a full CRC* which gave a vastly increased capability to control the movement of all aircraft in III and IV Corps. The CRC was operated by USAF personnel with VNAF personnel in a training status.

A CRP was established in the highlands at Pleiku. This station was operated by the VNAF with USAF personnel providing on-the-job assistance. With VNAF exclusively operating this site, we hoped to learn how quickly the VNAF would be able to operate the entire tactical air control system without direct USAF support.

A CRC was also installed at Danang. "Monkey Mountain," as the site was called, was operated by the USAF with Vietnamese Air Force people again in a training status. This station provided coverage to all of I Corps and linked with the CRP at Pleiku. By the end of the year, aircraft for the first time could be radar controlled in all areas of South Vietnam. This radar system remained throughout the war and was later enlarged by the establishment of two additional CRPs.

The Army had resisted establishing the tactical air control system because of the argument about the control and employment of helicopters. [12] At the time, U.S. Army and Marine helicopters operating in the four corps areas were assigned to the operational control of the U.S. Army advisor of the respective ARVN corps commander. The missions were conducted in the corps areas without reference to VNAF and FARM GATE. The Air Force had argued the need for bringing all air operations in South Vietnam under a centralized control system. Establishing the air control system provided the mechanism for directing and controlling air operations regardless of the type of aircraft and the service to which those aircraft belonged.

In 1962, the Howze Board (named after Lieutenant General Hamilton H. Howze, who chaired the Board) was created in response to Secretary McNamara's directive of 19 April 1962 to develop "a plan for implementing fresh and perhaps unorthodox concepts which will give us a significant

---

*CRC—control and reporting center: A subordinate air control element of the tactical air control center from which radar control and warning operations are conducted.

Aerial view of Monkey Mountain near Danang Air Base, South Vietnam.

increase in mobility."[13] This Army board proposed the creation of a number of air assault divisions with organic helicopters for airlift, fire support, and reconnaissance. Fixed wing assault transports were also proposed as part of the force structure. The Army was most anxious to test these new units in South Vietnam. Since it was argued that these units were in direct support of the ground battle, they shouldn't come under the control of the Air Force in a theater of operation.

Shortly thereafter, the Air Force established the Disosway Board (named after Lieutenant General Gabriel P. Disosway, chairman of the Board) as a counterpart to the Howze Board, and its mission was to explore established doctrine and techniques as a means of increasing the mobility of Army forces. This board concluded that a combination of C–130s, fighters, and a limited number of helicopters for the movement of combat troops and supplies forward of the C–130 assault air strips was a better solution to the problem.[14] The Disosway Board's fundamental point was the need to design the force so that it had the capability to fight across the spectrum of limited, non-nuclear war. The assault helicopter force proposed by the Howze Board had limited utility in the opinion of the Disosway Board, since it couldn't function in a high threat environment because of vulnerability to enemy fighters, AAA, and SAMs. It was further contended that limited funds wouldn't allow the development of such highly specialized forces that could be used only for a particular type of war.

EVOLUTION OF SYSTEM FOR AIR SUPPORT

This background had a direct influence in the early problems of air-ground operations in Vietnam. Even though operations were relatively small during the 1962–1964 period as compared to the later campaigns,

255

the problems were aggravated by the inability of the South Vietnamese to stop the take-over of the country by the North Vietnamese and Viet Cong. The loss of five helicopters and damage to nine others during a helicopter assault at Ap Bac in January 1963 drove home the need for better coordination between air and ground forces. Because of another operation directed by the Vietnamese Joint General Staff, there were no fixed wing aircraft available to support the helicopter assault. The ARVN 7th Division commander, as well as the U.S. advisor, decided to go ahead with the operations and to employ helicopter gunships to suppress ground fire in the landing zone. The Viet Cong had reconnoitered the probable landing sites and had a battalion of troops dug in waiting for the assault. The helicopters were brought under intense ground fire with the attendant losses. Fixed wing aircraft were brought into the action later, but too late, The operation failed with 65 ARVN and 3 U.S. advisors killed. Admiral Felt, CINCPAC, stated the operation should not have been conducted without fixed wing aircraft support, and he directed that all future helicopter assaults be so supported. [15] Later, the point was made again by General Wheeler, Chairman of the JCS, that helicopter gunships were not a substitute for fixed wing aircraft.

NORTH AFRICA—THE CLASSROOM FOR VIETNAM

The organization for air-ground operations in Vietnam had its genesis in North Africa and was modified in the battle for Europe. Essentially, the interface between air-ground fighting forces was at the tactical air force-field army level. It was at this level that commanders made the major decisions on the hour-by-hour changes in the battle. Above this level, commanders concerned themselves with longer range issues and the general strategy for the campaign. It was at the tactical air force/army level that the campaign plans were translated into specific actions to achieve the objectives of the campaign.

In North Africa, an air support command was the initial organization that functioned with an army formation, either a field army or a corps. In the fall of 1942, the XII Air Support Command was the air organization that worked with II Corps. Following the Casablanca Conference in January 1943, General Eisenhower, the Commander-in-Chief Allied Forces Northwest Africa, established the Northwest African Tactical Air Force. It worked with the Allied 18th Army Group which controlled British forces in the north, French corps in the center, and the U.S. II Corps on the southern flank. [16] The XII Air Support Command was unique during this period of development of air commands designed to work with ground units.

By the time of the Normandy invasion, numbered tactical air commands had been created to work with each of the field armies, and numbered air forces with army groups. As the Allied forces became established on the continent, the 9th U.S. Air Force, commanded by General Vandenberg, worked with the U.S. 12th Army Group that

General Bradley commanded. General Vandenberg had three tactical air commands (TACs), the IX, XIX, and XXIX. Each of these commands worked with the field armies under General Bradley, the 1st, 3rd, and 9th. Thus, the IX TAC was a partner of the 1st Army; XIX TAC of the 3rd Army; and the XXIX TAC of the 9th Army.

EUROPE—REFINEMENT OF THE SYSTEM

General Vandenberg shifted forces among the tactical air commands as the campaign changed. When General Patton's 3rd Army had the mission of driving to the Rhine, Vandenberg augmented Weyland's XIX TAC with forces from Quesada's IX TAC and Nugent's XXIX TAC. The flexibility of the command and control system allowed Vandenberg to be responsive to strategic decisions.

The British had a similar relationship between the 21st Army Group and the 2nd Tactical Air Force commanded by Air Marshal Coningham. Below the tactical air force level, the British had an air group which was approximately the size of a U.S. tactical air command. Each of the British and Canadian field armies had an air group partner. These similar organizations made it easier to pass operational control of air units between the U.S. 9th Air Force and the British 2nd Tactical Air Force (TAF). For example, during the period when Montgomery had priority for a drive on the northern flank, units of the IX TAC were temporarily under the control of the 2nd TAF.

Each tactical air command operated a Joint Operations Center (JOC) to provide command and control. The JOC was the operational heart of the

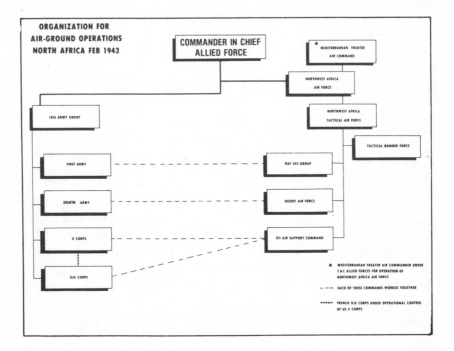

command where the staff scheduled all missions, selected weapons loads, and determined level of effort. The JOC was run by the Combat Operations Officer. He was an experienced combat fighter pilot who usually was a group commander before being assigned to the JOC. The JOC usually had a senior Army officer, Naval liaison officer (if the tactical situation so required), and various intelligence officers who closely followed and posted the current air and ground situation. Although the other services were represented, the JOC was the command facility of the tactical air commander—it was not a joint facility in the sense of being jointly directed by an air and ground commander.

The JOC contained a Tactical Air Control Center (TACC)* within seeing and hearing of the Combat Operations Officer. This center executed the JOC decisions, controlling all aircraft to the target. The TACC had sub-elements whose number depended upon the geographical area and the density of the air traffic. Normally, the TACC had two CRCs and three CRPs with an indefinite number of forward director posts (FDPs) deployed to fill in the spaces in the high and low altitude radar coverage. This tactical control system was conceived, designed, and operated for both offensive and defensive air operations. Each tactical air command was responsible for all offensive and defensive air missions within its geographical area to the range of its radars. This organization was also used during the Korean conflict.

Fifth Air Force was responsible for air operations in Korea and made only minor changes to the system. Instead of having a tactical air command parallel to a field army, a numbered air force now performed that function. Actually, except for the change in name, there was no difference between the WWII and Korean War organizations. Fifth Air Force provided the same support for 8th Army in Korea as IX TAC did for 1st Army in Europe.

ADAPTING THE AIR—GROUND OPERATIONS SYSTEM TO VIETNAM

As the ARVN forces expanded, the need for an air organization similar to that at the close of the Korean War was obvious. It was difficult to create the necessary organization because of U.S. policy which did not allow our forces to fight, but the command and control organization had to be tailored to the Vietnamese situation.

In Vietnam the Air Force added a new element to the system, the Air Support Operations Center (ASOC), later redesignated the Direct Air Support Center (DASC). DASCs were non-existent at the Army corps level during World War II and Korea, but the functions performed by the DASC were handled by an air liaison officer who maintained contact with the JOC. These air liaison officers did not have as much authority as the

---

*TACC—Tactical Air Control Center: The principal center from which air operations are controlled. It is the senior radar control facility in the tactical air control system.

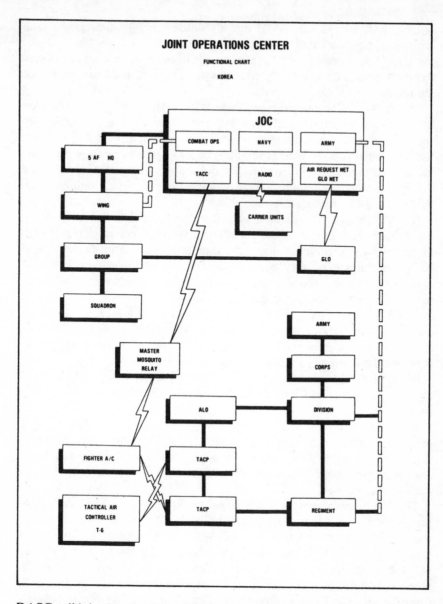

JOINT OPERATIONS CENTER

FUNCTIONAL CHART

KOREA

JOC

COMBAT OPS NAVY ARMY

5 AF   HQ

TACC RADIO AIR REQUEST NET
GLO NET

WING

CARRIER UNITS

GROUP GLO

SQUADRON

MASTER
MOSQUITO
RELAY

ARMY

CORPS

ALO DIVISION

FIGHTER A/C TACP

TACTICAL AIR
CONTROLLER
T-6 TACP REGIMENT

DASCs did in South Vietnam. They represented the tactical air force commander at the corps level and advised the Army corps commander on the employment of airpower. Because the air liaison officer acted in an "advisory" capacity only, the Army processed all field requests for "immediate" and "pre-planned" air strikes. Each request had to be laboriously channeled and approved through each higher echelon of the command structure until it reached the field army level. So, a battalion request had to be approved at regimental, division, and corps level before being forwarded to the field army. During this process, the air liaison officer, at division level, would directly notify the JOC through his own communications net that an "immediate" was on its way. This allowed the JOC enough lead time to divert aircraft from other missions or launch those on ground alert. This, of course, was done with the assumption that the request would be approved.

During the Italian campaign in World War II, Brigadier General Gordon P. Saville, Commander, XII Air Support Command, experimented at corps level with an organization similar to the DASC.[17] For various reasons, however, the Air Force did not adopt the organization for the subsequent campaigns in Europe, nor did it appear as part of air doctrine during or immediately following the Korean War. It wasn't until the early '60s that the DASC began to take shape.

South Vietnam was divided into four corps areas, and each corps commander had almost absolute authority within his area. Even though the corps were technically under the direction of the Joint General Staff (JGS), the highest military body in the South Vietnamese armed forces, corps commanders reported directly to the President and were responsive to the JGS only when they considered it desirable.

The Vietnamese assigned all military forces in a corps zone under the command authority of the corps commander. Even air units in each of the corps zones came under the control of the corps commander; he considered these units as being his and not for the use of adjacent corps commanders. USAF air doctrine, on the other hand, placed air units under the command and control of the senior air headquarters in a theater of war.

In establishing the air-ground control system, repeated attempts were made to create the system in accordance with U.S. Air Force doctrine. Experience in North Africa showed that parceling out air forces to ground commanders is ineffective. Centralized control provides the most efficient use of air resources.

The Vietnamese organization's complete inflexibility created additional problems. On the other hand, FARM GATE forces under the control of the 2nd Air Division commander responded in all corps areas. Even with a force of only two fighter squadrons, the 2nd Air Division commander was able to deploy a Direct Air Strike Team (DAST) to any corps area in which a battle was planned. These small forces responded to directions from the 2nd Air Division Operations Center at Tan Son Nhut. Because VNAF forces were under the operational control of the corps commander, they were seldom employed outside of the corps zone. This arrangement prevailed throughout the war, even during the final offensive of 1975.

The shortage of air units in 1963–1964 demanded more flexibility, not less, yet the Vietnamese organization denied this flexibility. This Vietnamese organization corresponded to the U.S. Army position on the control of helicopters in which they were under the U.S. corps advisor who employed them in accordance with the desires of the corps commander. In effect, this compartmented operations in each corps area and was

260

contrary to all fundamental principles of the employment of airpower. The need to integrate air doctrine with ground doctrine was again shown in 1964 during Operation DESERT STRIKE, a joint Army/Air Force exercise conducted in southern California. It was the largest joint military exercise conducted since World War II and second in size only to the North Carolina maneuvers in November 1941.

Out of DESERT STRIKE evolved the Air Support Operations Center (ASOC)* at Army corps level. Introduced into Vietnam shortly thereafter, the ASOC closely followed the functions developed during the exercise. Essentially, the ASOC enlarged the duties of the air liaison officer by providing him with specialists for the fighter, reconnaissance, airlift, and intelligence roles. Instead of being just an advisor to the corps commander, he now had more of an operational responsibility for the employment of air sorties allocated to the corps. This role represented a significant change in the perception of how airpower should be employed in support of ground forces.

In World WAR II and Korea there had been very limited decentralization of authority below the tactical air force level. The decisions were made at the field army/tactical air force level and not at the corps level. The air liaison officer had no authority to divert aircraft, nor was he given an allocation of sorties to be used by the corps commander. The corps commander received his air support through "preplanned" requests approved by the field army commander. It wasn't until Operation DESERT STRIKE that the need for more flexibility in the use of close air support at corps level was fully appreciated.

The establishing of the ASOC, or DASC, was a direct response to this need. The fluidity of the ground battle within a corps area often made it necessary for the ASOC to divert strike aircraft from preplanned targets in support of ground units. This gave the corps commander some flexibility to change the importance of targets at any given time or to support the ground unit which needed direct air support the most.

Early in Vietnam we helped the VNAF establish an ASOC at each corps headquarters. A VNAF officer ran the ASOC with a USAF officer assisting and advising him. The ASOC worked with the corps Tactical Operations Center (TOC) responsible for the ground effort. On preplanned missions the ASOC would estimate the air forces required to support the planned ground operations. This was done independently of the TOC. In the meantime, ARVN divisions within the corps areas would submit to the TOC what they thought would be required in the way of air support. These requests were then consolidated at the TOC which would then validate the requested air support with the ASOC. After the two agencies had completed their review the corps commander approved or disap-

---

*ASOC—air support operations center: A facility of the tactical air force commander located near an Army corps headquarters to advise and assist the corps commander in requesting and using air support.

proved and sent the request to the JGS. Usually the VNAF representative had very little impact since the senior TOC officer considered VNAF air units to be merely another element of the corps. The Joint Operations Center was located at JGS. Only a few airmen were assigned to this facility, and because of the small size of the VNAF, there was a lack of experienced officers for assignment. As a consequence, airmen had limited influence in the decision process. The JOC determined how many aircraft would be assigned to the mission and what the armament load would be. In almost all cases the JOC followed the request submitted by the corps commander. The JOC sent final request to the Air Operations Center for execution.

The Air Operations Center was jointly manned with USAF and VNAF personnel. The director was a VNAF officer with a USAF officer as his deputy. Again, at this stage of the war U.S. forces were not overtly involved, and their role was that of training the VNAF. The Air Operations Center developed the fragmentary (frag) order that directed the units to fly the mission. A copy of the frag order went to the ASOCs and all elements of the tactical air control system. The senior element of the control system was the Tactical Air Control Center located within the Air Operations Center. This was consistent with the organization developed in World War II and used in Korea, since the Air Operations Center had the same functions as the JOC in World War II and Korea. There was, however, one very fundamental difference: At the JOC, the senior Army representative presented Army requests for air support after they had been evaluated by the field army—the combat operations officer then determined what forces and armament would be needed to produce the desired effect requested by the field army commander. This was centralized control of air resources. Hence, the USAF/Army system of air-ground operations was fundamentally different from the Vietnamese system. In the Vietnamese system, the division of airpower into corps areas was a serious error that limited the capability of the VNAF to support its army. Without a centralized combined command structure, there was no way to adequately employ the very limited resources of the VNAF.

THE BATTLE CHANGES—NEW DEMANDS

By the end of 1964, the North Vietnamese had made a decision to escalate the war in South Vietnam. [18] Engagements were now approaching that of a battalion size. Heretofore, most of the contacts were at platoon and company levels with some attacks on strategic hamlets above company size. The hostile troops were armed with better weapons, and for the first time regular North Vietnamese troops were identified from battalion and in some cases larger units. From these activities, it was evident that the North Vietnamese had embarked on a sustained military effort to capture South Vietnam. [19]

With the appearance of regular North Vietnamese troops in South Vietnam, aircraft encountered heavier ground fire in all corps areas with a sharp increase in I and II Corps. Whereas most of the ground fire had been 30 caliber, 50 caliber became more frequent. This added firepower required a change to higher operating altitudes for most aircraft, because the effective range of 30 caliber weapons was about 1,400 feet while that of 50 caliber was approximately 3,500 feet. Aircraft operating below 2,000 feet expected to encounter some ground fire. In approaching most of the advanced or assault airfields, aircraft could expect spasmodic ground fire while on final approach. This was also true after dark on the approaches to most of the larger airfields such as Tan Son Nhut, Bien Hoa, and Danang.

Aircraft losses increased significantly during this time. The T-28 had reached the end of its useful life and was difficult to operate where ground fire was present. Structural failures and losses forced the replacement of these aircraft with A-1Es* for the air commando squadrons and A-1Hs for the VNAF. The shortage of A-1s made it imperative that we obtain a replacement aircraft. The U.S. Air Force was pressing, again, for the introduction of jet aircraft since their survivability was better and they could be logistically supported in a more efficient manner.

The B-26, which had done a good job, was no longer structurally sound. When a wing buckled on a B-26 during a glide bomb run, the Air Force decided to pull the remaining B-26s out of combat.[20] The Air Force made a number of studies to determine what could be done to increase the life of the B-26 and concluded that the price was prohibitive

The A-1E became the workhorse of the Air Commando fighter force, replacing the T-28. It carried a large payload and could stay over a target area for hours.

---

*A-1E was a multiplaced model whereas the A-1H was a single seat version.

for the few years of added life expectancy and the few aircraft available. With the withdrawal of the B–26 from Vietnam, the Commander-in-Chief of the Pacific Air Forces again proposed the introduction of two squadrons of B–57s. These aircraft were based in the Philippines and were scheduled to return to the states to be assigned to the reserve forces. The request was again turned down for the same reason—it would be an apparent violation of the Geneva Accord. The Air Force hoped to get additional A–1Hs from the Navy, although that service was having similar difficulties maintaining its inventory.

During this period, the frequency of ground contacts increased, and the number of helicopter assaults increased significantly. U.S. Army aircraft in South Vietnam constituted 47% of all aircraft. The 2nd Air Division had 117 aircraft of which approximately 50 were combat aircraft, the remainder being O–1s, SC–47s, and C–123s. The VNAF had about 170 aircraft. With so many U.S. Army helicopters conducting missions throughout most of the corps areas, the issue of who would control these helicopters became critical. The increased ground fire, probability of collision, and likelihood of mutual interference required more positive control of these aircraft. In previous wars, all combat aircraft in a theater came under the control of the air component commander. In Unified Action Armed Forces, the Air Force is charged with "the preparation of the air forces necessary for the effective prosecution of war . . . for the conduct of sustained combat operations in the air . . . to gain and maintain general air supremacy, to defeat enemy air forces, to control vital air areas, and to establish local air superiority . . ."[21] But precedent and this directive didn't seem to answer all of the questions about what to do with helicopters in Vietnam. The issue was control of U.S. Army helicopters since VNAF helicopters were already under the control of the tactical air control system even though they were assigned to the various corps and employed by the corps commanders.

The Army declined to place helicopters under control of the 2nd Air Division commander, again arguing that these aircraft were essentially part of the ground forces, like jeeps and artillery, and therefore operational control should remain with the senior U.S. Army advisor to the corps commander. The Air Force replied that it was not trying to dictate how the helicopters would be employed, but it felt obligated to prevent helicopter missions from going into areas already under heavy attack by other aircraft and to provide adequate cover and suppression of ground fire for helicopter air assaults. The Air Force component commander had the responsibility for coordinating all air operations including helicopter assaults. It was essential, therefore, that helicopters be controlled by the Air Operations Center.

Realizing the necessity of a coordinated effort, Military Assistance Command Vietnam (MACV) headquarters directed that the U.S. Army Aviation Operations Center colocate with the combined USAF/VNAF AOC and that the commander of Army aviation coordinate all helicopter assault operations with the 2nd Air Division commander.[22] Based on extensive experience from previous wars, the Air Force knew the vulnerability of airborne assaults, and helicopters were even more vulnerable to ground fire than the C-47 transports used for airborne assaults in World War II and Korea. Thus, Harkins' directive of 18 August 1962 resolved that all heliborne assaults would be escorted by fixed wing aircraft and that concentrated air attack would be conducted prior to the assaults to suppress any ground fire or to prepare areas where it was highly probable.

NO BATTLE LINES—FACS COME OF AGE

With the enemy infiltrating throughout the country, except for certain areas where there were few civilians the problem of preventing or minimizing civilian casualties was extremely critical. Obviously the bombing of innocent civilians, aside from being inhumanly wrong, would quite defeat our purpose—to convince the civilian population to help the government eradicate the NVA and VC. The enemy was aware of the favorable propaganda he could generate by enticing attacks in areas where civilian casualties were bound to happen; it was virtually impossible to root the enemy out of villages and other areas, particularly in the delta, if attacks were not made with great precision.

Because there were no front lines except for the 17th parallel which arbitrarily separated South Vietnam from North Vietnam, the enemy was apt to be anywhere; this was a distinguishing characteristic of the war as compared to World War II and Korea. In those wars, once the aircraft passed the "bomb line,"* the crew could assume that anything that moved was directly associated with support of the enemy's fighting force

---

*bomb line: Imaginary line beyond which aircraft could deliver ordnance without coordinating with the ground forces.

and was a legitimate target. Towns and villages were struck when the enemy used them for bivouac of troops, supply points, or staging for further attacks. When Intelligence indicated military activities in these villages, they were brought under attack. Civilians were not the target, of course, and if civilian casualties did occur, they were a collateral effect of the attacks against the military target.

In Vietnam, though, all villages and towns were in the combat zone. We had no way of telling whether there were enemy forces in the villages unless the villagers were willing to come forward and report their presence. In some hard core North Vietnamese Army (NVA) and Viet Cong (VC) areas like War Zones C and D, the U Minh Forest, and Bong Son Plain, civilians were warned to stay out; those who remained or filtered in were considered hostile and were brought under attack by ground or air firepower.

To minimize attacks against civilians, Forward Air Controllers became the fundamental means by which all strikes were controlled. With the deployment of the first FARM GATE detachment in 1961, the policy that strike aircraft would be under the control of a Forward Air Controller became firm.[23] All FARM GATE aircraft were controlled by VNAF FACs. The shortage of VNAF pilots and the lack of trained FACs, however, created an unsatisfactory condition.

As FARM GATE aircraft assumed more missions and as the demand for close air support increased, the need for USAF FACs was apparent. The USAF had phased down some of its forward air control structure at the end of the Korean War; FAC duties were handled by periodically rotating fighter pilots from squadrons to Army divisions. These FACs were on duty with Army divisions during field exercises and maneuvers. Considering the frequency of exercise, this was a satisfactory method of training air and ground forces. Such a program, however, did not have a broad enough base to meet a large wartime requirement. For conventional war we considered that there would be sufficient time to expand the inventory of FACs to correspond with the number of Army divisions being deployed.

Vietnam presented an unforecast demand for FACs; we hadn't supposed they would be needed for the control of all strikes irrespective of target location. Close air support doctrine dictated that these pilots would be needed only for the control of strikes that were in close proximity to friendly troops. Once strikes were scheduled beyond a line in advance of ground forces such as a bomb line, there would be no need for a FAC except to locate a target. The basic requirement in World War II and Korea was to employ FACs in the close air support role.

With no formal line of engagement of troops in South Vietnam, the role of the Forward Air Controller took on a vastly different character. With the requirement to have a FAC control all air strikes, an expanded force was needed and time was of the essence. The inability of VNAF to satisfactorily do the job resulted in the deployment of the USAF 19th

Tactical Air Support Squadron in June 1963.[24] This unit had 22 L–19s and 44 pilots; its role from the outset was not only control of air strikes, but development of information about the enemy through daily visual reconnaissance (VR).

During the period 1963–1964, Forward Air Controllers were a prime source of intelligence about the enemy. FAC procedures closely followed a RAND* study which suggested that the assignment of FACs to each of

SOUTH VIETNAM'S
PROVINCES

*RAND—A California corporation that has had a contract with USAF since 1948 to undertake a variety of studies concerning strategy, force development, employment, and design of weapon systems.

the 44 provinces would provide a good source of current intelligence that could be used for air and ground action.

By April of 1965, with the introduction of major U.S. ground and air units, the FAC structure was expanded to four 0–1 squadrons. Our policy then assigned a FAC to each province headquarters and major ground force unit down to battalion level. All Tactical Air Support Squadrons (TASSs) were assigned to the 507th Tactical Air Control Group.

Two categories of FACs met the expanded demand. The first category were fighter pilots who served in fighter units for the first six months of their tour in Vietnam. After this period, they were given a short course in the functions and duties of a FAC and then checked out in an 0–1, 0–2, or OV–10 depending upon the type of aircraft assigned to a given TASS. Upon completion of the course, these fighter pilots were assigned to U.S., Korean, and ARVN ground force combat units.

These pilots were familiar with the conduct of close air support strikes. Since they worked with ground combat units, it was essential also that they have a good rapport with the ground commanders and that these commanders have confidence in them. The Air Force assumed that this relationship could come about more easily if a fighter pilot who had been flying close air support missions were selected for FAC duty with Army units. Since the FAC was an extension of the command system, and since he was the direct representative of the tactical air force commander, he advised and assisted the ground forces commanders in determining their need for air support requirements. This policy prevailed throughout the war.

The second category of FACs included those assigned to each of the province headquarters and to special units. These pilots came from all types of organizations in the Air Force. Some came from Air Training Command, Air Staff, Military Airlift Command, Air Defense Command, as well as all the other commands of the Air Force. They were given FAC training at Hurlburt Field, Florida, and then assigned to a Tactical Air Support Squadron in South Vietnam.

These pilots worked daily with the province chiefs. By flying over the province day after day, the pilot got to know the province in detail and easily noticed changes in the activities of villages which would often indicate the presence of enemy elements. Based on this information, patrols were dispatched to confirm the intelligence. The FAC was usually overhead to provide direct assistance or request air support. Many times the province FAC would recommend targets to the province chief who would go to the corps headquarters to get a strike approved.

As in Korea, the Vietnamese terrain seriously restricted the utility of FACs on the ground. The airborne FAC was the only effective way of controlling a strike. In Korea the T–6 was used for this mission. These FACs were called MOSQUITOES, and flew close to the ground battle

O–1 "Bird Dog" over a Special Forces camp in South Vietnam.

line.[25] They received their missions from the TACC and were in contact with the ground units receiving the close air support. Targets were marked with 2.75 inch rockets, and the MOSQUITO pilot controlled the fighters making the strike.

We had a similar system in South Vietnam. The Air Force FAC assigned to an Army battalion would be over the target area prior to the scheduled time of the mission. Being familiar with all aspects of the battalion's current operation, he didn't require orientation once the action started. After taking off from their home base, the fighters would contact the nearest tactical air control facility which would direct them to the location of the FAC. When the fighters established radio contact, the FAC would provide a description of the target, attack heading, artillery information, and location of a recovery area in the event a fighter was damaged. The fighters then reported their armament on board, and the flight was ready to begin the attack.

Before he allowed the fighters to attack, the FAC had to determine the exact position of friendly forces. He would request the ground commander to indicate the forward line of U.S. forces with smoke markers. Once this line was identified and the fighters confirmed visual contact with the smoke, the FAC would make his run into the target, mark it with a smoke rocket, and clear the fighters for attack. Using the smoke as a reference point, the FAC would relay adjustments to the fighters until they had expended their ordnance on the target.

269

Depending on the type and amount of ground fire being received and the position of friendly troops, the FAC called for multiple passes on the target from appropriate directions. Circling near or over the target and keeping it and the fighters in sight at all times, he would often hold up the attack and mark remaining areas of the target. The greatest percentage of targets in South Vietnam were not visible to the fighter pilot because of terrain, jungle cover, or speed of the aircraft; usually it was a combination of all three. In most instances the fighter pilots never actually saw the specific target because it was hidden in the dense vegetation of the jungle. These men had to rely almost entirely on the eyes of the FAC to get their ordnance on the target. At all times, the FAC was the final air authority on whether or not the strike would continue. He was, in fact, the local air commander for the conduct of air operations, and his authority was recognized by the ground force commander and flight leader alike.

JETS ARE APPROVED

As 1964 came to a close, the military and political situation had deteriorated to the point that a major decision to introduce U.S. combat forces had to be made if South Vietnam were to survive. North Vietnam had embarked on a full scale military campaign to eliminate all South Vietnamese resistance. Instead of isolated small scale actions by VC troops, battalion size engagements were taking place, and North Vietnamese troops were more in evidence, particularly in I and III Corps and along the tri-border area in II Corps. More airpower was needed to cover these areas, but the VNAF was unable to meet the increasing requests. The graduated increase in the use of airpower was ineffective against enemy forces and lines of communication in Laos and did not convince the North Vietnamese that further military actions against South Vietnam would lead to larger scale military action by U.S. forces. With the enemy attacks on Pleiku in early February 1965, 2nd Air Division was given authority to employ B–57s in South Vietnam.[26] It had been a long uphill struggle to get a modern jet aircraft into South Vietnam to replace the worn out T–28s and B–26s. Although the A–1s were in relatively good condition, there were insufficient numbers of them to support the enlarged war. Furthermore, these aircraft didn't have the performance to cope with the increase in anti-aircraft fire, nor the range to operate between corps areas when fully loaded. As the North Vietnamese expanded ground operations, the quality and quantity of their automatic weapons went up sharply. Although not comparable to ground fire experienced in the upper route packages in North Vietnam, it had greatly increased over that in 1961–1963.

With the introduction of the B–57s, some other restrictions were lifted. These aircraft could operate in support of troops in contact with the enemy without the requirement to have a Vietnamese observer on board. Additionally, for the first time U.S. markings were authorized on combat aircraft delivering air-to-ground munitions. Until this time, FARM GATE aircraft had carried VNAF markings even though they were flown exclusively by USAF pilots.

With the introduction of jet aircraft into South Vietnam, a major expansion of airfields and runways was needed. Only three airfields in South Vietnam were capable of handling jet aircraft and these were badly in need of major repair and runway extension. Danang had been a major French airfield for I Corps, and although it needed major expansion, it could support F-100s. Bien Hoa was marginal, and to support sustained jet operation, concrete runways were needed to replace the asphalt strip used by T-28s, B-26s, and A-1s. The third airfield, Ton Son Nhut AB in Saigon, was already handling jet commercial traffic. However, the hard wear and tear of the heavy commercial traffic meant the runway would soon have to be replaced if combat jet aircraft were to be located there.

The decision in March 1965 to deploy U.S. Marine and Army units to Vietnam accelerated the engineering effort to increase the capability of the three major airfields.[27] Furthermore, we started plans to build new airfields at Cam Ranh Bay, Phan Rang, Phu Cat, Tuy Hoa, and Chu Lai, as well as to expand the strip at Pleiku for the emergency recovery of F-100s and F-4s. With the expected buildup of forces, we decided to construct two 10,000-foot runways for each of the major jet airfields. In building to this objective, temporary runways of 10,000 feet were constructed first to meet the operational ready dates of the incoming units. As the temporary runways were finished, along with supporting structures, work was immediately started on the permanent, parallel 10,000-foot concrete runways. When the first concrete runways were finished, traffic was shifted from the temporary runways to the new ones, and work was started on the second concrete runways. Before the end of 1967, the runway construction program was completed.

By the end of 1965, U.S. strength had built to more than 184,000 men.[28] The forces under the control of 2nd Air Division had been expanding at a rapid rate, and by now the headquarters should have equaled that of a numbered air force. Not having sufficient staff officers, the Air Division had to depend on 13th Air Force in the Philippines to provide temporary personnel to help handle the workload. This was certainly not a satisfactory arrangement, but was consistent with the policy at the time of not building up headquarters in Vietnam.

All forces assigned to South Vietnam came under MACV headquarters which was now established as a sub-unified command under the Commander-in-Chief, Pacific (CINCPAC). Under this arrangement, MACV could not employ its air units in areas not under the jurisdiction of COMUSMACV without the approval of CINCPAC. Thus, CINCPAC was directly responsible for the war effort north of Route Package I and in northern Laos.

THE AIR-GROUND OPERATIONS SYSTEM EXPANDS

By November 1965, the need for a complete tactical air force command and control system was recognized. With 500 combat aircraft now under the control of 2nd Air Division, an air-ground system like the one developed in World War II and Korea was needed. The South Vietnamese Air Force was no longer the prime agency for air operations; the USAF had taken over the main task of supporting all ground operations. Even though command and control facilities were jointly manned, USAF personnel ran the activities with the VNAF members controlling only a limited part of the operation.

As U.S. ground forces were introduced into II and III Corps, an organization analogous to a corps headquarters was needed to direct the subordinate divisions. Normally, each three divisions would require a corps headquarters, but a corps headquarters would have been confused with the four South Vietnamese Army corps, and since there was no combined command, separate South Vietnamese and U.S. headquarters were dictated. This situation led to the creation of I and II Field Force Vietnam (FFV) headquarters. These headquarters were equivalent to a standard U.S. Army corps and had the same functions. In addition, each of these headquarters had the flexibility to expand to accommodate a field army.[29]

Inasmuch as South Vietnamese and U.S. forces did not operate under a combined command structure, separate air-ground networks were also necessary. To support South Vietnamese forces, a DASC was collocated with each of the ARVN corps headquarters. The DASCs were jointly manned. Although the director was a VNAF officer who had a USAF officer as deputy, the DASC was essentially run by the USAF. The DASC handled all requests for air support generated within the ARVN corps area. Missions allocated to the DASC were employed against preplanned targets, but they could be altered at the direction of the corps TOC. Throughout the war this system never changed—the DASC always coordinated USAF close air support of ARVN ground forces. Close air support of U.S. troops was managed through a separate system. II FFV was collocated with the ARVN III Corps headquarters, and the DASC serving the corps also served II FFV; however, all U.S. air requests were handled separately. Poor Vietnamese security made it necessary to function in this manner. I FFV was located at Nha Trang and was responsible for all U.S. Army divisions operating in the II Corps area. The DASC for ARVN forces in II Corps was located at Pleiku with the corps headquarters. This arrangement, therefore, separated the U.S. I FFV headquarters from ARVN II Corps by some 150 miles, making it necessary, as well as prudent for security reasons, to have separate DASCs.

Additional DASCs did not complicate the conduct of air operations since allocation decisions were made at a higher level. COMUSMACV and the commander of 7th Air Force established the priority of the various corps and field force requests for air support.

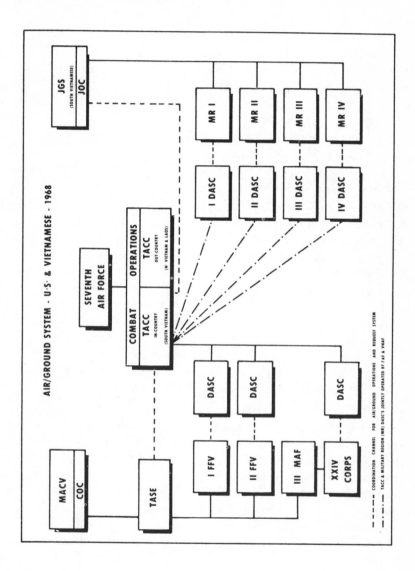

AIR/GROUND SYSTEM - U.S. & VIETNAMESE - 1968

- - - - - COORDINATION CHANNEL FOR AIR/GROUND OPERATIONS AND REQUEST SYSTEM
- · - · - TACC & MILITARY REGION (MR) DASC'S JOINTLY OPERATED BY 7AF & VNAF

2ND AIR DIVISION BECOMES 7th AIR FORCE.

As more U.S. units moved into Vietnam, 2nd Air Division expanded to become a numbered air force on 1 April 1966.[30] Three months later, on 1 July, I replaced Lieutenant General Moore as the unit's commander.

Until the Tet offensive of 1968, all of the air operations within South Vietnam were flown by units based in South Vietnam or on carriers in the Gulf of Tonkin. Although I had operational control of all USAF units in Thailand, I didn't employ them in South Vietnam until January 1968 because Thai leaders were concerned about the international political implications of allowing U.S. aircraft based in Thailand to provide regular close air support to ARVN divisions in South Vietnam. I did have authority to use Thailand based units if I considered the situation in South Vietnam critical, but I could do so only after coordinating with the U.S. Ambassador in Thailand. Beginning with the Tet offensive, I and other 7th Air Force commanders used Thailand based units as needed throughout the remainder of the war.

The command and control system, as designed during World War II and Korea and refined for Vietnam operations, facilitated the employment of airpower wherever it was needed. The 7th Air Force commander's available aircraft in South Vietnam and Thailand were based throughout the theater so that no target was ever more than a few minutes away from strike aircraft.[31] Additionally, KC–135 tankers made it feasible to use fighters from one end of the country to the other. It was this flexibility within the command and control system which permitted fighters from Cam Ranh Bay, for instance, to fly close air support missions in I, II, or III Corps, or if necessary to strike targets in the lower portion of North Vietnam. Also, the tankers provided even greater flexibility to shift the fighter force from close air support missions in South Vietnam to counter air missions in North Vietnam if the situation demanded it.

By early 1967, there were hundreds of thousands of troops in country. The 7th Air Force was well established by this time to support these U.S., ARVN, Korean, and Australian ground forces in all of the four corps areas. Centralized control of airpower was the only feasible means by which each of these ground forces could get air support when it needed it. If the air had been divided-up among these various forces, COMUSMACV would have been unable to concentrate the airpower of 7th AF where he wanted and needed it. With the control centralized, he was able to move around anywhere within his area of responsibility, concentrating firepower as needed. General Creighton Abrams, the Army Chief of Staff in the mid-seventies, and a previous COMUSMACV, put it this way:

> I'm talking about sheer power in terms of tonnage, bombs on
> the target, and that sort of thing, and rockets, because high

performance fixed-wing aircraft carry a much greater payload. And you can focus that very quickly. . . . I don't mean from the first brigade to the second brigade. I'm talking about going anywhere, instead of putting it in MR–4, you go to MR–1. You switch the whole faucet, and you do it in about 45 minutes. The whole control system and base system that supports that, there is nothing in the Army like it. There is nothing anywhere in the world like it.[32]

EMPLOYING THE FORCE

Because a formal line between two opposing ground forces did not exist, the techniques of applying airpower were under constant revision. In World War II and Korea only a small fraction of the sorties were devoted to immediate air strikes, although the fighters could be diverted to these missions from preplanned ones in a matter of minutes. The most effective use of airpower in close air support in both World War II and Korea was in a preplanned mission designed to break through enemy defenses or to stop a penetration. In these missions airpower could be massed, and the full shock of the attack exploited before the enemy could get reorganized.

It was at El Almein on 23 October 1942, that Montgomery's 8th Army launched an offensive to break the defenses of Rommel and roll up the Afrika Korps.[33] The Western Desert Air Force, having gained control of the air, concentrated on destroying supplies badly needed by Rommel. He had supplies for only a limited offensive and depended on capturing stocks to sustain his Panzer forces in a final thrust to take Cairo and the sorely needed oil of the Middle East. As Rommel's drive halted at the marshes of El Almein, airpower was turned to a series of sustained attacks against the dug-in forces. Montgomery launched his ground offensive on 24 October 1942 with more than 1,000 aircraft creating a way for his armor to move through the German defenses. Although Rommel was able to elude the 8th Army and withdraw his battered forces into

P–40s taxiing up to the take-off line at an airfield near Surg-El-Arab in North Africa.

275

Libya, the Western Desert Air Force took a heavy toll of his retreating forces.

The use of massed airpower to support a breakthrough then became a standard procedure. Massed air support was invaluable on 6 May 1943 when a combined air and ground offensive defeated von Armin's army, which was trying to make its final stand before Tunis in North Africa.[34] Later, more than 3,000 aircraft, the largest number used in World War II, supported the breakout of Patton's 3rd Army from the Brittany peninsula. Periodically throughout World War II, airpower was massed in a similar manner: at Caen, Eschweiler, Aachen, and all major offensives by Allied armies. When concentrated in this way, tactical airpower had the greatest potential for helping to break the back of the enemy.

The Korean War produced few circumstances in which airpower could be used in concentrated close air support. Only at Pusan and during the stabilization of the line in 1952 was airpower employed in such a manner. The absence of open terrain for tank warfare and the strong defensive positions provided by the mountains tended to reduce the potential for a decisive offensive by either force. Further, the United Nations objective of accepting a return to the pre-hostility positions along the 38th parallel made a decisive offensive unnecessary. As a result, United Nations forces conducted spoiling attacks, and since these attacks had such limited objectives, there wasn't the need for massed close air support.

Close air support in Korea set the pattern for South Vietnam. The FAC was the key element in Korea, as he was in Vietnam, because the targets were relatively small and very close to the position of friendly forces. The demand for close air support in both wars tended to be scattered across

B–24 "Liberators" of the 467th Bomb Group dropping tons of bombs on enemy installations in France.

the front with some targets more active than others. With these small targets, the number of aircraft in a strike was usually from four to eight unless the strike itself revealed an enemy concentration requiring more aircraft.

Preplanned missions historically have been the most productive since there is better integration of the air and ground effort in accordance with a specific plan of action. In addition, pre-briefed pilots have a better understanding of the target area and the scheme of maneuver to be employed by the ground forces. Other advantages stem from the opportunity to select the best timing and the best weapon for the target, and the least vulnerable approach and withdrawal routes from the target. These factors have always meant that preplanned missions were best for both the air and ground commanders.

In the 1965–1968 period, about 65% to 70% of the 7th Air Force strength was employed on preplanned missions. The Army conducted search and destroy operations like ATTLEBORO, JUNCTION CITY, and CEDAR FALLS, and the air support for these operations had to be planned in detail. Planning for such division and multi-division operations covered a period of weeks, but this detailed planning enabled airpower to be most effectively worked into the scheme of maneuver.

Once the operation started, fighters were scheduled against a number of preselected targets in the assault area such as dug-in enemy troops, suspected concentrations of troops, routes of approach and withdrawal, and areas in which suspected automatic weapons posed a threat to helicopter assaults and the subsequent support of the troops. The destruction of these preselected targets provided maximum security for the assault force during the critical phases of landing, consolidation, and movement from the objective area. Additionally, fighters were scheduled into the area at frequent intervals to take care of targets of opportunity. These preplanned strikes provided air support for all aspects of the operation and ensured airpower was available to handle those parts of the action in which artillery and organic weapons were insufficient.

The remaining 30% of the air effort, not committed to preplanned missions, was used for "immediates" or troops-in-contact situations. All ground operations were designed to seek out the North Vietnamese and VC and to force an engagement in which our superior firepower, particularly airpower, could be employed. It was our policy that after contact with the enemy was established, our ground forces would pull back a sufficient distance to allow artillery and airpower to be used without restraint. Then the Army would follow up these attacks with reaction forces.

When the ground forces made contact, we diverted fighters from preplanned missions of lesser priority, or aircraft held on ground alert were scrambled. COMUSMACV had the flexibility to divert fighters from any part of the country to a troops-in-contact situation. He, therefore, had at his disposal the forces necessary to mount a superiority of firepower against the enemy no matter where the enemy elected to stand and fight.

On any given day in South Vietnam, 7th Air Force flew about 300 preplanned sorties, the Marines in I Corps another 200, and the VNAF 100. The number of aircraft on ground alert varied according to the number of ground contacts expected throughout the country. On an average, 40 aircraft were held on alert, and were scrambled approximately three to four times a day. At the end of a typical day, we had flown between 750 and 800 sorties in support of the ground forces. This effort was sustained day in and day out. We also had the potential to surge to a much higher rate for short periods. During the 1968 Tet offensive and the peak period of the assault at Khe Sanh, for example, the sortie rate jumped from 1.2 per day per aircraft to 1.8.

Because the tactical air control system spanned the country, all combat aircraft were under positive control. The TACC had a minute-by-minute display of all missions, including the ordnance of any given formation. If a ground contact suddenly developed, the combat operations officer could immediately divert fighters to the scene. The TACC, which controlled the fighters through subordinate elements of the system, directed the new mission and provided vectors for the fighters to the target area.

While the TACC issued these instructions, ground alert forces were being launched and backup forces brought to alert status to replace those launched. In a matter of minutes, we could apply a major air effort to any battle area. For a divert, it took an average of 15 to 20 minutes to get fighters into the area and in contact with the FAC. Then the time it took to put munitions on the target was determined by the speed with which the FAC could get clearance from the ground force commander and set up the pattern for attack. If the FAC had received prior clearance, it required only a matter of minutes to mark the target. To save time, the FAC began briefing the fighter pilots while they were still some distance from the target. All that remained for the fighters to do was to visually acquire the target.

Ground alert aircraft normally took 35 to 40 minutes to get bombs on the target.[35] This compares with 40 minutes in the Korean War and about 45 minutes in World War II. When a faster reaction time was needed, either diverted aircraft or airborne alert aircraft would provide it, but most targets could wait the 40–45 minutes necessary to get alert aircraft to the target. Usually a ground force commander took longer than this to decide to call for air support rather than handle the situation with organic weapons or artillery.

ADVANTAGES OF AIRCRAFT ORDNANCE OVER ARTILLERY

When a contact was in the jungle, air strikes were more effective than artillery or mortars because the latter two munitions didn't have enough penetrating power to get below the trees. These shells frequently detonated prior to hitting the ground. Delayed action rounds helped, but these munitions simply didn't have enough destructive power for the heavy rain forest.

Fighters with 500- or 750-pound bombs, on the other hand, were effective against troop concentrations in the jungle and in fortified bunkers. Bombs had the capacity to penetrate the jungle foliage and have the desired destructive effect. Understandably, when fighters were diverted, they didn't always have the optimum weapons for the target. A fighter diverted from a preplanned target and carrying general purpose bombs, for instance, may have been more effective had it been carrying a load of CBUs (cluster bomb units). We recognized this loss of effectiveness, but the need to get munitions on the target quickly often outweighed our wish to hit each target with the optimum munition. Follow-on strike aircraft then carried the munition most suitable for the target.

Alert aircraft, usually in flights of two to four, were loaded with a combination of general purpose bombs, CBUs, rockets, and napalm. Not all aircraft in the flight had the same mixture of weapons, but in the formation there would be sufficient flexibility to take care of any type of target. Many targets required a number of different munitions. For example, for assaulting a fortified position, general purpose bombs were needed to break open the fortification. However, once the enemy troops were in the open, either CBUs or napalm was more effective. Thus, alert flights needed mixed loads.

Air alerts are the most expensive missions. Aircraft dispatched without a target loiter in the vicinity of the ground force until a target appears. While waiting for a target the fighter consumes fuel, and after a time he may have to depart without expending ordnance. We used air alerts in situations in which the vulnerability of the ground operation was so great that aerial firepower had to be available immediately. Although providing air alerts is an established procedure for airborne or amphibious operations, they were not used extensively in Vietnam since there were few circumstances that demanded them. Where there was a need for airpower in such situations, fighters were held for a specified time, and if a target had not developed they were passed to the control of a FAC who either gave them a target or cleared them for a strike into hard core Viet Cong areas not requiring a FAC for control. Alternate targets in these areas were assigned by the TACC and passed to the fighters by the CRC or CRP which was monitoring the progress of the fighters in the area.

DAYLIGHT FIGHTER TACTICS

Most day missions were flown by flights of two to four aircraft. We always had the ability to increase the size of the force, but most targets didn't require it. Furthermore, the enemy's defenses were limited during the 1965–1968 period, allowing for multiple passes on the target. Multiple passes provided maximum weapons effectiveness and superior target coverage. This was just the opposite of target conditions in North

Vietnam in which anti-aircraft defenses limited the number of passes to one, and all munitions were delivered on that single pass. When targets in South Vietnam required more than four aircraft, another flight would be scheduled some 10–15 minutes later.

In South Vietnam most of the anti-aircraft fire came from automatic weapons, which meant that most of our aircraft losses occurred below 2,000 feet. When ground forces were involved and needed the support, pilots pressed their attacks as low as possible to get the job done. There were, however, occasions when friendly ground forces were not actively involved in the target area or even scheduled to enter the area after an air attack. In those cases, the minimum pull-out altitude for the fighters was raised to 3,500 feet. We simply did not want to risk the life of a pilot and the loss of an aircraft by over-exposure in the danger zone when no friendly ground forces were involved. The pilots, of course, didn't like this, always wanting to go as low as possible for better accuracy. In the Korean War, we had a similar policy once the battle lines had stabilized. U.S. ground forces in Korea were directed not to engage in offensive operations that could produce high casualty figures like those that resulted from the battles for Triangle Hill and Sniper Ridge. Because of this policy, minimum pull-out altitudes in Korea were set at 3,000 feet. [36]

I enforced the altitude limitations in Vietnam most vigorously in IV Corps because the enemy had relatively unrestricted fields of fire there and pilots were tempted by the flat rice fields to go lower. Furthermore, ground contacts in IV Corps were not pursued to the same extent as in the other corps areas. Terrain was a major obstacle to the exploiting of a contact; the many canals and swamps made it difficult to get our troops into the desired tactical positions. Additionally, enemy gunners in IV Corps used spider holes that were hard to detect. They would hide in these holes and pop up to fire at the fighters pulling off a target. The gunners were such small targets that they didn't warrant risking the loss of a fighter unless friendly troops were sweeping the area. Thus for several reasons I was quite uncompromising with my pilots about the altitude restrictions in IV Corps.

NIGHT AND BAD WEATHER—SPECIAL TECHNIQUES

The North Vietnamese did much of their fighting at night, a tactic which tended to reduce the effectiveness of our airpower and to counterbalance the numerical superiority in men and artillery we had. We found it difficult to pinpoint the attacking troops and even more difficult to use large numbers of aircraft. To offset this disadvantage, we developed techniques using AC–47s and AC–130s to illuminate the target. These aircraft patrolled each of the corps areas throughout the night. They were in contact with the TACC, the DASC, and various isolated outposts along the Cambodian and Laotian borders, always ready to provide support at a moment's notice.

During the 1962–1965 period, the AC–47 was used to strike, as well as illuminate, attacking enemy forces. For example, as an attack started against a Civil Irregular Defense Group (CIDG) camp, the AC–47 pilot would ask the camp commander for information on the direction of the attack and the estimated distance of enemy troops from the camp's perimeter. Since the AC–47 was not equipped with electronic sensors and had to depend on the eyes of the crew, camp personnel would light smoke pots in the form of a "T" with its base pointing in the direction of the attack. Crossbars were placed across the stem of the "T" to indicate the distance to the attackers. Based on this information, the AC–47 would illuminate the appropriate area and fire on the enemy with its side-firing machine guns. During these preliminary steps, the crew contacted the nearest radar facility to request fighters for follow-up strikes.

When approaching the target area and in contact with the AC–47, the fighters would request illumination of their target. By the time they reached the target, it would be lit up like a ballpark on Friday night—the

The aging Gooney Bird has been used for just about everything. Here, members of the 4th ACS load their AC–47 in preparation for a night mission over South Vietnam.

fighters would then make their firing runs until the flares ran out or the enemy ceased to attack. If the enemy persisted, a second AC–47 would be scrambled to keep the target illuminated. These tactics were very effective in beating off attacks against isolated villages and outposts throughout South Vietnam. In fact, the North Vietnamese and Viet Cong frequently broke off their attacks as soon as the AC–47 began dropping flares.

With the introduction of the AC–130 and its sophisticated sensor system late in 1968,* nighttime air support improved measurably.[37] The aircraft proved the best weapon system we had for target acquisition. With its low light TV or infrared sensor, the AC–130 could precisely detect enemy movements and immediately put firepower on the enemy. Also, the AC–130 was able to mark targets more accurately than the AC–47 for follow-on attacks by fighters. By the time of the 1972 offensive, AC–130s were delivering direct fire against enemy tanks at Quang Tri, Kontum, and An Loc. Furthermore, using laser designators, these aircraft illuminated targets for our fighters to attack with laser bombs. Enemy night attacks no longer posed the serious threat they had during the early 1960s.

Air cover of outlying areas during bad weather required different measures than night attacks using flares. During periods of good weather, fighters under the control of MSQ radars located in each of the corps areas were flown over all the outposts and known routes of infiltration. We carefully positioned the MSQs to permit full coverage of all South Vietnam and most of the panhandle of Laos and Route Package I. Radar beacons were installed in our fighters to facilitate tracking by the MSQ. The fighter would fly at a constant altitude and speed, and the MSQ operator would plot the track and determine the bomb release point for the specific target and type of bomb. This procedure helped our night and bad weather operations. Each MSQ site had on file all the presurveyed targets within range of the site. With this information the site was able to control fighters or B–52s without any appreciable advance notice regardless of atmospheric conditions. If the enemy launched an attack against one of the surveyed targets during night or bad weather, the TACC would launch the fighter directly to the control of the MSQ. A stream of fighters could be directed to the target area to deliver ordnance as directed by the MSQ controllers. During the southwest monsoon (May to October), MSQ was the only way some of the border camps could get timely air support because of the weather.

The use of MSQ was also the primary means for directing B–52 attacks after 1965. Most of the B–52 strikes were made from high altitude, and most targets could not be seen on the B–52 radar. Since MSQ equipment

---

*Actual combat evaluation of the prototype was flown in late 1967 and again from February-November 1968.

provided greater accuracy than we could obtain by having the B–52s use offset aim points, the MSQ was used almost exclusively.

B–52s AND THE GROUND BATTLE

As I explained in the preceding chapter, we targeted B–52s differently from the tactical forces. Each of the field forces and the ARVN corps nominated targets to MACV. Seventh Air Force also nominated targets based on intelligence from FACs, reconnaissance, and reports from the DASCs and air liaison officers. Then COMUSMACV selected the targets using the recommendations submitted by the two field forces and 7th Air Force. [38] As a consequence, we applied the B–52 effort on a priority basis to close air support targets within South Vietnam. This departure from fundamental air doctrine reflected the primary interest of COMUSMACV in the ground war and the Secretary of Defense's guidance that gave priority to the war in South Vietnam.

In the Korean War we observed different priorities for the employment of the B–29 bomber force. For the most part, the B–29s were used in the counter-air and interdiction roles. On occasion, however, during critical situations we used the full force for close air support. One particular instance was during the 8th Army's withdrawal from North Korea. Pounding the advancing enemy forces, the bombers made a significant contribution to finally stabilizing the front.

From the beginning of B–52 bombing on 18 June 1965, the number of sorties continued to climb until the Tet offensive of 1968 when the number reached 60 sorties per day in South Vietnam. This was double the rate that these aircraft were flying in 1966. COMUSMACV used these sorties to strike suspected concentrations of North Vietnamese and Viet Cong troops.

Because of the extremely rugged terrain in many areas, our ground forces were either not available or incapable of maneuvering. B–52 strikes in some respects, then, became a substitute for ground force operations.

The tactical air control system controlled all B–52 strikes against targets in South Vietnam. Because of the lack of SAM and MIG threats in South Vietnam during this period, it wasn't necessary to provide ECM and fighter support for the B–52s. However, when strikes were scheduled into the northern part of I Corps where there was an active SAM and a latent MIG threat, EB–66s, F–105s, Wild Weasels and fighter cover were provided. These missions placed a strain on the supporting forces because of their concurrent commitment to all strikes going into North Vietnam. Nevertheless, the effectiveness of these tactics was evident in that not a single B–52 was lost to enemy action during this time period. It must be recognized, however, that the SAM threat near the DMZ was very limited during this time when compared to the deployments the North Vietnamese made in support of the 1972 Easter offensive.

Many times the enemy tried to place a SAM near the DMZ that could be brought to bear against the B–52s. Most of the time, however, these

Before they entered the air effort over North Vietnam, B–52 Stratofortresses were bombing suspected and known enemy concentrations along the Ho Chi Minh Trail and in South Vietnam.

movements were noticed by reconnaissance aircraft and high speed F–4 FACs. Once a site was detected, 7th Air Force would saturate the target with as many as 30 or 40 fighter strikes, continuing as long as new targets were discovered in the surrounding area. This fast reaction had the effect of deterring the North Vietnamese from establishing a SAM ring in the DMZ similar to that around Vinh in 1968. Whether SAMs were actually fired against B–52s near the DMZ is an open question. The North Vietnamese fired some rockets which had a trail like that of a SAM, making it easy to mistake such a rocket for a SAM. On the other hand, B–52 ECM operators and EB–66 and Wild Weasel crews near the DMZ reported électronic signals from Fan Song tracking radars, so it must be presumed that some SAMs were probably fired at B–52s. However, never were such missiles closer than a few miles to the B–52.

At that time in the war, the downing of a B–52 would have been a psychological plum for the propaganda machine of the North Vietnamese. The JCS and all other headquarters were especially sensitive to this possibility. For this reason, protection of the B–52s had the highest priority in the commitment of the tactical air force. There was a constant shortage of Wild Weasel forces, especially when the weather allowed the F–105s to fly two strike missions a day in the Hanoi delta. During the northeast monsoon, the demands on these forces abated, and protecting B–52 strikes was no problem.

NAVY AND MARINE AIR

Of course, USAF fighters and bombers were not the only aircraft employed in South Vietnam. Until July of 1966, the U.S. Navy maintained a carrier off the coast of IV Corps. Known as DIXIE Station, its aircraft were employed in III and IV Corps. Unlike the B–52s, these aircraft came under the control of the 7th Air Force commander, and therefore operated under the same tactical control procedures as USAF fighters. In August of 1966, because the demands for operations against North Vietnam were increasing rapidly, the Navy at the direction of CINCPAC moved the DIXIE Station carrier north to join the carriers at YANKEE Station in the Gulf of Tonkin.[39] From that time most Navy missions were directed against targets in North Vietnam or along the road networks in Laos. On occasion, however, we used Navy aircraft in South Vietnam to augment the 7th Air Force effort. They flew predominantly in I Corps since it was closer to YANKEE Station and because I Corps had a greater demand for close air support.

We used a large number of the Navy's sorties just above the demilitarized zone where the enemy's entrenched artillery was a problem for U.S. bases and outposts along the southern edge of the DMZ. From YANKEE Station these aircraft would check in with the CRC at Danang for assignment of tentative targets and FACs. As the fighters approached the target area, our ABCCC aircraft, HILLSBORO, would assume control of the flights and make the final target selection and FAC assignment.

With the introduction of Marine ground units into South Vietnam in March 1965, elements of a Marine air wing (MAW) were also deployed.[40] The Third Marine Amphibious Force (III MAF) reached full strength by early 1966. The 1st MAW provided support for its divisions. Until the decision in February 1968 to make the Deputy Commander for Air Operations, MACV, the single manager for the employment of airpower in South Vietnam, Marine air was used almost entirely in I Corps in direct support of the 1st and 3rd Marine Divisions.

The Marines established their own tactical air control system in I Corps which controlled all of their air operations. The DASC located with each of the divisions, and the requirements for close air support missions for the following day were sent from the DASC to the Marine Air Wing TACC. III MAF did not evaluate the requests for the air support, nor determine what the priority for support would be. Instead, the Marine tactical air control system scheduled all in-commission aircraft into each of the division areas on a planned flow, a costly way to manage air resources for sustained operations of an air-ground campaign. The Marine system was designed for amphibious operations where the lack of supporting artillery required airpower overhead at all times. In this operation, where attaining a beachhead is critical, the use of airpower in this manner can be justified. However, it is highly expensive to keep aircraft overhead at all times throughout the day and during critical periods at night when there are no targets. Furthermore, even if the Marine ground situation does not require the aircraft that are on air alert, they are already committed and it is difficult to move them to where they might be needed more.

Marine divisions are not designed for sustained ground operations and therefore do not have heavy or numerous artillery. Thus, the Marines depend upon planes and naval guns to provide heavy firepower for the amphibious assault. In the normal scheme of joint operational plans, once the Marines have secured a beachhead, Army units take over the ground campaign and press the offensive. In Vietnam, the 1st and 3rd Marine Divisions in I Corps settled down to sustained ground combat, with the Army providing them the heavy artillery augmentation. Even so, the Marines employed aviation as though they were still conducting an amphibious operation. The 1st Marine Air Wing divided its aircraft between the two Marine divisions and, irrespective of the ground situation, scheduled these aircraft into their areas in a steady stream. In the event of a major engagement by one division, the air wing would "surge" to meet the additional requirements rather than change to any significant degree the number of aircraft and sorties committed to the other division.

This type of arrangement for sustained land warfare prevents the application of airpower where it can have the greatest effect on the ground battle. Dividing the available airpower between ground units does not allow it to be concentrated on a decisive part of the overall theater campaign. This concept of employing airpower caused us to suffer terrific losses during 1942–43 in the North African campaign. The same issue surfaced in the Korean War when, at the conclusion of the Inchon landing in September 1950, General Almond, Commanding General of the X Corps, proposed the permanent assignment of the 1st Marine Air Wing to the operational control of the corps.[41] Stratemeyer and Weyland resisted such an assignment, articulating again the arguments that finally prevailed in World War II. And I used the same rationale against the arrangements in Vietnam during the 1965–1968 campaign. Further, the debacle of the South Vietnamese defeat during the 1975 North Vietnamese offensive was rooted to some extent in the same fundamental problem: VNAF resources were divided between the corps commanders, and no single air authority looked at the whole of the battle to determine where massed airpower could do the most good in disrupting and breaking the offensive.

Seventh Air Force, during 1965–1967, had little influence on how much of the Marine air capability could be used to support other ground force units. The Marine Air Wing made available daily to 7th Air Force those sorties not used for support of the Marine divisions. The Marine Air Wing decided the number of sorties to be made available. Released sorties usually consisted of A–6s and some F–4s for interdiction in Route Package I or Laos. The Marines provided no significant amount of sorties for use by 7th Air Force for close air support in South Vietnam during this period.

U.S. Marine A–6 "Intruder" loaded to the hilt and ready for take off from Danang Air Base, South Vietnam.

Besides the Marine units, there were three ARVN divisions and one U.S. Army division in I Corps prior to the 1968 Tet offensive. All of these divisions normally required air support during engagements with the enemy, and it was essential that we apply the available airpower to the divisions doing the fighting. It would have been much more effective, when there was no significant ground action, to use aircraft on the lines of communications in Laos or the lower Route Packages where there were some significant targets.

This issue of efficient centralized control of airpower resulted in the decision to give the Deputy Commander for Air Operations, MACV, responsibility for "single management" of the fixed wing aircraft in South Vietnam. With the buildup of enemy forces in I Corps in 1968 and the movement of more U.S. Army units into the area, a change in the control of air operations was inevitable. The demands of the situation were such that COMUSMACV had to be able to turn to a single air commander for advice on when and where to apply the assigned airpower. Thus, the Marine system for amphibious operations gave way to the Air Force system of centralized control for supporting a sustained air-ground campaign.

SUMMARY

Whereas the war in South Vietnam was initially viewed as a counterinsurgency, it was soon apparent that the North Vietnamese were employing forces similar in firepower, mobility, and strength to the units that assaulted Dien Bien Phu. The airpower conceived for "wars of liberation" was totally inadequate for this level of conflict. Thus, we introduced a tactical air force with a complete tactical air control system and an air-ground operations system to blunt the enemy's attacks and halt his offensive operations.

In expanding the air-ground operations system in Vietnam, we encountered the same arguments about control of airpower that came up in

World War II and Korea. The absence of a unified theater command and the Vietnamese command relationships aggravated the problem and initially had a significant effect on the employment of the force. With the experiences gained in combat, these issues began to fade and by the time of the surge of Khe Sanh, General Westmoreland decided to support my position on the control of all tactical aviation including Marine aviation. The employment of airpower in furtherance of the ground campaign assumed a role more consistent with the experiences of World War II and Korea.

CHAPTER VII

FOOTNOTES

[1] George S. Eckhardt, Command and Control, 1950–1969, U.S. Army Vietnam Studies (Washington, D.C.: Government Printing Office, 1974), p. 12.

[2] Illustrated History of the United States Air Force in Southeast Asia, Office of Air Force History, July 1974, p. 5.

[3] Working Paper for CORONA HARVEST, USAF Activities in Southeast Asia, 1954–1964 (Maxwell AFB, AL: Air University, January 1973), p. 5.

[4] Maxwell D. Taylor, The Uncertain Trumpet, (New York: Harper & Brothers, 1959), p. 63.

[5] Working Paper for CORONA HARVEST Report, USAF Activities in Southeast Asia, 1954–1964, vol 2, book 1 (Maxwell AFB, AL: Department of the Air Force, 1970), p. 4–54.

[6] CORONA HARVEST, Command and Control, book 1, p. 1–22.

[7] John J. Tolson, Airmobility, U.S. Army Vietnam Studies (Washington, D.C.: Government Printing Office, 1973), p. 5.

[8] CORONA HARVEST, USAF Activities, book 1, p. 4–62.

[9] CORONA HARVEST, USAF Activities, book 2, p. 6–23.

[10] Wesley F. Craven and James L. Cate, eds., The Army Air Forces in World War II, vol. 3: Europe: Argument to V-E Day (Chicago: The University of Chicago Press, 1951), p. 460.

[11] William C. Westmoreland, A Soldier Reports (Garden City, N.Y.: Doubleday & Company, 1976), p. 86.

[12] CORONA HARVEST, USAF Activities, book 1, pp. 4–63, 4–70.

[13] Tolson, Airmobility, p. 19.

[14] Memoranda, Lieutenant General Gabriel P. Disosway, for Chief of Staff, U.S. Air Force, subject: Comments on Report of Army Tactical Mobility Requirements Board, 14 August 1962, and attachments.

[15] CORONA HARVEST, USAF Activities, book 1, pp. 4-65, 4-67.

[16] Craven and Cate, Army Air Forces in World War II, vol 2: Europe: Torch to Pointblank, pp. 157-165.

[17] Ibid., p. 161.

[18] Working Paper for CORONA HARVEST Report, Out-Country Air Operations, Southeast Asia, 1 January 1965-31 March 1968, book 1, p. 16.

[19] Lyndon B. Johnson, The Vantage Point (New York: Popular Library, 1971), p. 121.

[20] CORONA HARVEST, USAF Activities, book 1, p. 1-21.

[21] Unified Action Armed Forces (UNAAF), Joint Chiefs of Staff Publication 2 (JCS Pub 2) (Washington, D.C., 1974), p. 28.

[22] Eckhardt, Command and Control, p. 38.

[23] CORONA HARVEST, USAF Activities, pp. 2-19, 2-20.

[24] Ibid., vol 2, book 2, p. 6-34.

[25] United States Air Force Operations in the Korean Conflict, 1 July 1952-27 July 1953, No. 127 (Washington, D.C.: Department of the Air Force, 1956), p. 215.

[26] U.S.G. Sharp and William C. Westmoreland, Report on the War in Vietnam (Washington, D.C.: Government Printing Office, 1968), p. 98.

[27] Johnson, Vantage Point, p. 138.

[28] Sharp and Westmoreland, Report, p. 100.

[29] Westmoreland, A Soldier Reports, p. 155.

[30] Eckhardt, Command and Control, p. 64.

[31] William W. Momyer, The Vietnamese Air Force, 1951-1975: An Analysis of Its Role in Combat, USAF Southeast Asia Monograph Series, vol 3, monograph 4 (Washington, D.C.: Goverment Printing Office, 1977), p. 33.

[32] Creighton W. Abrams, Hearings Before the Committee on Armed Services, House of Representatives, on Cost Escalation in Defense Procurement Contracts and Military Posture, 93rd Cong., 1st sess., part 1, p. 967.

[33] The Lord Tedder, With Prejudice, The War Memoirs of Marshal of the Royal Air Force (Boston: Little, Brown and Company, 1966) p. 357.

[34] *The Army Air Forces in World War II*, vol. 2, *Europe: Torch to Pointblank*, p. 204.

[35] CORONA HARVEST, *Command and Control*, book 2, part 2, p. III–4–20.

[36] *Korean Conflict*, no 127, p. 152.

[37] Jack S. Ballard, "The United States Air Force in Southeast Asia, Development and Employment of Fixed-Wing Gunships, 1962–1971," Office of Air Force History, January 1974, p. 135.

[38] Westmoreland, *A Soldier Reports*, p. 76.

[39] CORONA HARVEST, *Out-Country*, book 1, p. 31.

[40] Sharp and Westmoreland, *Report*, p. 109.

[41] *Korean Conflict*, No. 72, p. 202.

CHAPTER VIII

BLUNTING THE ATTACK WITH AIRPOWER
(JUNCTION CITY, KHE SANH, TET, EASTER OFFENSIVE)

Once a week COMUSMACV reviewed the past week's combat actions and projected operations for the weeks ahead. These reviews used intelligence reports as bases for studying the enemy forces' locations and capabilities, and produced the best strategy for drawing the enemy into an extended engagement. Attending these meetings were COMUSMACV staff officers, the Deputy Commanding General of USARV, and the 7th Air Force commander. Field commanders and the Commanding General, III MAF, came to these strategy sessions when a particular operation was under review for their commands.

Routinely, I would present a review of air operations in North and South Vietnam. Although COMUSMACV had no responsibilities for operations in North Vietnam above the 18th parallel, all of us needed to know what the North Vietnamese and Viet Cong were up to on the ground and in the air. I would also assess the out-of-country operations, including Laos, then detail my recommendations for the coming week's interdiction. Although allocations varied widely, we usually put about 60% of our air effort into the interdiction task. The remainder would go to pre-planned close air support missions.

"Immediates"* would get thirty percent of our effort. These are targets we found out about through ground troop contact. They were usually very productive missions since there was no question about the location of the enemy. We would even shift part of our pre-planned air effort to worthwhile targets coming to us as immediate air requests. The important consideration here is the need to exploit a ground force contact that forces the enemy into the open where airpower can be most effective.

---

*immediates: A request for an air strike on a target that could not be identified sufficiently in advance to permit detailed mission coordination and planning.

This, of course, was not always possible, for contacts in the jungle were difficult to exploit. The enemy could disperse and avoid most of our munitions.

AIR OPERATION—JUNCTION CITY

All "search and destroy" operations were designed to find the enemy, fix his position, and destroy him. This fundamental kind of ground warfare is a tactic with one objective—to draw the enemy into a position where superiority in artillery and airpower can inflict a high rate of casualties. Search and destroy operations ranged from small company engagements to large multi-division actions.

Operation JUNCTION CITY represented a multi-division action. It took place in War Zone C, approximately 30 miles north-northwest of Saigon, embracing an area of some 1,000 square miles. The French did not penetrate this area during the Indochina War. It had become the most impenetrable enemy base area in South Vietnam. In late October 1966, three regiments of the Viet Cong 9th Division and the 101st North Vietnamese Regiment deployed into Tay Ninh province.[1] The 9th was probably the best unit in the Viet Cong Army and compared favorably with battle-tested divisions in the North Vietnamese Army. War Zone C was considered the site of COSVN, the political and military headquarters for all enemy activities in South Vietnam. For various reasons, the French did not attempt to clear the enemy from this zone during the Indochina War and until JUNCTION CITY, neither had we.

Over the years, having been left alone, the Viet Cong built in this dense jungle area a self-contained base complete with supply depot, hospital unit, troop housing, and training facilities. Most of the base, except for the training area, was underground in a well-developed bunker system. Some of their bunkers were fortified with concrete. Because of the heavy jungle, it was virtually secure from aerial reconnaissance. The only way to determine the extent of the enemy supplies and forces was to penetrate on the ground. However, because of the size of the area and the number of troops there, a reconnaissance-size force would surely be annihilated. Something on a much larger scale was necessary.

Most of the area along the Cambodian border was hostile to South Vietnamese forces, and, except for outposts, the enemy controlled much of the terrain. North of Tay Ninh City and west of An Loc, which was to be the site of a major battle in the 1972 Easter offensive, the jungle and large rubber plantations provided excellent cover for the undetected movement of forces. Ambushes were frequent in the area because of the nearness of the roads to the jungle. Furthermore, intelligence about the enemy was difficult to obtain. People in the area were not dependable. Although rubber plantations in the region continued to operate throughout the war, their owners paid taxes to both the enemy and the South Vietnamese. So they did not help us, either.

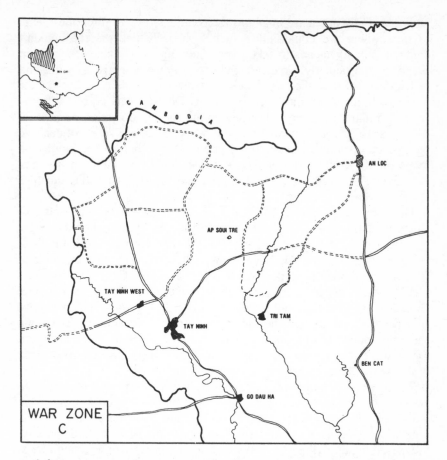

WAR ZONE
C

A large amount of enemy supplies for III and IV Corps in this period, 1967, came by boat to Sihanoukville and were then transported by trucks to base areas along the South Vietnamese-Cambodian border. From there, the supplies were filtered into other base areas within South Vietnam. War Zone C was one of the principal destinations of supplies.

Cambodia's "neutrality" kept us from getting those supplies. The evidence was convincing that a Chinese trucking firm in Phnom Penh moved the North Vietnamese supplies to base camps along the border. Ostensibly, supplies coming into Sihanoukville were for the Cambodian Army, but it was evident even to the untrained eye that such quantities of arms were far above the requirements of a Cambodian Army that was not engaged in combat nor even beginning its later expansion.

JUNCTION CITY was in the planning stage for many months. It would be a multi-division operation to find, engage, and destroy the enemy along with his supplies and equipment. There was no intent to hold the area. Once War Zone C was cleared, the enemy would see that similar attacks could make the area untenable. Also, by neutralizing the area, a major threat to Saigon would be eliminated. The main objective,

295

however, as was the case in all of the operations, was to force the enemy to fight where we could inflict such high casualties that he couldn't continue military means to take over South Vietnam. Our superiority in firepower, particularly airpower, gave us a great potential to withstand heavier combat than the enemy could tolerate.

We concluded that most of the 9th VC Division was currently in the zone. February was selected as the month for the operation because of the dry season; if the enemy could be provoked into a series of decisive battles, there would remain three good months before the southwest monsoon. Of course, the monsoon would make operations in the area very difficult because of the soft ground and flooding—negating much of the effectiveness of armor.

The operation began on 22 February, employing the 1st, 4th, and 25th Infantry Divisions; the 173rd Airborne Brigade; the 11th Armored Cavalry Regiment; the 196th Light Infantry Brigade; and several South Vietnamese units.[2] Also, three South Vietnamese divisions and other elements of Vietnamese forces were positioned near Tay Ninh City, An Loc, and other cities to keep enemy forces from reinforcing. The 173rd Airborne Brigade conducted the only parachute assault of the war. The concept of the operation was to drop the 2nd Battalion of the Airborne Brigade along the Cambodian border as a blocking force, while the 1st and 25th Divisions drove up from the south. Air support was planned on a larger scale than it had been for any previous operation.

The general scheme of employment was to force the enemy into pockets where airpower and artillery could be concentrated. From what we expected our ground forces to do and our estimate of the enemy's reaction, we decided to have about 200 close air support sorties a day. We were already flying over 300 sorties a day throughout Vietnam. Thus, with a requirement for more than 300 sorties, we would employ aircraft from Thailand (after an agreement was reached with the Ambassador). As a final measure, we would direct aircraft from TF–77 to the battle.

We didn't need to augment the FACs; each battalion had two, plus the brigade and division FACs. Most of the pre-planned strike missions were flights of two aircraft with control being exercised by PARIS CRC, located in Saigon next to the 7th Air Force headquarters. If an engagement developed elsewhere, pre-planned missions would be diverted to handle requirements over and above the alert force which numbered about 40 to 50 aircraft.

Prior to D-Day, B–52s would bomb COSVN's probable location (which was to remain elusive throughout the war). Many times Westmoreland requested fighter strikes against a suspected location of COSVN, and on a few occasions he personally scheduled the B–52s against these suspected sites. Before scheduling such strikes, because of difficulty in locating the suspected area of the headquarters, RF–4s were dispatched to reconnoiter the position. Most of the time the intelligence information was very scanty, and reconnaissance, both visual and photographic, failed

F-100 delivering its ordnance in support of ground operations during Operation "JUNCTION CITY".

to reveal the location of COSVN. Nevertheless, such information could not be dismissed since even an outside chance of knocking out COSVN was worth the effort.

Even though JUNCTION CITY continued until May, it actually was a series of three distinct operations. The first phase, which went to the middle of March, was very disappointing. The enemy avoided battle, and the more than 35,000 ground troops had little to show for their effort. Very few of the enemy were killed, although considerable quantities of rice were taken by Major General William E. DePuy's "Big Red One" (1st Infantry Division). We also captured large quantities of ammunition and small arms.

The enemy apparently had advance intelligence of the operation and elected not to make a fight during the initial phase of the assault. After the first day, when 7th Air Force flew more than 180 sorties, we recommended the effort be phased back to approximately 100 sorties, with the understanding that sorties would be increased to the pre-planned level if an enemy contact developed.

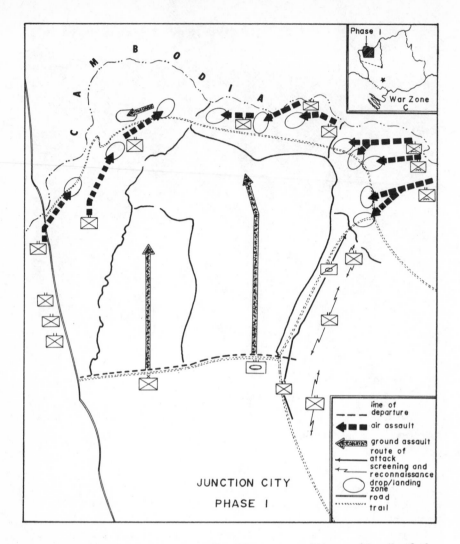

Phase I

War Zone C

line of departure
air assault
ground assault
route of attack
screening and reconnaissance
drop/landing zone
road
trail

JUNCTION CITY

PHASE I

   The airborne assault went off without a problem, with all of the
paratroopers hitting within the landing zone. We dropped 845 paratroopers
of the 173rd Airborne Brigade. Our F–100s had attacked the area prior to
the jump to suppress potential ground fire. Even with this effort, the
enemy continued to fire spasmodically at the C–130s air-dropping supplies
throughout the day. Although there was no 50 caliber ground fire
reported, smaller caliber fire was encountered throughout the day. It was
very difficult to eliminate this type of fire, for it came from troops literally
strapped in trees and other protected areas.
   In addition to the large close air support effort, a major part of the
airlift force was engaged in the positioning of troops at advanced airfields
that had been constructed for the operation. The C–130s and C–123s
became the life line for the support of our troops deployed in the

298

operation. On the first day, more than 40 C–130s were committed to drop paratroopers and resupply our forces. This represented about 45% of the C–130 force in South Vietnam. Our fighters suppressed ground fire directed at these transports. FACs, who were overhead throughout the day and night, called in strikes as required.

As the operation moved into the second phase, we had flown more than 1,500 pre-planned sorties and over 400 immediates. This seems a large number of immediates, but these were the missions we were after—to provoke the enemy to fight so we could use the firepower of the fighter rather than expose troops on the ground.

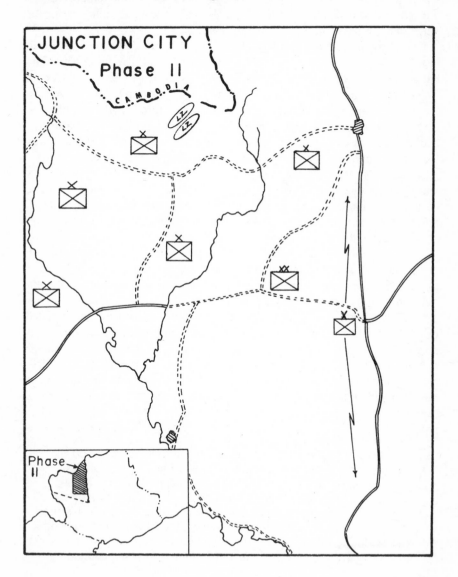

In phase two of the operation, which continued through 15 April, the enemy elected to fight in regiment-size forces. The Army's practice was to establish a series of interlocking fire support bases so that artillery was in position to cover the troops no matter what part of the operational area they were patrolling. In advance of patrolling, a number of landing zones were selected and detailed surveys made of probable enemy gunners' locations. Then helicopters and the artillery pieces were brought in to expand the base. The North Vietnamese and Viet Cong had also reconnoitered most of the potential helicopter landing zones and were particularly adroit at attacking us from hidden positions. Some of the sharpest fights of the operation were in defense of fire support bases located on tops of hills where the jungle was cleared to make room for helicopter landings and a defense perimeter.

One of the largest battles of JUNCTION CITY took place on 21 March in the rubber plantation country 18 miles northeast of Tay Ninh City. The battle of Suoi Tre involved more than 2,500 troops of the Viet Cong's 272nd Regiment. On the 20th of March, 450 U.S. troops were flown into the Suoi Tre fire support base (FSB). Additional forces had been lifted in earlier and were deployed a half mile south to protect the base. An additional force of two companies of armor and mechanized infantry was located a couple of miles away, separated from the FSB by heavy jungle.

At 0631 on the morning of the 21st, an enemy force of some six battalions assaulted the FSB.[3] They penetrated the outer defenses and were threatening the inner lines; the need for air support was immediate. An O–1 FAC aircraft with two pilots was scrambled from Dau Tieng, only a few minutes flying time away. As the FAC put in the first strikes along the tree line, the enemy opened up with heavy machine gun fire, knocking down the O–1 and killing the pilots. By this time the situation on the ground had become desperate with the FSB about to be completely overrun.

Two additional FACs were scrambled while 7th Air Force launched more F–100s from Bien Hoa and F–4s from Cam Ranh Bay, both bases only a few minutes flying time from the battle. By 0900 more than 85 fighters had been committed, delivering a variety of munitions from 750-pound bombs to 20mm cannon fire. A combination of air support and the timely arrival of armor broke the attack. Captain Sager, one of the FACs, reported:

> By that time, the ceiling was so low that we were picking up intense fire. We only had about a thousand foot clearance off the ground, and I'd run out of smoke rockets. I had a flight of F–4Cs come in—three F–4Cs—talked them in underneath the overcast. In fact, one time I had to go up above the overcast and pick up the fighters and bring them back down on my wing, which is pretty hard to do in an aircraft like mine. We got the F–4s under and had them orbiting down there. Visibility was only about two miles and, boy, I thought it was an outstanding job just maintaining

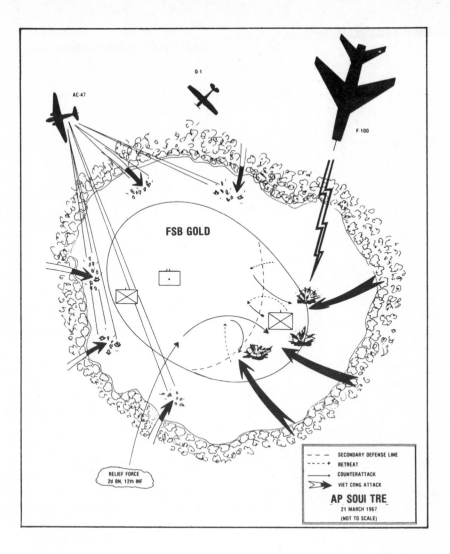

FSB GOLD

AC-47

0-1

F-100

SECONDARY DEFENSE LINE
RETREAT
COUNTERATTACK
VIET CONG ATTACK

RELIEF FORCE
2d BN, 12th INF

AP SOUI TRE
21 MARCH 1967
(NOT TO SCALE)

contact. As I said, I was out of smoke rockets by this time, so I pulled the grenades mounted under the wing which ejected white smoke, and I circled over the target. I told the fighters to strafe beneath me and try not to pull up into me. They did, and it was a fine job. More than 600 enemy troops were killed in the battle. [4]

JUNCTION CITY came to a close by the middle of May. More than 5,000 sorties had been flown in support of the operation. F–100s and F–4Cs provided most of the close air support strikes. AC–47s kept the area illuminated at night and controlled the fighter strikes. During the day, FACs did a superlative job of controlling air strikes in an area filled with helicopters, C–130s, and strike aircraft. Along with the FACs and fighter strikes, forward logistics support by C–130s and C–123s, air resupply by

301

Army helicopters, and determined fighting by troops on the ground inflicted a major defeat on the VC's 9th Division. The 9th retreated into the Cambodian sanctuary, no longer an effective combat unit. MACV reported more than 2,700 enemy troops killed, approximately 500 weapons captured, and 800 tons of rice taken.[5] The operation was not without loss to U.S. and South Vietnamese forces—289 killed.

JUNCTION CITY was the largest operation of the war in South Vietnam. During this period, 1967, a series of operations across the country had left the North Vietnamese forces seriously weakened. Wherever the enemy had elected to fight, the combined arms of infantry, artillery, and airpower had inflicted on him a decisive defeat. If the war continued in this vein, it was apparent that it would be only a matter of time until the enemy threat would no longer be a significant factor in the future of South Vietnam.

ENEMY PROBES FOR AN ATTACK

But the enemy's strategy began to change by the end of the summer. Intelligence indicated that he was building toward a decisive victory in 1968, yet any assessment of the relative capability of ground forces indicated that U.S. and South Vietnamese forces backed up with airpower had sufficient strength to defeat any move the North Vietnamese might make to take over large populated areas, particularly the northern part of the country.

With the airpower available to MACV, the potential of the North Vietnamese to turn the tide of battle was indeed slim. Wherever there was a concentration of troops to assault a hamlet, outpost, CIDG camp, battalion fire support base, key bridge defense point, or provincial headquarters, the rapid application of airpower, both day and night, invariably defeated or repelled the enemy. To weaken the impact of our airpower as much as possible, the enemy consistently tried to stage his attacks at night or in bad weather. When he couldn't follow this strategy, he would try to make contact with our forces, fighting as close in as possible so that our airpower couldn't be used for fear of causing casualties to our own forces.

Neither of the enemy's tactics was very successful. The use of light-producing aircraft and radar-controlled strikes provided the needed capability to accurately strike the enemy in darkness and bad weather. Furthermore, in some cases fighters strafed within 75 feet of friendly troops. With cluster bomb units (CBUs), strikes were often delivered within 100 feet; with bombs it depended upon the size of the blast. During one engagement in JUNCTION CITY when the enemy had ambushed a column moving up to an assault position, CBU strikes were put in so close to friendly forces that a few minor casualties resulted. When the division commander was advised that strikes this close were unsafe, he opted to have them anyway. Without our air strikes, his column would be overrun before relief could be brought in, and the casualties inflicted by the enemy would far exceed those of the "short rounds" from our own air strike. The air strike was delivered and the enemy assault was broken, permitting the column to continue its advance.

With the movement of large bodies of troops into I Corps in September 1967, and the shelling of all the northern bases along the DMZ, the enemy strategy was starting to unfold. From captured documents, the Allies learned that the coming offensive was to be divided into three phases. The first phase would consist of sharp but intensive attacks, mounted from sanctuaries, to draw U.S. troops away from population centers. The second phase was to consist of attacks throughout the country so ARVN would lose control of the people and the people would rise to the support of the National Liberation Front. The third phase would be a ground battle in a northern province favorable to a major victory.

The opening phase of the strategy was soon felt by the Marines and ARVN in I Corps. During 1967, a series of fire support bases manned by Marine and South Vietnamese troops had been established below the DMZ. These 175mm-equipped artillery bases were capable of bringing interlocking fire against enemy troops moving through the DMZ deploying for an attack against the main line of resistance at Dong Ha. At the same time, the exposed position of these fire support bases—especially at Con Thien and Gio Linh—made them vulnerable to artillery fire from north of the Ben Hai River. Realizing their vulnerability, the Marines had requested all the assistance they could get from 7th Air Force to destroy these enemy artillery positions.

Concurrent with the mounting threat to these forward bases, the enemy began the movement of the 304th and 325C Divisions into the vicinity of Khe Sanh. The terrain around Khe Sanh was favorable for concealing large numbers of troops. From these movements, it appeared that the enemy was positioning his forces for a frontal assault against Dong Ha, Quang Tri, and Hue, with a flanking movement through Khe Sanh. To execute such an attack, the enemy was thought to believe it necessary to neutralize Gio Linh, Con Thien, and Khe Sanh. By doing so, he would protect his rear positions and extended logistics lines.

As the threat developed across most of I Corps, I proposed to COMUSMACV that an intensified air campaign be initiated to disrupt and suppress the artillery attacks north of the DMZ and to locate enemy troops that would be staging for the expected attacks. He agreed, and the operation, named NEUTRALIZE, would embrace Khe Sanh, all of the remainder of I Corps, and the area above the DMZ formerly called TALLY HO. [6]

We formed a combined intelligence center at 7th Air Force to oversee operation NEUTRALIZE. The center was augmented with intelligence specialists from the United States and from MACV. The center's product was a compilation of targets that would be struck by the 7th Air Force, Marine, and Navy aircraft under my control for the duration of the operation. High level interest was generating in Washington as to what

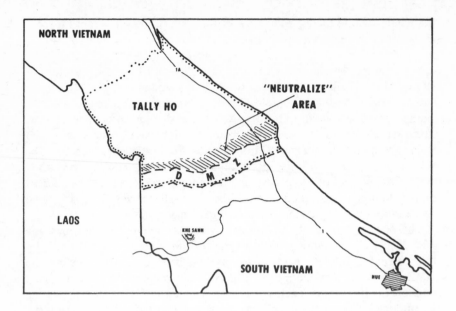

airpower could do to break up the intensive shellings. By September, for example, the Marines were taking over 1,000 rounds per day against the northern fire support bases.[7]

To make the strike effort as effective as possible, targets had to be pinpointed and carefully analyzed. Because of excellent camouflaging and well dug-in positions, the enemy guns were very difficult to locate. In some cases, they were actually kept in caves, rolled out to be fired, and then rolled back in. For these types of targets, counter-mortar radar and other electronic intelligence gathering techniques were not proving very effective. Our best results came from the photo interpreters, and we photographed on a large scale. The RF–4s had to make repeated runs at two to three hundred feet over heavily defended areas where 37mm and 57mm anti-aircraft guns were always active, and from time to time a SAM was sneaked into firing position. The reconnaissance pilots flew their runs at about 580 knots without being able to jink the aircraft. Jinking distorted the photographic image, so the run over target had to be particularly stable.

Using the intelligence data, we struck suspected artillery and troop positions day and night. We located an ABCCC below the DMZ to control all strikes in the area. Aircraft reported to PANAMA, the CRC at Danang. From there they were turned over to the ABCCC. The massing of airpower (fighters, bombers, and reconnaissance) finally broke the siege of these northern bases which had been under intensive attack for more than 49 days. It was the constant pounding of airpower that the enemy had not foreseen when planning this offensive to resemble his successful assault against Dien Bien Phu. The French Air Force had been depleted and totally incapable of mounting an effort in any way comparable to that in NEUTRALIZE.

During NEUTRALIZE, we flew more than 3,000 tactical and 820 B–52 sorties.[8] It was, indeed, a massive effort, and it relieved pressure on the northern two provinces. Consequently, these areas were in much better shape to withstand the assaults that were to develop during the Tet offensive. If the regular North Vietnamese troops had been able to roll up Con Thien, Gio Linh, and Dong Ha, there would have been little to stop them before reaching Hue. The enemy's strategy for the Tet offensive initially appeared to call for a decisive battle in Quang Tri province as a follow-up to a breakdown in the South Vietnamese government and its ability to effectually continue the war.

As the battle for the outposts subsided—even though the assault failed—the enemy continued to occupy U.S. and ARVN troops. Then attacks intensified around Khe Sanh where two enemy divisions, the 304th and the 325C, were positively identified.

KHE SANH—AIRPOWER IN THE FOREFRONT

Khe Sanh was significant only as a political symbol. Situated in a remote part of I Corps near the Laotian border, Khe Sanh was surrounded by rugged mountains blanketed by dense jungle. It controlled no major line of communication, except for the interdicted Highway #9, nor did its terrain hold any advantage for ground warfare. Over the years, the CIDG camp at Khe Sanh had become a jumping off point for long range patrols. These patrols were used to probe the axis of enemy troop movements from Laos into I Corps and the northern part of II Corps. Tchepone was the hub of lines of communication in southern Laos, and from this hub men and supplies were filtered into the Khe Sanh area and then into Quang Tri, Hue, and the A Shau valley. Thus, we saw Khe Sanh as useful principally for intelligence-gathering.

To support the CIDG patrols in 1961 and 1962, a detachment of 0–1s was stationed at Khe Sanh. Two FACs provided most of the support for ground patrols. Patrols reported their observations to the FACs who relayed the information to the CIDG command post. The command post would then request air strikes from I Corps headquarters at Danang. Those activities provided substantial intelligence on the enemy's movement both in the western part of South Vietnam and along the trail network leading into the area.

The major land route to Khe Sanh began at Dong Ha, passed through Tchepone, and terminated at Savannakhet. The Viet Cong had kept most of the road's bridges blown, and U.S. forces thought it too costly to maintain the bridges because the VC and North Vietnamese could knock them out at will. We held out no hope of reinforcing the Khe Sanh overland movement.

For all practical purposes, Khe Sanh was totally dependent upon air support for its existence. By the fall of 1967, enemy activity around Khe

Sanh forced us to decide whether the base should be evacuated or defended. We thought of Dien Bien Phu and its isolation but decided we could do the job with intensive air support.

Defending Khe Sanh would not be easy, for this part of Vietnam suffers from both the southwest and northeast monsoons. During January and February, normally good weather months in Laos, Khe Sanh can expect low clouds in the morning with visibility of one to two miles and a ceiling of 500 to 1,000 feet. By 1000 to 1100 hours, visibility increases to a couple of miles and the clouds start to break; over the next four hours, conditions improve. By 1600 hours in the afternoon, however, the weather again deteriorates.

With these weather conditions, air operations take place during a relatively short period daily when visibility is good enough for air strikes and parachute deliveries. Too, we knew the airfield wouldn't remain operational because of the large amount of artillery and rockets the enemy could deliver from concealed firing positions on all sides of Khe Sanh. We planned to use the runway as long as we could. We would fly a large number of sorties during the good weather periods and use MSQ to drop supplies during bad weather. Because of Khe Sanh's restricted runway approach, incoming aircraft were vulnerable to anti-aircraft weapons that could actually be fired downward at them when they were approaching or taking off. When the enemy's shooting made landing too dangerous, we would parachute supplies in.

The enemy had in the Khe Sanh area two divisions of 15,000 to 20,000 troops, with another division in the vicinity.[9] The base couldn't support a sufficient amount of artillery for its own defense because of the additional load that would be placed on the airlift force to bring in ammunition. The only artillery pieces that could cover Khe Sanh were the Army's 175mm

Supporting the Marines at Khe Sanh, a USAF C–130 lands at the battered airfield bringing much-needed supplies.

guns located at the Rock Pile Camp Carroll. So the enemy held every advantage save airpower—shades of Dien Bien Phu.

In early January the Marines increased their forces at Khe Sanh to two battalions of the 26th Marine Regiment. A third battalion was flown into the base on the 16th of January.[10] By this time President Johnson had taken a personal interest in the buildup of enemy forces around Khe Sanh and was concerned about our ability to hold the base. Obviously, he had Dien Bien Phu on his mind and the political consequences the French suffered from the military defeat. During his Christmas visit to Cam Ranh Bay, the President brought up the question of defending Khe Sanh, and I reassured him that with the massive use of airpower, the base could be defended.

Seventh Air Force was authorized to use whatever aircraft it considered necessary for the task. We would use aircraft from TF–77 and Thailand along with all of our available aircraft in South Vietnam. We put into effect a plan to employ 400 strike sorties a day. Given the unfavorable weather conditions, we would have to use radar control extensively to prevent mutual interference and to enhance safety among the many aircraft.

HILLSBORO, the ABCCC we used for controlling strikes along the LOCs in Laos, would be our main control agency. We reinforced the ABCCC crews with command personnel from 7th Air Force, giving us a balanced airborne command post able to make decisions without clearance from other headquarters. The area surrounding Khe Sanh was divided into sectors with FACs assigned to each one. The FACs gave us continuity of intelligence and detailed familiarity with the terrain and enemy locations. Khe Sanh couldn't be viewed in isolation to the interdiction campaign, for most of the enemy's logistics would flow through the Laotian network. With the enemy's large expenditure of shells, which reached 1,307 rounds of mortar, rocket, and artillery fire on 23 February,[11] replacement supplies had to move through the Laotian road network. To put as much pressure on the logistical system as feasible, 7th Air Force stepped up its offensive against these roads. The enemy's need of greater quantities of material came from several influences. He supplied North Vietnamese divisions now operating in South Vietnam; he introduced more conventional artillery; he suffered heavy losses in CEDAR FALLS, JUNCTION CITY, and other engagements. So we saw more trucks than ever in the resupply network.

During December, January, and February, we flew over 20,000 sorties against the LOCs in Laos and destroyed more than 3,000 trucks.[12] We felt that any reduction in enemy logistics at this time would have not only an effect on the battle for Khe Sanh, but also an influence on the strategy of the forthcoming Tet offensive. Khe Sanh was as much a psychological battle as a military engagement. All of these complications of Khe Sanh made it imperative that the enemy not only be defeated militarily, but psychologically: An unequivocal setback was essential to neutralize the political offensive against the South Vietnamese and U.S. home fronts.

The final defenses at Khe Sanh were the main base: Hills 861, 558, 881 south, and 950.* All patrols ceased to operate after the 21st of January. An additional Marine and Vietnamese Ranger battalion was flown in to complete the buildup for the expected battle; the total strength was about 6,000 men. The concrete bunker command post that was developed for the FACs in 1962 became the command post for the 26th Marine Regiment. Fighting bunkers and a trench network surrounded the base. Because of the plateau on which the airfield was located, the trench network was very close to the runway on three sides. The fourth side was not trenched because the approach end of the runway was at the edge of a sharp drop to the river valley below, making an enemy assault from this direction unlikely.

For our air strikes to be effective, we would have to make them as close to our own trenches as possible. Since there were no troops outside of the base or the hill outposts, air strikes could be brought in very close to the defended positions without endangering our own forces. We planned to deliver most ordnance close to the base perimeter and make selected strikes against the primary approaches to the base. These approaches were seeded with seismic and acoustic sensors that had been

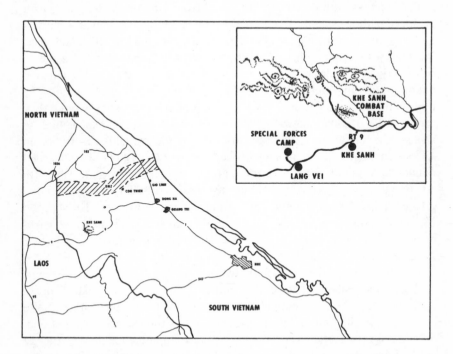

---

*See Nalty, Airpower and The Fight for Khe Sanh, Office of Air Force History, 1973, for detailed discussion.

proven on the road network in Laos. An EC–121 was airborne to relay the sensor information to Dong Ha. The information then went to the ABCCC who would direct strike aircraft into the area.

The problem of air controlling became acute. The Marines had maintained that this was a Marine air-ground team operation and that all air used for close air support should come under their control. Furthermore, a circle had been drawn around Khe Sanh, and it was proposed to prohibit all but Marine air strikes within that circle. A reduced DASC called a "mini DASC" was established within the command post of the 26th Regiment, and the mini DASC was linked to the DASC with the 3rd Marine Division at Dong Ha. Communications were then established with the Marine TACC at Danang. With the magnitude of air traffic around Khe Sanh, the system was totally inadequate.

Because of these problems I told Westmoreland that centralized control of the air had become absolutely essential.[13] Without it, Khe Sanh could well be lost. Our discussion led Westmoreland to designate me, as the Deputy Commander for Air Operations, MACV, the single manager for all fixed wing aircraft in the theater.

On 18 January 1968, Westmoreland sent a message to Sharp which read as follows:

> The changing situation places a demand for greater organization and control of air resources and a premium on the need for rapid decision-making. It is no longer feasible nor prudent to restrict the employment of the total tactical air resources to given areas. I feel the utmost need for a more flexible posture to shift my air effort where it can best be used in the coming battles. Consequently, I am proposing to give my Air Deputy operational control of the 1st Marine Air Wing, less the helicopters.[14]

Sharp concurred after being briefed by Major General Gordon F. Blood, Deputy for Operations of 7th Air Force, and discussing the matter with Wheeler.

In addition to the fighter strikes, B–52s also made a significant impact on the enemy's efforts. The bombers were employed along the outer range of the main defenses where the enemy forces were expected to stage. The purpose of these raids was to prevent the enemy from organizing the attack as he had at Dien Bien Phu and at base camps in other parts of South Vietnam. We expected the enemy to dig tunnels up to the perimeter of defenses where his assault forces would attempt to force an opening that would be immediately exploited by troops advancing with heavy artillery support.

The tactics used at Dien Bien Phu were much in evidence at Khe Sanh: intensive artillery barrages followed by a probing in strength. Tanks had made an appearance for the first time in the war at Lang Vei, which was the last Special Forces camp on Route #9 before it entered Laos. As a prelude to the major assault on Khe Sanh, the North Vietnamese overran Lan Vei using nine PT–76 amphibious tanks.[15] Because of their use at Lang Vei, we assumed that significant numbers of tanks would be used to pave the way for the infantry in breaching Khe Sanh defenses. The use of B–52s against these suspected staging areas surrounding Khe Sanh was for the purpose of destroying these forces before they could move out. The fact that such an organized attack never developed against Khe Sanh can probably be attributed to the effectiveness of these missions coupled with all the other air strikes from the tactical air force.

Because of weather, 50% to 60% of the air attacks were under the control of MSQ. All B–52 attacks were controlled in this manner. With the vast experience 7th Air Force had acquired in MSQ attacks, there was a high level of confidence that these strikes could be delivered within 400 to 500 feet of friendly troops by fighters and within 1,000 to 1,500 feet by B–52s. There was much controversy on this issue when I proposed that the B–52s strike within 1,000 feet of the perimeter. III MAF was opposed because of the possibility of short rounds.[16] Westmoreland, however, felt the B–52s could be employed as I suggested. SAC felt the attacks should come no closer than 3,000 feet for the same reasons advanced by Lieutenant General Robert E. Cushman, Jr., Commanding General of III MAF. MACV finally settled on 3,000 feet with the understanding that if the enemy attack developed in force, the B–52s would bomb within 1,000 feet of friendly positions. The enemy attacks never developed to more than a large scale probing maneuver, so no urgent tactical requirement developed for moving the B–52 attacks closer than 3,000 feet from the dug-in Marines.

The AC–47 also contributed to the air effort. These aircraft were used throughout the night to keep the area illuminated. As was the case with other outposts, this tactic helped to deter attack. If the attack developed anyway, enemy losses would be comparable to a daylight assault because of the absence of concealment that darkness provides. Aircraft were periodically directed to the AC–47 which acted as a FAC for controlling the strike. By striking within the area throughout the night, we created a further deterrent to a night assault.

On any given day after the siege began, 7th Air Force managed approximately 350 tactical fighters, 60 B–52s (even though MACV was responsible for targeting, the 7th Air Force command and control system actually controlled the strike and the decision to strike), 12 to 15 C–123s/C–130s, 10 RF–4s and 30 0–1/0–2s.[17] Numerous helicopter sorties were coordinated, but we didn't control these aircraft.

There is little doubt that airpower was decisive at Khe Sanh, although the Marines and ARVN Rangers fought courageously under trying conditions. The enemy at Dien Bien Phu was able to move without much fear of the French Air Force, but circumstances differed at Khe Sanh. Airpower was pounding him day and night. We never let up through the long siege from January through March. We flew more than 22,000 sorties and dropped 82,000 tons of bombs. General Wheeler, Chairman of the

JCS, estimated the number of enemy casualties to be 10,000. Moreover, the 304th and 325C Divisions were left unfit for further combat. [18]

Khe Sanh was probably the turning point in the enemy's strategy for Tet. If Khe Sanh had fallen, the regular NVA troops would have moved against the major cities that were initially assaulted by VC local forces. The fact that there were no significant actions by regular forces indicated the enemy backed away from a combined military-political offensive and settled primarily on a political offensive designed to undermine the South Vietnamese' confidence in their government and to create more dissension on the U.S. home front.

PRELUDE TO TET

By the time of the Tet offensive in January of 1968, the air war in the North had scored a major victory over the North Vietnamese Air Force. It was essentially a defeated air force and was withdrawn from battle to be retrofitted and prepared to resume the fight at a later date. The airfields of China provided a sanctuary for the recuperation of the force.

Along with the counter-air compaign, interdiction of the major LOCs in the northern routes also had been effective. All the main bridges were down, and most of the marshalling yards were blocked. A single through-line was kept open at great expense in repair crews. Troops trying to get to the front took three months as compared to the earlier periods when it took only three or four weeks. The traffic on the LOCs in Laos was the heaviest of the war with the number of trucks destroyed or damaged averaging 45 to 50 per night. [19] Logistical requirements were spiralling upward. The entire enemy logistical structure was under severe strain to support the increased level of the major engagements in the south. This was evident by the increasing amount of supplies that had to enter the pipeline in order to get enough at the other end to sustain the enemy's increased combat operations. At no previous time in the war had the enemy expended such large quantities of rockets, mortars, and artillery ammunition as he did at Cam Lo, Con Thien, Gio Linh, Khe Sanh, and Dak To.

Stocks for these operations had been accumulated over a long period of time, perhaps a year. The enemy's low level of activity until late summer of 1967 may have been a deliberate strategy to avoid consuming precious stocks for the Tet offensive. However, these stocks were not enough and had to be replenished. The level of fighting early in the Tet offensive caused rapid consumption of the stocks. The relatively short length of the offensive was probably attributable to a combination of the resistance encountered from the U.S. and ARVN ground forces and airpower, and the rapidly dwindling supplies.

The critical year for the North Vietnamese was 1968. If the war continued, it would be only a matter of time until the air offensive expanded, and with this expansion the price for continuing the conflict would become prohibitive. Furthermore, the South Vietnamese armed forces, under the umbrella of U.S. and VNAF airpower, would enjoy superiority in battle throughout all corners of the country. Time was not running in the enemy's favor. He needed to create a dramatic change in the political situation in South Vietnam which would lead to United States disengagement because of lack of support on the home front. Widespread demonstrations against the war were taking place throughout the U.S., and a major military and political offensive could reinforce those arguing that the U.S. should get out of the war.

While the pressure against Khe Sanh increased, the NVA 1st Division launched an attack against Dak To in the central highlands. These actions were designed to pin down ARVN and U.S. main forces. The same strategy was implemented in the northern provinces with sustained attacks against the fire support bases north of Dong Ha. Only in III and IV Corps was the enemy relatively quiet. The large operations, CEDAR FALLS and JUNCTION CITY, had probably persuaded him to withhold the 7th Division, especially since the 9th NVA Division had been practically decimated in JUNCTION CITY and was probably in Cambodia being reformed.

The enemy was taking a severe beating from each of these actions. His operation at Khe Sanh hadn't developed as planned; the northern FSBs hadn't fallen to open the way to a frontal assault on Dong Ha, Quang Tri City, Hue, and eventually Danang; and Dak To had held off a sustained attack by three regiments, thus blocking the gate to Kontum and Pleiku. The objective of attaining a major military victory like Dien Bien Phu seemed remote.

On the other hand, the opportunity for a major political victory was still at hand, although it would be costly in human life. In the long term, however, it could provide the ingredient for an eventual military defeat of South Vietnam. The evidence seems to indicate that the Tet offensive was designed for its political effect, with whatever military successes it realized being only a by-product.

The basis for this assessment is the fact that the local Viet Cong forces launched the offensive against all the major population centers of South Vietnam. Only in the case of the attack at Hue were regular main force units employed. Accompanying the local forces were political cadres who were to take over the administration of the cities and villages. The regular forces were held in reserve and apparently were to be committed only if there were a disintegration in ARVN regular, RF, and PF* organizations. If the defenders didn't stand and fight, and the NVA anticipated they wouldn't, the ARVN regular divisions would rise up against the U.S. and their own government and demand the withdrawal of all U.S. forces.

---

*RF–PF: Regional Forces and Popular Forces. The Popular Forces were part time soldiers for local defense of hamlets and villages, while the Regional Forces might be employed throughout a province.

The North Vietnamese had planned the Tet offensive for many months. Men and equipment had been infiltrated into all of the major cities over a long period of time. Uniforms, ammunition, and weapons were sneaked in with all types of disguises. Weapons were hidden in cemeteries, and small arms ammunition was hidden in produce that came from the delta and Dalat. Trojan horses were everywhere.

As in previous years, a standdown of U.S. and South Vietnamese forces was being discussed for the coming Tet holidays. Seventh Air Force had always opposed the Christmas, New Years, and Tet standdowns because of the conclusive evidence that the enemy took advantage of them. Large amounts of troops and supplies had been brought south during these periods. Shortly before Tet, North Vietnam proposed a truce covering the period 27 January to 3 February; the U.S. and South Vietnam announced a 36-hour truce to be in effect from 29–31 January. Although the interdiction of the LOCs in Laos continued, all operations in North Vietnam were halted except for Route Package I. The enemy selected and rushed his war materials southward. Because of the unrestricted movement allowed the enemy during these "truces," the net result was a reduction in the effectiveness of the interdiction campaign. During the six-day standdown for Tet in 1967, the enemy moved an estimated 45,000 tons into Route Package I.[20]

Just prior to the new year MACV alerted all forces to the enemy buildup but did not specify the magnitude or intensity of the expected offensive. The forces in Vietnam were not placed on alert but on an increased readiness posture. I put 7th Air Force on alert,[21] however, because my staff believed that major attacks were imminent throughout the country and that the forces at all air bases and installations in Southeast Asia should be at their battle stations. Consequently, we were already in "condition red" when the Tet offensive began.

TET—THE OFFENSIVE BEGINS

Duplicity of the enemy at Tet had occurred in 1953 during the French-Indochina War, so the tactic was not a new one. The attack was uncoordinated during the initial phase. Danang and Pleiku were among other cities that were struck 24 hours ahead of the main offensive. Then on the evening of the 30th and the early morning hours of the 31st, simultaneous mortar, rocket, and sapper attacks were made against 36 of the 44 provincial capitals, 64 of 242 district capitals, and 23 airfields; they were followed by the main assault forces.

The burden of defending the district and provincial capitals fell on the shoulders of the local RF and PF forces. A significant portion of ARVN forces were on leave, and it was some time before regular South Vietnamese Army troops were effective. The RF and PF militia stood and fought gallantly. If these forces had not held, the outcome of the battle might have been much different. When assaulting these population centers, the enemy's tactic was to get into the heart of the city so that airpower couldn't be used without destroying homes and churches, thus killing innocent civilians. If we did bomb in support of friendly forces, the resultant effect on the civilians would tend to make them join the VC in the uprising. Air support was, therefore, a delicate matter.

Although the attacks spread throughout the country, the 7th Air Force base structure remained secure, and operations were suspended at only two bases—Tan Son Nhut and Bien Hoa. The enemy made a concerted effort to overrun these two bases. Heavy barrages of mortars and rockets were laid down against aircraft, flight line facilities, and housing. Some aircraft were destroyed or damaged. Because of the red alert condition, however, the defense of these two bases was greatly enhanced. All outposts had been reinforced, and all fighting positions were manned. Thus, when the enemy followed-up the mortar and rocket fire with assault troops, the base security forces were ready. These men held off the enemy at both bases through the night of the 31st and into the next day when U.S. and ARVN ground forces entered the battle. Army helicopter gunships in conjunction with the AC–47s provided the fire support. Helicopters delivered attacks within a few feet of the bunkers where security forces were fighting for their lives. The attacks at both bases were contained, and enemy sapper teams never reached the flight lines. Aircraft were flying missions the following day.

The flexibility of centralized control was never more dramatically demonstrated than when fighters and airlift forces were shifted from area to area to meet the enemy. Throughout the offensive the command and control system remained intact, and no unit was out of contact with the headquarters. As the attack developed, FACs were launched all over Vietnam to cover the battle areas. Already airborne AC–47s, while illuminating the battlefields for the fighters, were also providing valuable information to the tactical air control system. In a short time, follow-up came through the normal command and control network.

Except for the area around Hue and Saigon, the offensive had run its course within a few days. In these two cities, however, the enemy fought from house to house making it very difficult to put in air strikes without destroying a considerable number of structures. It was 7th Air Force policy not to approve strikes in the Saigon area unless there was a desperate situation. Even though FACs controlled the strikes, there was no way to avoid adverse reaction from the South Vietnamese people to our bombing of VC holed-up in houses and other buildings. Strikes in Saigon, therefore, were conducted mostly on the periphery of the city against the infiltration points. When it was necessary to support U.S. tanks and APCs, as in the Phu Tho race track area, we made strafing attacks because they gave us better accuracy.

As the Tet offensive mounted in intensity and the threat to I Corps became more pronounced, Westmoreland elected to reinforce the area with the 1st Air Cavalry Division, 101st Airborne Division, elements of the 82nd Airborne Division, and the 11th Armored Cavalry Regiment. We

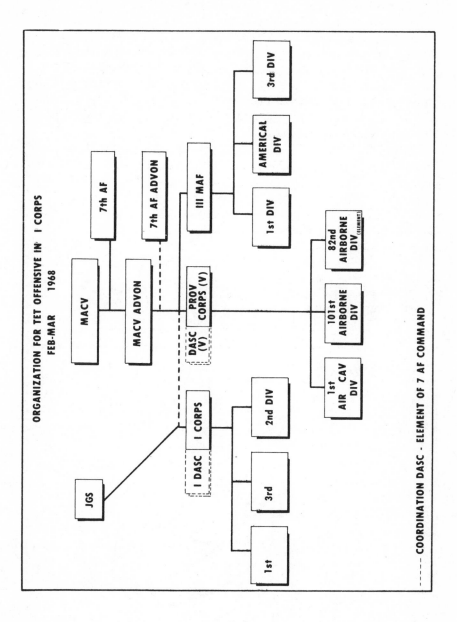

ORGANIZATION FOR TET OFFENSIVE IN I CORPS
FEB-MAR 1968

------- COORDINATION DASC - ELEMENT OF 7 AF COMMAND

315

needed a new organization to control these units. The III MAF, up until this time, had functioned basically as a corps of two divisions, but with the battle for I Corps reaching a peak, Westmoreland activated a MACV forward headquarters on 9 February and moved Abrams, his deputy, to Hue Phu Bai to assume command of this new headquarters.

Considering his additional responsibilities for advice and support of ARVN I Corps, Abrams was functioning as a field army commander. Because of the earlier decision not to establish a combined command of U.S. and Vietnamese units under a single U.S. commander, ARVN I Corps was technically not under the operational control of Abrams. However, Abrams exercised decisive influence with his advice to the I Corps commander on how best to employ the ARVN divisions in the combined counteroffensive.

Abrams' headquarters was augmented with an advanced command element of 7th Air Force. This command element was given discretionary authority for the employment of all airpower committed in the northern two provinces. It had no authority, however, with respect to operations in Route Package I and Laos, since these areas were still controlled from the 7th Air Force headquarters in Saigon. The 7th Air Force ADVON coordinated all air activities in the northern provinces, including I Corps DASC, and remained the paramount air organization working with ARVN and the Marine DASC.

The system in I Corps by mid-February 1968 appeared very similar to a field army-tactical air force relationship, except that the 7th Air Force ADVON was not staffed to perform as a numbered air force headquarters.

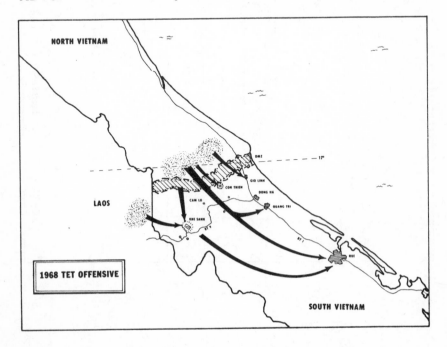

The ADVON commander did, however, advise on how the air should be used and where it could be most effective. Also, he had the power to make commitments in the name of the Commander of 7th Air Force, who was the final authority for the northern area in accordance with the overall priority established by Westmoreland for all of the combat activities within South Vietnam.

With the new organization established, procedures for requesting and approving air support were revised. The TASE commander, located in 7th Air Force headquarters, coordinated with MACV and established a priority between the two field forces and III MAF on requests for close air support. These headquarters were then notified of the decisions, and missions were laid on for the next day's operations. Each of the DASCs, including the Marine TACC and DASC, was notified of the missions by 7th Air Force, and the fragmentary order was issued to all flying units except the Marines. The order for them went to the Marine TACC which issued the order to the air group.

For the first time in the war, all of the fixed wing aircraft came under the control of the air component commander, making it feasible for him to respond with airpower anywhere in the theater and to whatever priority COMUSMACV established. Helicopters, on the other hand, were not brought under the control of the air component commander except during the invasion of and subsequent operations in Cambodia in April 1971. In South Vietnam, the helicopters assigned to the 1st Aviation Brigade were further allocated to the control of the field force commanders based on the requests submitted to MACV. The decision as to which field force would receive the most helicopter support was dependent upon the mission for the week assigned that field force commander. The field force did the basic planning for the helicopter employment and, when U.S. helicopters were employed in support of operations within the ARVN corps, passed control to the senior Army liaison officer to the ARVN corps.

The helicopters assigned to the 1st Air Cavalry and 101st Airborne Divisions were used according to the planning guidance of the field force commander. However, the complexity of helicopter assaults demanded greater centralization of control as the war continued. Furthermore, the increase in the intensity and quality of enemy ground fire made coordination between the Air Force and Army even more important. Little of the planning could be done below the field force level because of the number of units involved and the need for close integration of all elements of the assault forces. If SA–7s had been used in the 1965–1968 period, as they were in the 1972 Easter offensive, greater support by 7th Air Force would have been essential to suppress such defenses. Whether the helicopter assault operations conducted in 1965–1968 would have been feasible in the 1975 offensive is a basic question for the future of such forces. In a war in which large numbers of SAMs, radar directed AAA, and high performing fighter aircraft are used, it is doubtful that such air assault operations can be conducted without excessive losses.*

Hue presented a most difficult problem. The enemy made a concerted effort to capture the city using regular NVA forces. The forces apparently staged in the A Shau valley and moved into position for the attack during a period of bad weather. The initial assault force was estimated to be eight battalions of NVA/VC troops. The battle was reminiscent of the house to house fighting in World War II and the related problems of providing close air support. Additionally, the weather was very poor at this time of year, being the period of the northeast monsoon. Most of the time it was quite similar to the weather experienced at Khe Sanh, except there were more extended periods of heavy fog. This made it extremely difficult to use fighters and armed helicopters. Furthermore, MSQ was not a solution because the enemy was imbedded in the city and even more indiscriminate damage would have been done by these attacks. Additionally, we had learned from experience at Casino during the Italian campaign in World War II that heavy bombing could actually enhance the enemy's defenses by creating rubble that would slow the advance of friendly forces. Although the situation differed, there were some examples of this problem as the enemy retreated into the inner fortress of the city. As the battle mounted, the enemy brought in the 5th NVA Division and the VC 416th Battalion. It was apparent that he was trying to salvage at least one military victory out of the offensive. Between breaks in the weather, hundreds of sorties were flown in support of the Marines and the 1st Air Cavalry Division.

Finally, on the 25th of February the battle for the inner city was over. Tet was an extremely costly offensive in terms of the men that were lost by the enemy. The enemy had lost 5,000 troops at Hue and another 3,000 in northern I Corps.[22] Some 84,000 of the 200,000 troops in South Vietnam at the onset of the offensive were used. It was estimated that by the end of February 45,000 troops were killed and another 24,000 wounded.[23] By any standard of measurement this was a major military defeat. The North Vietnamese would need almost three years to prepare for another offensive of such magnitude, and they could do it then only because of the bombing halt in North Vietnam that provided secure supply points above the DMZ.

The expected effect on the South Vietnamese people didn't materialize. They, in fact, stood behind the government; and instead of collapsing, the government actually became stronger. There were none of the uprisings that the North Vietnamese expected, and not a single province fell to the enemy. Politically, although the offensive failed in its effect upon the South Vietnamese people, it succeeded in the effect that the North Vietnamese hoped to achieve on the U.S. home front.

---

*See my testimony (1973) to the Senate Armed Services Subcommittee on Tactical Airpower for a detailed discussion of this matter.

Disillusionment with the war was rife in the United States, and confidence in the policy of continuing the war was shaken. Instead of being able to follow-up the Tet offensive with a major military effort in South Vietnam and an all-out bombing campaign in the north, which would have been consistent with fundamental principles for employing military power, the President was compelled to suspend the bombing and step down as a candidate for reelection. Although the North Vietnamese could not win on the field of battle, they had won a resounding psychological victory.

Airpower met all expectations throughout the offensive. More than 16,000 sorties were flown from 30 January to 25 February 1968 in support of U.S., ARVN, Australian, and Korean ground forces. Close air support was flown under demanding conditions of weather and troop location. The airlift forces of some 280 aircraft moved over 12,200 troops at crucial periods of the battle. If these troops had not been deployed to reinforce threatened areas, some cities might have been temporarily lost to the enemy. Centralized control made airpower responsive to the threat.

As the crisis in I Corps abated, COMUSMACV decided to deactivate MACV Forward on 10 March 1968, and he assigned all of those forces to III MAF.[24] This entailed adding more Army officers to the III MAF staff. III MAF now functioned in the same manner as MACV Forward, controlling the employment of all U.S. ground forces located in I Corps. The Provisional Corps (V), which was activated on 10 March 1968 (redesignated XXIV Corps on 12 August 1968), assumed operational control of the U.S. Army divisions in Quang Tri and Thuy Thien province, with the divisions in the other parts of I Corps reporting to III MAF.[25]

When Westmoreland made the decision to inactivate MACV Forward and activate PROV Corps (V), I determined that a 7th Air Force advanced headquarters was no longer needed and a more appropriate organization to work with the new corps was a DASC. Consequently, DASC(V) was activated 10 March 1968, to take care of the air support needs of the Corps, and 7th Air Force Advance was inactivated that same date. With these changes in organization, the control of air operations, however, continued to be centralized in 7th Air Force headquarters, and the basic principles for coordinating the air effort with that of the ground forces remained intact.

I continued to be the single manager for air. The pre-planned air requests were consolidated at III MAF and then forwarded to the TASE. The system from that point on functioned the same as prior to the reorganization. An attempt was made to combine I Corps DASC and a new DASC adjacent to the III MAF HQ, similar to the arrangement of II FFV and ARVN III Corps at Bien Hoa. However, the ARVN corps commander opposed the consolidation fearing he would lose control of the VNAF units under his command. As a consequence, the split location and assignment of the two DASCs created problems of coordination. Still

319

we made all of the decisions on where the next day's air effort would go at the 7th Air Force-MACV (TASE) level.

With the conclusion of the Tet offensive and the halt of the bombing, the war in South Vietnam became one of limited engagements somewhat analogous to the campaigns in early 1966. Probing operations were conducted throughout the country with a few sharp engagements in the highlands. For the most part, the North Vietnamese were fully committed to a political offensive to force the withdrawal of U.S. forces and to achieve a temporary settlement that would provide the best possible posture for resumption of a full military and political offensive in the future.

PREPARING FOR WITHDRAWAL

On the U.S. home front, President Nixon announced a schedule of planned withdrawals of ground forces as the talks in Paris proceeded. By the spring of 1969, it was apparent, however, that the North Vietnamese had no real intent of ceasing military operations in South Vietnam. Their objective, stated many times in captured documents, was the domination of all of Vietnam, and whatever force was required to achieve this objective would be used. The evidence of the continued buildup of military power was reflected in the traffic on the lines of communication in Laos. At a time when the U.S. was withdrawing forces as announced, the flow of trucks and supplies on the Ho Chi Minh network was reaching a new high. Some of the flow can be attributed to the need to replace the losses in the Tet offensive, but the magnitude was over and above such an explanation. If there had been a sincere interest in bringing military operations to a standstill, there would not have been a requirement for such large amounts of materiel. In October of 1970, it was estimated that 60,000 tons of supplies would be put into the system in Laos and South Vietnam during the 1970–1971 dry season.

Our excursion into Cambodia in the spring and summer of 1970, and the invasion of Laos in February 1971, were designed to reduce the stocks that were flowing to enemy base camps along the borders. For future operations, these base camps, as they had done since 1965, would support the main offensive from I to IV Corps. With more sophisticated weapons on both sides and the higher consumption rates of munitions, it would take a much longer time to build up the level of stocks needed for an offensive by the 200,000 (13 divisions) troops now in South Vietnam. Following the bombing halt, the bulk of the North Vietnamese Army had been deployed into and along the borders of South Vietnam.

The invasion by U.S. and South Vietnamese forces into Cambodia in 1970 and by South Vietnamese forces into Laos in 1971 provided the South Vietnamese government additional time to develop its military forces. Large quantities of supplies were destroyed, and the concentration of enemy forces along the borders was temporarily disrupted. These operations also facilitated the withdrawal of U.S. ground forces without undue exposure when the fighting strength was constantly going down, a time of maximum danger to our troops. The invasion of Laos, however, was a disappointment in many respects. It was hoped that ARVN would be able to straddle the main junction of Highways #9 and #92 which was the hub of the enemy's logistical system for all of the northern provinces, the central highlands, and the northern part of III Corps. This junction had long been a strategic target for strangling the logistical flow into South Vietnam.

EXCURSION INTO LAOS

The planning for LAM SON 719* had been underway for a number of months. There were divided opinions within the U.S. military and political structure as to the capability of ARVN to march into the teeth of hard core North Vietnamese troops securing the Laotian panhandle. The strength of this enemy force had been estimated at 25,000 to 30,000 with considerable anti-aircraft defenses and some evidence of SAMs in the vicinity. Following the bombing halt, anti-aircraft fire had continuously expanded in southern and central Laos. Whereas it had been feasible to operate FACs and strike aircraft at two to three thousand feet, it would be necessary to move air operations up to higher altitudes. FACs could not function in some areas along the LOCs unless they were in high speed aircraft. The AC–130s had been under increasing fire and were now reaching the upper altitude limitation of their on-board weapons. In many cases, these AC–130s were moved to safer zones, and laser bombing F–4s took over the tasks where heavy anti-aircraft fire was suspected.

It was against this background that considerable differences of opinion existed concerning the conduct of LAM SON 719. ARVN had made significant progress, but there would be no U.S. troops fighting alongside them when they met the North Vietnamese. They would be fighting in very difficult terrain against well-entrenched troops and some heavy firepower. Furthermore, it could be expected that the enemy would use tanks against the ARVN armor moving along Route #9 to join up with the forces lifted in by helicopter. Seventh Air Force plans provided for the extensive use of airpower to "soften" the landing zones and to support the subsequent landings. We believed that helicopters would be extremely vulnerable to the heavy anti-aircraft fire and that the only way they could survive would be to use large quantities of fighters, bombing and strafing the target area before and during the helicopter assault. Even then, it would be most difficult to prevent determined fire from being brought to bear against the helicopter forces as they hovered over the landing zones.

Based on experience in South Vietnam, the XXIV Corps didn't expect the anti-aircraft fire to be as heavy as predicted by 7th Air Force and believed supporting the FSBs both north and south of Route #9 leading

*LAM SON 719—ARVN code name for the invasion of Laos. LAM SON was used as a common designation for all ARVN operations.

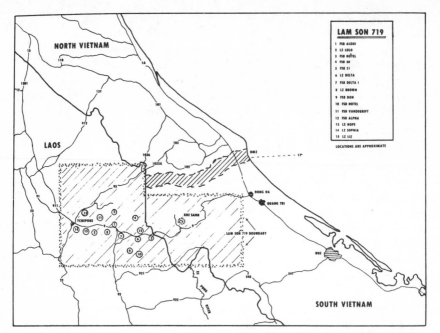

to Tchepone would not be unduly hazardous.[26] XXIV Corps felt that the 15 fire support bases to be established would protect the armored forces moving along Route #9, secure the road for logistical support of the troops to be deployed around Tchepone, deny the high ground to the North Vietnamese, and interlock artillery support for all the troops in Laos.

Because of the disagreements over the vulnerability of helicopters, XXIV Corps decided that most of the fire support for the troop-lifting helicopters would be provided by Army helicopter gunships. Seventh Air Force would provide some limited support, but the bulk of the air firepower was to come from these gunships. Seventh Air Force objected to giving so little close air support and stated it was prepared to provide large quantities of fighters wherever the operation demanded.[27] Further complicating the erroneous assessment of the ability of helicopters to function in a moderately heavy defense (no fighters or SAMs employed by North Vietnam) was the command structure. The operation was a South Vietnamese action with the U.S. providing only support. No U.S. troops would be allowed to accompany ARVN, and as a result the flow of battle information, requests for support, and rapid command decisions would be lacking. There would be no single authority in the combat zone to make decisions, and it would be sometime after the initial invasion before all ground and air headquarters were located together at Khe Sanh to thrash out decisions.

The entire air assault and continued air operations should have been under a single commander, 7th Air Force. The vulnerability of the helicopters, difficulties of support, and the need for thorough integration of all aspects of air operations demanded such a structure. In all previous wars, an airborne assault had been under the direction of an air commander until the troops were on the ground, and even then a single air commander provided the detailed air support (firepower and logistical) for that ground commander. A helicopter assault is still an airborne operation. To make it succeed demands a continuous stream of fighter cover taking the place of artillery. To deliver such firepower, there must be the ability to shoot and bomb regardless of the weapons used by the enemy. Helicopters were not able to cope with the firepower the North Vietnamese brought to bear against the landing zones.

As the enemy brought the fire support bases under intensive fire, the losses became prohibitive and the ability to reinforce was insufficient. As the armored forces bogged down on Route #9, some 20 kilometers inside of Laos, the pressure against the northern fire support bases became too much for ARVN rangers; with the T–54 tanks making their appearance, the entire ARVN force was in jeopardy. Even with the use of B–52s and the extensive use of fighters, the situation deteriorated rapidly. It was only through the effort of FACs, bringing in a stream of fighters, that the North Vietnamese tanks were stopped some five kilometers from the retreating column of ARVN armor trying to make it back into South Vietnam.[28]

LAM SON 719 was indeed a costly operation to the South Vietnamese and U.S. helicopter forces. It brought home quite convincingly that helicopter assaults have the same limitations and vulnerabilities as did the airborne assaults conducted in World War II and Korea*. From U.S. and British experience in North Africa, Sicily, Italy, and Europe, we know these operations were costly in men and equipment. The employment of such forces requires almost complete air superiority and the ability to maintain a stream of fighters overhead throughout the initial phases and until such forces can link up with an advancing column on the ground. Until such a link-up, the force is vulnerable to an armor attack. For airborne troops to survive such an assault, airpower must provide the heavy firepower until the soldier again has his own organic support.

During the South Vietnamese assault at Landing Zone Lolo, which was about half way to Tchepone, the enemy made an all-out effort to defeat the landing. The assault began on the 2nd of March with eight B–52s striking south of the landing zone. Most of the initial suppression of enemy fire around the landing zone, however, was attempted with helicopter gunships. For fighter strikes, forward air controllers were dependent upon the ground forces to determine where the ordnance should be delivered. These requests were slow developing. On the morning of the 3rd, only six fighter strikes were put in around the landing zone and the alternate landing zone. Additional sorties that dropped anti-

*See Vietnam Studies, Airmobility 1961–1971, by Lt Gen John J. Tolson, Department of the Army, Washington, D.C., 1973, for a different point of view.

personnel weapons were used around the perimeter of the landing zone. After four of nineteen helicopters were shot down and many others hit, the assault was suspended. We sent in more fighters and the enemy fought off another landing attempt. We flew still more fighter strikes. By 1600 hours, enemy positions were beaten down to such an extent that the operation was resumed. All troops were on the ground by 1830 hours. More than 40 helicopters were employed in this one assault. Almost all of them took hits—20 were shot down, and seven more were totally destroyed.[29]

The enemy's efforts against Lolo were characteristic of the tactics he used against the other fire support bases which led to the eventual abandonment of the objective to take Tchepone. The plan was to remain in Laos until the wet season in late April and destroy large stocks of supplies the North Vietnamese had been building for the future. With the withdrawal of U.S. forces, disrupting the enemy's supply system in Laos would provide much-needed time for the South Vietnamese to improve its military, economic, and political systems.

Even with the early withdrawal from Laos, both sides lost heavily in men and equipment. Over 100 enemy tanks were destroyed, mostly by fighters. For the first time laser weapons were used against tanks in combat. ARVN entered Laos with 71 tanks and 127 armored personnel carriers (APCs) and was able to get out with only 22 tanks and 54 APCs. The enemy lost over 13,000 men while ARVN lost more than 2,500. Because of accountability procedures, the exact number of helicopters destroyed is difficult to determine. We estimated the losses at 200 of more than 600 helicopters used.[30] The Army contends the losses were much less. Seventh Air Force flew more than 8,000 tactical sorties with a loss of seven aircraft.

LAM SON 719 presented the first real challenge to air mobile operations. The problem is the amount of airpower that must be employed to create a favorable environment for the use of such assault forces as LAM SON 719. Up to this time the South Vietnam theater of operations hadn't tested air assault operations. In LAM SON 719, the ground fire was not as intense as in the 1972 offensive, nor had the SA-7 Strella* been employed yet. (As with any new weapon, however, a countermeasure is always developed; SAMs were neutralized in North Vietnam, so they could also have been managed when employed in the south against helicopters and other slow flying aircraft). Still, LAM SON 719 was too costly because of weak planning that produced inadequate tactical air support.

WITHDRAWAL—AN UNEASY TIME

In 1971, in spite of the intransigence of the North Vietnamese, the United States was disengaging from the war. There was no longer any

---

*SA–7 Strella: Russia-made, hand-held, shoulder-fired, anti-aircraft missile.

desire to make the sacrifices required to bring about a military solution. Within South Vietnam, most of the main air bases had been turned over to VNAF. They had been rapidly expanded in anticipation of continued fighting before a final peace agreement was reached. The number of aircraft that 7th Air Force had in South Vietnam was down to a squadron of A–37s at Bien Hoa and a detachment of F–4s at Danang. The total strength of the force was 350 aircraft, mostly at bases in Thailand. Air strikes were still being conducted against logistical targets in North Vietnam, mainly Route Packages I and II, but on a limited basis. They were insufficient to reduce or even impair the rapid enemy buildup taking place above the DMZ. In view of the total disregard of the bombing halt being demonstrated by the North Vietnamese, 7th Air Force made repeated requests for the air campaign against all of North Vietnam to be resumed at once. The North Vietnamese were thought to have 13 divisions in South Vietnam. They were the best of the enemy's total force. Seven other divisions were held in reserve in the central and northern part of the homeland. The road network, all the way from the Chinese border to the DMZ, was in a good state of repair and able to support heavy truck traffic during all weather conditions. And the rail network had been repaired and all major lines were open. With these secure logistic lines, the North Vietnamese were in a position to employ large tank forces backed up by artillery and protected by a SAM blanket.

Never had the enemy been in a more favorable position to employ the modern weapons provided by the Soviet Union. Seven AAA regiments of their Soviet-built anti-aircraft defenses were deployed in South Vietnam, with another eight regiments deployed along the borders ready to move as needed. Three more SAM regiments were located along the DMZ and Laos. SAMs had never been positioned so far south. These weapons' firepower closely compared to some of the heavily defended targets in the Hanoi delta.

With the bombing halt in 1968, we expected the enemy to move their defenses farther south to complement the troops being readied to invade South Vietnam. The northern fire support bases at Con Thien and Gio Linh would probably fall if a major offensive were launched to take Quang Tri. Our massive employment of airpower prior to the 1968 Tet offensive had disrupted their plans to take these northern provinces, but with most of the U.S. airpower now gone and with the carrier force down to two, the North Vietnamese could test the South Vietnamese Army's will to fight without U.S. troops beside them. Further, the enemy apparently believed riots would occur on the home front if we sent our troops back into battle and our airplanes back into North Vietnam.

Given the upper hand, the North Vietnamese thought they could take Quang Tri and Thuy Thien provinces in the north, occupy most of Kontum in the highlands, and place their troops in outskirts of Tay Ninh City. Based on what was happening back home, theirs was a reasonable assessment of our situation. If the offensive developed as expected, the

North Vietnamese would be in an excellent position to sue for a cease-fire; if they were exceptionally successful, they could push on to a complete military victory.

EASTER OFFENSIVE—A TEST

On 30 March 1972, the enemy launched the offensive in three of the four Corps. The main thrust was in I Corps with the apparent objective of capturing the two northern provinces. More than 40,000 troops spearheaded the assault that was supported by artillery and rocket regiments, 400 armored vehicles, and SA–7s and SA–2s.[31] As expected, the main thrust was against the forward fire support bases at Gio Linh and Con Thien. The ARVN 3rd Division was overrun, and withdrew in such disorder that it would not be capable of fighting for months. As the enemy advanced toward Quang Tri, the U.S. began redeploying air units back to Thailand. Seventh Air Force was being rapidly increased to a thousand aircraft, and the carrier force was reaching the highest level of the war—five on the line at YANKEE Station.

SA–7s took their first toll of U.S. and Vietnamese aircraft on 1 May. No longer was it feasible to operate below 10,000 feet without using countermeasures. Fighters with their high speeds had little problem with the SA–7s, but 0–2s, OV–10s, A–1s, C–130s, and helicopters were severely restricted even with the use of countermeasures.

As the FACs were forced to the higher altitudes, their ability to locate targets and control strikes became more restricted. Conditions were even more severe than in Korea when the MOSQUITO FACs in T–6s were forced to fly above 6,000 feet to stay out of the anti-aircraft fire protecting Chinese ground forces. The technique of using high speed FACs, first used with F–100s, was employed at Quang Tri with the FACs in the back seats of F–4s.[32] Because of the speed and maneuverability of the F–4, it could operate within the enemy's defenses without unacceptable risk. FACs operating around Quang Tri used the same techniques and procedures developed in the air campaigns of the 1960s. The bombing altitude of the fighters was about the same as for those earlier operations in North Vietnam, as opposed to earlier times in South Vietnam when bombs were being released at four to five thousand feet.

For the first time, the enemy employed tanks (T–54s and PT–76s) in South Vietnam in quantity. The only tanks we had encountered before were PT–76s at Lang Vei, where only nine were used to lead the enemy's assault. Now they were being used in quantity. As the tanks moved into the open, our fighters and AC–130s rapidly knocked them out of action. Two hundred sixty-seven tanks were destroyed in the course of the offensive. Topography played an important role in our successes. Eastern Quang Tri is flat land, so the enemy's tanks were exposed when moving through there. Additionally, the flat land made easier our delivery of laser and general purpose bombs.

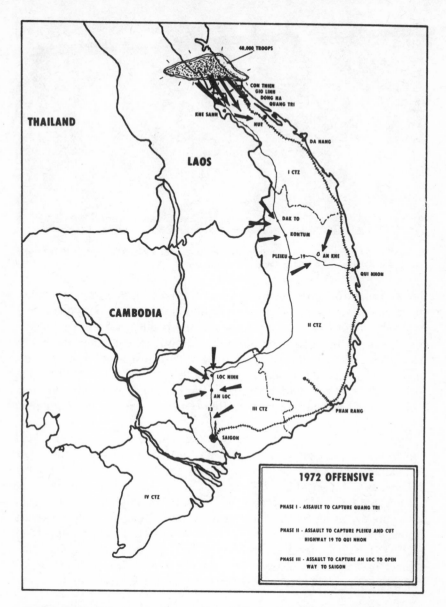

THAILAND

LAOS

CAMBODIA

40,000 TROOPS

CON THIEN
GIO LINH
DONG HA
QUANG TRI

KHE SANH

HUE

DA NANG

I CTZ

DAK TO

KONTUM

PLEIKU 19 AN KHE

QUI NHON

II CTZ

LOC NINH

AN LOC

13 III CTZ

PHAN RANG

SAIGON

IV CTZ

1972 OFFENSIVE

PHASE I - ASSAULT TO CAPTURE QUANG TRI

PHASE II - ASSAULT TO CAPTURE PLEIKU AND CUT
HIGHWAY 19 TO QUI NHON

PHASE III - ASSAULT TO CAPTURE AN LOC TO OPEN
WAY TO SAIGON

LAM SON 719 was quite another matter. We found getting to their tanks extremely difficult there. They enjoyed much better concealment. Exposing themselves for only brief periods, they hardly gave our fighters the time needed to find and destroy the tanks. Lasers were sometimes effective against concealed tanks, but more of the tanks were destroyed with 500–pound bombs used in conventional attacks.

The North Vietnamese used 122mm and 130mm guns to support tanks in the DMZ. These long range guns were difficult to locate. They were

A USAF A-37 supporting ground operations in South Vietnam. Aircraft is from the 604th Special Operations Squadron.

dug in, firing from concealed positions. When we located them, however, the laser weapons were accurate enough to destroy these positions with a minimum number of passes. Contrary to the popular notion that concentrated artillery barrages came from guns placed side by side, the North Vietnamese shot at their targets from widely dispersed positions. By timing their shots, they could get the results they wanted without putting their artillery in one vulnerable mass. We had no more success locating their artillery positions north of the Ben Hai River than we had during the Tet offensive.

When the 3rd ARVN Division broke and retreated, we couldn't hold Dong Ha. As the enemy moved forward in Quang Tri, we increased air sorties hoping to blunt the attack. By this time 7th Air Force was flying 207 sorties a day in Military Region I. Fighters from Thailand staged at Danang and flew another sortie before returning to base. We used tankers for fighters staging at the more distant Bien Hoa. B–52s were flying around the clock. The VNAF averaged 45 sorties a day in support of ARVN. These sorties were scheduled by 7th Air Force in the same manner as before the phase down. Enemy ground fire remained heavy throughout the offensive; we lost ten U.S. and six VNAF aircraft to SA–7s.

Quang Tri fell by the 30th of April, and the Allies left large quantities of supplies for the enemy. With the loss of Quang Tri, on 8 May the President announced his decision to increase the bombing of North Vietnam and to mine Haiphong. The combination of this announcement plus the constant pounding that the North Vietnamese were taking from airpower brought the enemy's offensive to a temporary halt.

In June the ARVN moved to break the stalemate. Supported by around-the-clock bombing, they counterattacked and gradually moved back into Quang Tri. The Allied victory in Quang Tri was complete on the 16th of September.[33] Strategists credit airpower with the decisive role in recapturing Quang Tri. Reinforcing 7th Air Force were 162 F–4s, 12 F–105s, and Task Force 77's combat planes to do the tactical job. Too, the B–52s flew more than 2,724 sorties from 30 March to 30 June.

When the attack developed in MR I, the enemy had already been probing the approaches to Kontum. Intelligence sources detected the 320th Division in the area around Dak To. We were to see a familiar pattern of attack against Kontum. They used tanks and heavy shelling prior to the infantry assault. If the enemy made it to Pleiku, he would cut across the waist of Vietnam along Highway #19. Already, sapper attacks had effectively cut this principal supply route from the port city of Qui Nhon. Indeed, the enemy had interdicted Highway #19 throughout the war. Even at the height of U.S. activity in Vietnam, supply convoys had to be escorted. Although Highway #14 from Ban Me Thuot was an alternate route, Qui Nhon was the major supply center for all of the central highlands, and Route #19 was the key for the support of forces at the end of the line—Pleiku and Kontum.

As the logistics stopped moving on Route #19, the 7th Air Force airlift force picked up the task of supporting these beleaguered ARVN forces. Landings were made under hazardous conditions. The VNAF had insufficient airlift capability to meet the needs of more than two divisions of troops deployed in defense of the area. Our C–130s had to air-drop more than 2,000 tons of ammunition and food to ARVN troops after most of the airfield at Kontum had been overrun. Fortunately, the Allies recaptured the airfield on the 8th of June, and we resumed landing to off-load supplies.

The weather around Kontum during this time of the year is changing from the northeast to the southwest monsoon. Consequently, many of the air strikes had to be delivered under MSQ control. Although no losses were reported to missiles, the enemy used SA–7s against strike and airlift aircraft.

Our F–4s used the LORAN method to bomb targets around Kontum. A LORAN F–4 would lead a flight of four aircraft in which the wingmen and element leader not having LORAN would drop their bombs on a signal from the lead F–4. This method is similar to MSQ formation bombing, and even though the preferred technique is to break the bombing formation into a smaller size for better accuracy, the F–4s using LORAN did drop bombs at Kontum within 500 feet of friendly troops.[34] In fact, aerial photographs of LORAN bombing in Route Package I against the Quang Khe ferry showed some bomb plots within 125 feet of the aiming point.

In defense of Kontum, 7th Air Force averaged 137 sorties a day, and the VNAF 33. Kontum held. Even so, as the enemy defenses had driven helicopters and lower performing aircraft out of the Quang Tri battle, the same conditions prevailed at Kontum. These lower performing aircraft were severely restricted in the missions that could be performed and the

329

conditions in the target areas. By the time Kontum and Pleiku came under attack, however, the enemy had increased the level of anti-aircraft defenses from those that prevailed in the 1968 Tet offensive. SA–7s coupled with 37mm and 57mm made speeds greater than 400 knots essential for survival. Although our aircraft used countermeasures, rapid maneuvers and high speeds were the best defense against the SA–7s.

With battles raging at Quang Tri and Kontum, the enemy threw out the third prong of the offensive at a little district capital in the middle of the rubber plantations near the Cambodian border—An Loc. On 9 April the assault began with 25 tanks leading the way.[35] The area surrounding An Loc provided excellent cover for the tanks. Supporting the assault were large quantities of artillery, SA–7s, and anti-aircraft guns. Again, the tactics were similar to those used at Kontum and Quang Tri. The enemy's apparent objective was to lay siege to An Loc, capture it, seal off Tay Ninh City, and then move against Saigon. For years the enemy had used base camps across the border in Cambodia to supply and support forces in III and IV Corps. The support for the forces assaulting An Loc apparently came from these same base camps.

As the assault proceeded, Highway #13, the main line of communication to Tay Ninh City and Saigon, was interdicted. The ARVN attempted to reinforce An Loc with armored forces but was beaten back. Each time an attempt was made to push through an armored column, the North Vietnamese, from ambush positions along the rubber plantations and heavy jungle, were able to force the column to give up the advance. Close air support strikes were not effective because of jungle concealment. The enemy did not expose his forces where a direct attack could be made with fighters.

We saw the enemy tighten his ring around An Loc, making air supply the only way to support ARVN defenders. At first we tried using CH–47 helicopters to deliver supplies, but anti-aircraft fire soon made it infeasible to use helicopters for the task.[36] Because the enemy's firing positions were along the tree line, it was difficult to attack them. Experienced tacticians knew the VC and North Vietnamese liked to place gunners with light, hand-held, automatic weapons high in the trees where they would have an unobstructed field of fire covering helicopter landing areas.

With the supply situation becoming severe, VNAF C–123s made low altitude parachute drops. But enemy fire was so intense these aircraft had to be pulled out after the first three weeks of the siege. With the withdrawal of the C–123s, 7th Air Force took over the air resupply of the troops at An Loc using one of the most effective methods of dropping supplies into a small area—the container delivery system (CDS).* This

---

*CDS—container delivery system: Supplies were not actually para-dropped in containers. The items dropped were bundles—a ton of food, fuel, or ammunition—covered with a shroud and lashed to a wooden pallet. For details, see Nalty, Air Power and the Fight for Khe Sanh (Office of Air Force History, 1973), pp. 46–47.

Where there's a will there's a way. Coming in low and pulling up hard, a USAF C-130 delivers its much-needed cargo to isolated ARVN forces engaged in the battle of An Loc, July 1972.

system had been used by C-123s and C-130s for the support of troops where there were no airfields or where the airfield was of insufficient length. We at first made air drops from 500-600 feet, but the ground fire was so heavy, battle damage to the C-130s was unacceptable. Consequently, we tried high altitude drops under radar control. (Most all of the air drops at Khe Sanh were made at the lower altitude in spite of the ground fire.) On 18 April, the enemy shot down a C-130 making a CDS air drop. Shortly thereafter, daylight C-130 missions were suspended.[37] Still another C-130 was shot down, and more than 37 aircraft were damaged. Furthermore, because of rigging problems, delivery accuracies were not up to standards. After the Army improved the rigging and packing of chutes, MSQ drops achieved a recovery rate of about 85%. Because of the close proximity of the enemy to the drop zone, troops recovering bundles were often subjected to a heavy barrage of mortars and automatic weapons fire, and any bundles that landed outside the drop zone were up for grabs by either side.

With countermeasures installed in the C-130s to defeat the SA-7s, MSQ control was the best way to drop supplies from altitudes above 10,000 feet. The drops used high velocity parachutes that had a rate of descent of 105 to 120 feet per second. With these high velocity chutes, friendly forces began to recover 96% of the bundles. The airlift force demonstrated a magnificent determination to keep the garrison supplied with food, ammunition, and weapons.

In all, three C-130s were lost, but An Loc was saved because the men on the ground were kept supplied to continue the fight. Because of the SA-7s and the very small drop zone, the operating conditions at An Loc were even tougher than Khe Sanh. The jungle gave the enemy better coverage at An Loc than Khe Sanh. During the campaign from 9 April through 10 May, the airlift force flew 448 missions and air dropped almost 3,700 tons of supplies.[38] Early in the campaign, the drop zone was only 200 feet by 200 feet. By the end of the battle, however, the drop zone had been enlarged to 800 feet by 1,600 feet. The enlarged drop zone represented the improved condition of the defending troops who were successfully pushing the enemy back into the jungle.

Close air support, as it had been a decisive factor in holding the northern two provinces and Kontum, played an equally decisive role at An Loc. There is little doubt that the battle would have been lost without the day and night support flown by fighters and the AC-130 and AC-119 gunships. However, the saturation of the target area with so many aircraft caused problems similar to those at Khe Sanh. The target area was so small and so close to friendly troops FACs had to direct strikes with great exactness. Even though there were sufficient FACs to control the strikes, only so many aircraft could be controlled at a time. The area at Khe Sanh was by contrast much larger, so it was feasible to have two or three strikes going on there simultaneously.

On a typical day, 185 strikes were flown in defense of An Loc. There sorties were flown for the most part by the aircraft at Bien Hoa and those that staged through the bases from Thailand, mostly F-4s. The VNAF flew 41 sorties per day. SA-7s and anti-aircraft fire forced the A-37s to operate at a much higher altitude than the F-4s. Consequently, the burden fell on the 7th Air Force higher performing fighters to provide most of the support. B-52s averaged 11 sorties a day, and they were primarily targeted to areas where it was thought the enemy was staging and reforming for attack.

Although the siege was broken by the end of June, the enemy made one final effort to overrun the village with an attack across the airfield. Fighters caught the troops in the open and decimated most of the attacking force. With this last assault, the enemy withdrew most of his forces into base camps in Cambodia and along the border, bringing the siege and most of the Easter offensive except in Quang Tri Province to a close.

THE ASSESSMENT

The North Vietnamese were probably surprised at the reaction the Easter offensive produced. With the U.S. withdrawing, they probably thought the U.S. public wouldn't permit a bombing campaign against their homeland. The fact that the U.S. suspended the peace talks on 4 May as the offensive was in full swing must have also been cause for concern among the North Vietnamese leadership. Surely their miscalculations on the employment of U.S. airpower, both in South Vietnam and against the homeland, were two most significant factors in their turn-around in attitude about the negotiations.

The fight put up by ARVN was probably assessed with some mixed emotions. The First Division had fought tenaciously and with considerable professionalism at Quang Tri. Even though the upper third of Quang Tri province was lost, the enemy's apparent objective of capturing Quang Tri and Thua Thien provinces had failed. The ARVN 3rd Division's ineffectiveness tells us that only a portion of the 12 ARVN divisions were battle-ready. But the stiff defenses at An Loc and Kontum still came as a surprise to the North Vietnamese. From previous probing operations in the highlands and along the Cambodian border, the North Vietnamese thought these two points would fall, particularly with a large amount of artillery fire and the wide use of tanks. At An Loc alone, there were between 1,000 and 2,000 artillery rounds fired from 25 April to 1 May.[39] The holding of these two strategic points stabilized the relative positions of Allied forces in the event a cease-fire took place. Only in Quang Tri, therefore, did the North Vietnamese help themselves significantly toward winning the politico-military struggle for Vietnam.

The U.S. continued withdrawing ground forces, and by August there were no U.S. ground combat forces in Vietnam for the first time in seven years. Our aircraft continued bombing North Vietnam while the President intensified efforts to reach a cease-fire. When on 23 October we were making progress at the peace table, we stopped bombing above the 20th parallel. Hopes were high that a cease-fire was at hand, but again, a cease-fire was not to be. On the 18th of December, the 11-day, all-out bombing offensive was launched against the enemy to bring about the final agreement for a ceasefire on 23 January 1973.

403-892 O - 83 - 23

CHAPTER VIII

FOOTNOTES

[1] U.S.G. Sharp and William C. Westmoreland, Report on the War in Vietnam, 1964–1968 (Washington, D. C.: Government Printing Office, 1968), p. 129.

[2] Ibid., p. 284.

[3] Bernard W. Rogers, Cedar Falls-Junction City: A Turning Point, U.S. Army Vietnam Studies (Washington, D. C.: Government Printing Office, 1974), p. 137.

[4] Project CHECO Report, Operation Junction City (Hq PACAF, Hickam AFB, HI, 1967), p. 26.

[5] Rogers, Junction City, pp. 149–151.

[6] Sharp and Westmoreland, Report, p. 143.

[7] Bernard C. Nalty, Air Power and the Fight for Khe Sanh (Washington, D. C.: Government Printing Office, 1973), p. 38.

[8] Illustrated History of USAF in Southeast Asia, Office of Air Force History (Conf.), p. 34. (draft).

[9] Nalty, Khe Sanh, p. 14.

[10] Ibid., p. 15.

[11] Sharp and Westmoreland, War in Vietnam, p. 164.

[12] Working Paper for CORONA HARVEST Report, Out-Country Air Operations, Southeast Asia, 1 January 1965–31 March 1968, book 2 (Maxwell AFB, AL: Department of the Air Force, 1971), p. 60.

[13] William C. Westmoreland, A Soldier Reports (Garden City, N. Y.: Doubleday & Company, 1976), p. 343.

[14] CORONA HARVEST Report, Command and Control, book 1, p. II–2–25.

[15] Sharp and Westmoreland, Report, p. 184.

[16] Nalty, Khe Sanh, p. 86.

[17] Ibid., p. 103.

[18] Congressional Testimony by General W. W. Momyer before the Senate Armed Services Subcommittee on Close Air Support, October-November 1971.

[19] CORONA HARVEST Report Out-Country, book 2, p. 60.

[20] CORONA HARVEST Report, Out-Country, book 1, p. 83.

[21] Illustrated History, p. 144.

[22] Sharp and Westmoreland, Report, p. 160.

[23] Ibid., p. 161.

[24] George S. Eckhardt, Command and Control, 1950–1969, U.S. Army Vietnam Studies (Washington, D. C.: Government Printing Office, 1974), pp. 74–75.

[25] Ibid., p. 74.

[26] CORONA HARVEST Final Report, USAF Operations in Laos, 1 January 1970-30 June 1971, pp. 205–206.

[27] Ibid., pp. 11–12.

[28] Ibid., p. 14.

[29] Ibid., p. 12.

[30] Ibid., pp. 206–214.

[31] John A. Doglione, et al., Airpower and the 1972 Spring Invasion, USAF Southeast Asia Monograph Series, vol 2, monograph 3, p. 4. See also William W. Momyer, The Vietnamese Air Force, 1951–1975: An Analysis of its Role in Combat, USAF Southeast Asia Monograph Series, vol. 3, monograph 4, p. 45.

[32] Momyer, The Vietnamese Air Force, p. 67.

[33] Ibid., p. 46.

[34] Ibid., p. 4.

[35] Ibid.

[36] Doglione, Airpower, p. 86.

[37] Ibid.

[38] Momyer, The Vietnamese Air Force, pp. 49–50.

[39] Ibid., p. 47.

CHAPTER IX

CONCLUSION

The war in South Vietnam presented difficult challenges to airmen at all levels. I've explained the perspectives that were important to me, as Seventh Air Force commander from 1 July 1966 until 1 August 1968, as I attempted to meet those challenges. *Strategy*: Our air strategy before 1972 was, of course, severely limited, but within the slight freedom allowed us we attempted to raise the cost of the enemy's aggression unacceptably high and to confront him with overwhelming firepower whenever he elected to join in battle. External restraints greatly reduced our ability to achieve the first of those objectives, but we did achieve the second. *Command and control*: I observed that for efficient command and control, a theater component commander must be in charge of his portion of the war. If the employment of an air force is to be sharply attuned to the realities of combat, the controlling headquarters should be within a few hundred miles of the battles. The further removed a headquarters is from the scene of combat, the greater the tendency to lose contact with hour-by-hour developments, to become excessively involved in the political chessboard aspects of the war while neglecting the realities upon which success in combat depends. *Counter air*: The contest for air superiority is the most important contest of all, for no other operations can be sustained if this battle is lost. To win it, we must have the best equipment, the best tactics, the freedom to use them, and the best pilots. We had the best pilots. Our experiences suggest that superiority in equipment and superiority in tactics must be viewed as two elusive goals to be constantly pursued, not as assumed conditions. We are not apt to have marked superiority in both equipment and tactics for an extended period; neither side is likely to corner the market on ingenuity for long. Because so much depends on this battle, because it is so fiercely contested, and because it is so readily affected by technology, tactics, and rules of engagement, this is the battle in which our airpower can most easily be crippled by external restraints. *Interdiction*: It's easy for laymen to build exaggerated conceptions of airpower's capabilities here. Airmen must work in percentages when conducting interdiction campaigns; to reduce the flow through an enemy's supply line to zero is virtually impossible so long as he is willing and able to pay an extravagant price in lost men and supplies. To reduce the flow as much as possible and to make his price painfully high, though, we must focus our campaign upon the most vital supply targets: factories, power plants, refineries, marshalling yards, and the transportation lines that carry bulk goods. To wait until he has disseminated his supplies among thousands of trucks, sampans, rafts, and bicycles, and then to send our multimillion-dollar aircraft after those individual vehicles—this is how to maximize our cost, not his.

Close air support: The tactical air control system was surely one of the unquestioned successes of our airpower in Vietnam. We continually refined our air-ground operations, and by the end of 1968 we had become so responsive to ground commanders' needs that the characteristic engagement was one in which our ground forces located the enemy and kept him in sight while waiting 30-40 minutes for the fighters to arrive. This kind of support was made possible by our central control of all in-country tactical air forces.

Finally, I want to end this book as I began it, by mentioning a few of the perspectives that didn't meet my criteria for extended discussion.

I have deliberately avoided the perspectives from which one would make judgments about the wisdom of our national commitment to maintaining an independent government in South Vietnam. I'm aware that future airmen could infer from the preceding pages that the fighting in Vietnam ended in 1972 and our side won. It didn't and they didn't. But U.S. airpower accomplished what the President asked of it in 1972. Beyond that fact the professional airman can say little without exceeding the limits of his professional expertise.

Another perspective I have avoided is the one which would allow me to protest the many restraints we imposed on our own airpower in Korea and Vietnam. Of course my bias in this matter has been clear: I deeply resented the proscription of attacks on North Vietnamese airfields, SAM and AAA sites, and other targets. Airmen are bound to resent such restraints; it is an ugly and bitter thing to hold a hand voluntarily behind one's back while being beaten or while watching one's friend being beaten. But self-imposed restraint has been a fact in all U.S. conflict since World War II, and obviously our hope in the age of nuclear and thermonuclear weapons is that some restraint will be exercised by all superpowers in all future conflicts. Thus, however the airman may feel about restraints, and I know how he will feel, his professional responsibility is to articulate the probable consequences of his alternative courses of action to his superiors and then to act as effectively as possible within the instructions he is given. For future airmen it's worth stressing here, too, that as technological advances make warfare ever more complex and tempt political leaders to exercise direct control at lower and lower levels of command (disturbing as this may be to subordinate commanders), an extremely high premium must be placed on the airman's ability to articulate options thoroughly and clearly for those leaders.

A final viewpoint deserving of mention is one oriented toward the future. I said in the Foreword that I wouldn't have us rely entirely on yesterday's ideas to fight tomorrow's wars. Ironically, that is what our airmen are most apt to do if they are *not* thoroughly conversant with airpower history. Our air leaders must look closely at their history to prepare themselves for the future. I will mention here only two of the kinds of trends that seem reasonable to extrapolate from my experiences and my study of other airmen's experiences in World War II, Korea, and Vietnam.

First, theater commanders will attach increasing value to airpower's flexibility if, as seems likely, political restraints continue to reduce their combat options. Airpower in Vietnam constituted a uniquely switchable faucet of firepower (to borrow a metaphor from General Creighton Abrams); its point of application could be shifted 450 miles and more in less than an hour. By the end of 1972 we could strike point targets in heavily defended zones, using only a few aircraft, with very high probability of success and very low probability of collateral damage. Technological developments will bring further improvements in speed of response, range, and ability to apply enormous amounts of firepower with great precision; all of these improvements can help airpower compensate for the limitations imposed upon combat commanders by economic, geographical, and political considerations.

Second, my sense of history leads me to expect another trend that may be seen as a corollary of the first: The increasing complexity of international politics and the unique flexibility offered by airpower will entice us again toward parceling our air forces for the winning of battles rather than unifying and focusing them for the winning of wars. Aware that our every move in a combat theater today sends ripples around the world, we are reluctant to act decisively. We prefer to make smaller decisions, win battles, and hope that the enemy will lose heart. And our airpower will permit us to win most battles. But that way leads to a series of Khe Sanhs and eventually in a free society to war-weariness and dissent. As an alternative to this approach, airpower offers the possibility of an early LINEBACKER II campaign (with the enforcing threat of subsequent LINEBACKERs, a threat that was conspicuously missing in 1975). Airpower can be strategically decisive if its application is intense, continuous, and focused on the enemy's vital systems.

In short, airpower can win battles, or it can win wars. All commanders since Pyrrhus have been tempted at one time or another to confuse the two, but few distinctions in war are more important. The future airman's right to insist that such distinctions be made is, I believe, one of the things our airmen purchased so dearly in Vietnam.

INDEX

Aachen, 276
Abrams, General Creighton, 274–275, 316
acoustic sensors, 308
Advanced Research Project Agency (ARPA), 249
advisor program, 247
advisors, 66, 73
advisory group: See, Air Advisory Group
agreement violations, 215
Afrika Korps, 43, 231, 275
AIM–7, 155, 157–158
AIM–9, 155, 158
Air Advisory Group, 248, 250
air alert, 279, 285
air attache, 85–86
airborne assault: See, airborne operation.
airborne command and control center (ABCCC), 86, 88, 95–96, 151, 154, 197, 202–203, 205–206, 218, 285, 304, 307
airborne operation, 81, 265, 279, 298, 323
air commando, 11, 252.
air commando squadron, 263
air component commander, 44, 46, 51–52, 54–58, 62, 65–66, 70, 72, 74, 81–82, 86, 90, 95, 98, 102, 264–265, 317
air cover 67, 73
aircraft:
  A–1E, 263, 270–271, 326
  A–1H, 263–264
  A–4
  A–6, 286
  A–7, 87, 241
  A–26, 204
  A–37, 325, 332
  AC–47, 280–282, 301, 310, 314
  AC–119, 213–214, 332
  AC–130, 211–214, 280, 282, 321, 326, 332
  AN–2, 248
  B–17, 43, 113, 125, 165
  B–24, 113, 125, 165
  B–25, 43, 95
  B–26, 11, 43, 55, 67, 95, 180, 204–205, 251–253, 263, 270–271
  B–29, 1, 55–56, 147–148, 168, 180, 283
  B–52, 21, 31–32, 87, 99, 101–106, 124, 130, 133, 145, 148, 177–178, 182–183, 214, 237–241, 282–285, 296, 305, 309–310, 323, 328–329, 332
  B–57, 17, 205, 212, 264, 270
  C–47, 248, 251–253, 265
  C–123, 204, 254, 264, 298, 301, 310, 330
  C–130, 202, 206, 255, 298–299, 301, 310, 326, 329, 331–332
  C–135, 154, 179
  CH–47, 330

342

Air Operations Center (AOC), 73, 86–87, 262, 265
air refueling, 32, 88, 91, 101, 105, 145, 155, 176, 218, 220, 223, 227–228, 231, 274, 328
Air Staff, 15, 70, 73, 77, 103, 251, 268
air superiority, 1, 5, 21, 40, 42–44, 57, 59, 83, 85, 111–113, 115, 117, 139, 148, 158–159,
    163, 167–168, 264, 275, 312, 323, 337
air support, 19, 59, 87, 214, 250, 256, 261, 268, 272, 277–278, 285, 296, 300, 305–306,
    314, 317–318, 323–324
Air Support Operations Center (ASOC), 258, 261
air supremacy, 264
air sweeps: See, fighter sweeps.
air-to-air, 1, 26, 42, 155, 157–158, 180
Air Training Command (ATC), 268
Air War College (AWC), 2
Alexander, General Harold L., 43, 50
Alexandria, 44
Allied armies, 40–41, 115, 164, 168
Allied Expeditionary Air Force (AEAF), 51–52, 56, 164
Allied invasion, Normandy: See, OVERLORD.
all-weather operations: See, weather operations.
Almond, General Edward M., 62, 286
ambassador, U.S., 82, 84–85, 87, 196, 228, 274, 296
amphibious operations, 59, 112, 279, 285, 287
An Loc, 214, 282, 294, 296, 330, 332–333
Anthis, Maj. Gen. Rollen H., 74–75, 81
anti-aircraft-artillery (AAA), 16, 20, 28, 32–33, 118–119, 123, 126, 133, 135–136, 177,
    179–181, 183, 192, 199–200, 205, 213, 217–218, 225–226, 234, 237, 240–241, 255, 280,
    304, 306, 317, 321, 325, 330
anti-tank defenses, 192
anti-tank missile, 31
anti-war demonstrations, 312, 325
Ap Bac, 256
Ap Soui Tre, 300
armed reconnaissance, 150, 163, 168, 183, 196, 200–201, 204, 216
armor, 31, 275, 300, 321–323, 326
armored personnel carriers, 314, 324
Army Air Forces Board, 1
army aviation, 265
Army Aviation Operations Center, 265
Army, U.S., 2, 6, 9, 14, 56, 59, 69–70, 74, 76, 107, 249, 271
Arnold, General "Hap", 48–49
Articles 17 and 19, 12
artillery, 59, 80, 88, 277–278, 280, 285–286, 294, 296, 300, 303–304, 306–307, 309, 311,
    322–323, 325–326, 330, 333
ARVN: See, South Vietnamese Army
A Shau Valley, 305, 318
Atlas mountains, 40
Atoll missile, 138, 143–144
atomic bomb, iv
ATTLEBORO, Operation, 277
attrition, 19, 170, 189
Australian forces, 274, 319

Bac Mai, 118, 231, 237
Bai Thuong, 93
Ban Karai Pass, 31, 193, 195, 200, 214
Ban Me Thuot, 329
Ban Raving Pass, 195, 217
"bar caps", 227
BAR LOCK, 231
barrage firing, 132, 179
BARREL ROLL, 86–87, 192, 196, 199–200

Chiefs of Staff, British, 45
China, 137, 140–141, 150, 183
Chinese Air Force, 22, 141
Chinese Army, 5, 24–25, 114, 170, 175
Chinese Communist airfields, 1, 115, 141, 146, 216, 311
Chinese Communist Offensive, 3, 167, 169
Chinese ground forces: See, Chinese Army.
Cho Do Island, 156
Christmas standdown: See, standdowns.
Chu Lai, 271
Churchhill, Sir Winston, 41, 45–46, 48
CINCPAC, 12–13, 19, 22, 27, 69–71, 76, 79, 87–88, 90, 91, 95, 99, 102, 104, 106–107, 144, 198–199, 233, 271
CINCPACAF, 23, 70, 77–78, 90, 99, 102, 106, 264
CINCPACFLT, 23, 71, 78, 90, 95, 99, 104, 106
"city busting", 46
civic action, 6, 70
civilian casualties: See, collatereal damage.
Civil Irregular Defense Group (CIDG), 281, 305
Clay, General Lucius B., 98
Clifford, Clark M., 28
close air support, 1, 5, 14, 17, 21, 32, 40, 42–43, 49, 55–59, 61–62, 67, 113, 117, 139, 163, 168–169, 203, 253, 261, 266, 268–269, 272, 274–276, 283, 285–286, 293, 296, 298, 301, 309, 317, 319, 322, 330, 332, 338
cluster bomb unit (CBU), 219, 226, 279, 302
Coastal Air Force (N. Africa), 41
collateral damage, 134–135, 176–177, 179, 188, 227, 265–266, 314, 339
COLLEGE EYE, 151, 154–155
Collins, General J. Lawton, 247
Cologne, 204
Combat Information Center (CIC), 154
Combat Reporting Center (CRC), 151, 154, 202, 254, 258, 279, 285, 304
Combat Reporting Point (CRP), 151, 155, 203, 251, 254, 258, 279
combined bomber offensive, 34, 45, 113
Combined Chiefs of Staff (CCS), 40, 43, 45, 48–49
command and control, 3, 39, 44–45, 52, 56, 78–79, 86, 95, 99, 107, 201, 251, 257–258, 260, 271–272, 274, 310, 314, 337
Commandant, Marine Corps, 14
Commander, Army Forces Far East, 53
Commander-In-Chief, Allied Forces Northwest Africa, 40
Commander-In-Chief, Far East (CINCFE), 52, 58, 62, 169
Commander-In-Chief, Mediterranean Theater: See, Wilson.
Commander-In-Chief, Pacific: See, CINCPAC.
Commander-In-Chief, Pacific Air Forces: See, CINCPACAF.
Commander-In-Chief, Pacific Fleet: See, CINCPACFLT.
Commander, U.S. Forces Southeast Asia (COMUSSEA), 69, 73
Commander, U.S. Military Assistance Command, Vietnam: See, COMUSMACV
Commanding General, 18th Army Group: See, Alexander.
COMMANDO HUNT, 211–212, 214
COMUSMACV, 6, 8, 30, 70, 73–74, 77, 82, 85, 87, 99, 101, 104, 198, 199, 271, 277, 283, 287, 293, 303, 317, 319
"condition red", 313
conference table, 6
Congress, U.S., 9
Coningham, Air Vice Marshall Mary, 42–43, 46–49, 52, 257
container delivery system (CDS), 330–332
Con Thien, 303, 305, 311, 325–326
control of the air: See, air superiority.
control of the sea, 57
conventional war, 2, 7, 22
convoy: See, truck convoys.

403-892 O - 83 - 24

operational control, 44–50, 52, 54–58, 61–62, 73, 77, 79, 83, 86, 90, 97–99, 101–103, 164, 254, 257, 265, 274, 309, 316, 319
organizations:
1st Air Cavalry Division, 314, 317–318
1st Air Commando Squadron, 17
1st Army (British), 40, 43
1st Aviation Brigade, 317
1st Division (ARVN), 333
1st Division (NVA), 312
1st Field Army, 257–258
1st Infantry Division, 296–297
1st Marine Air Wing, 285–286, 309
1st Marine Division, 285–286
2nd Air Division, 17, 67, 72, 74, 77–78, 80–83, 90–91, 99, 107, 264–265, 270–272, 274
2nd Air Division (ADVON), 21, 67, 71, 260
2nd Tactical Air Force (British), 46–47, 51–56, 165–166, 257
3rd Division (ARVN), 32, 326, 328, 333
3rd Field Army, 257, 276
3rd Marine Division, 285, 309
4th Infantry Division, 296
5th Air Force, 5, 54–55, 58–59, 61–62, 82, 90, 113–114, 117, 167–169, 175, 223, 258
5th Division (NVA), 318
7th Air Force, 17, 24, 26, 32–33, 83–87, 90, 92, 95–99, 102, 104, 106–107, 127, 139, 144, 154, 193, 196, 200, 203, 217–219, 226, 236, 240, 272, 274, 277, 283–284, 286, 296, 303, 307, 310, 314, 317, 319, 321, 325–326, 328–329, 332
7th Air Force (ADVON), 316, 319
7th Division (ARVN), 256
7th Division (NVA), 312
7th Fleet, 25, 78, 104
8th Air Force, 46, 101, 111–113
8th Army (British), 5, 24, 40, 42–43, 61–62, 168–170, 258, 275
8th Army (U.S.), 113, 283
9th Air Force, 46–47, 51–56, 95, 111, 165–166, 223, 256–257
9th Field Army, 259
9th Division (VC), 294, 296, 302, 312
11th Armored Cavalry Regiment, 296, 314
12th Army Group, 46–47, 50–51, 256
13th Air Force, 67, 70–71, 77–78, 80, 82, 271
15th Air Force, 47–48
18th Army Group, 43, 47, 50, 256
19th Tactical Air Support Squadron, 266–267
21st Army Group, 46–47, 50, 257
25th Infantry Division, p. 296
26th Marine Regiment, 307–309
33rd Fighter Group, 1
34th Tactical Group, 17
82nd Airborne Division, 314
101st Airborne Division, 314, 317
101st Regiment (NVA), 294
173rd Airborne Brigade, 296, 298
196th Light Infantry Brigade, 296
242 Group (RAF), 40, 43
272nd Regiment (VC), 300
304th Division (NVA), 303, 305, 311
320th Division (NVA), 329
325th Division (NVA), 303, 305, 311
355th Tactical Fighter Wing, 220
388th Tactical Fighter Wing, 220
416th Battalion (VC), 318
507th Tactical Air Control Group, 250, 268
4400th Combat Crew Training Squadron, 252

probability of kill (PK), 158, 236
prohibited zones, 20, 133, 234
propaganda, 179, 265, 284
propeller driven aircraft, 250–251
protective reaction strike, 30–31, 195, 217
Provisional Corps (V), 319
psychological warfare, 11, 177, 251
public opinion, 26, 30
Pusan, 167–168, 276
Pusan perimeter, 3, 5
Pyongyang, 59, 119, 168, 171, 180, 188
Pyrrhus, 339

Quang Khe, 193
Quang Khe Ferry, 329
Quang Lang, 215
Quang Tri, 214, 282, 303, 305, 312, 319, 325–326, 328–330, 333
Quang Tri Province, 215, 305, 325, 333
Quebec, 45
Queseda, Lt. Gen. Elwood R., 2, 257
Qui Nhon, 18, 118, 329
Q–34: See, drone.

radar beacons, 282
radar bombing, 33, 177, 179–180, 200, 218–219, 283, 302, 310, 329, 332
radar controlled strikes, 302, 307, 331
radar system, U.S. (Vietnam), 147, 151
radar homing and warning (RHAW), 127–128, 220
RAF, 40, 42, 45–46, 54
RAF Bomber Command, 48, 112, 204
railroads, 18, 24, 27, 32, 183, 325
rainy season, 192. See also, monsoon.
RANCH HAND, 68
RAND, 267
Readiness Command (REDCOM), 65
reconnaissance, 22, 28–30, 44, 67, 85, 101, 105, 123, 174, 186, 203, 205, 215, 217–218, 232–233, 235–236, 261, 283–284, 294, 296, 304
reconnaissance photos, 27, 218, 233, 235
recovery rates, 332
Red River delta, 156, 181, 183
refueling: See, air refueling.
Regensburg, 219
Regional and Popular Forces (RF-PF), 312–313
repair teams, 185, 311
rescue: See, search and rescue.
restricted zones, 20, 33, 133, 220, 234, 237
restrictions: See, rules of engagement.
resumption of bombing, 30, 33, 104, 212, 217, 236, 240
retaliatory strikes, 15, 17–18
retreat, 167
road network, 20, 188–190, 203, 205, 309, 325
rocket attacks, 306–307, 311, 313–314
rockets (V1 & V2), 113, 279
Rock Pile, 307
ROLLING THUNDER, 18–19, 23, 33, 79, 89–90, 97, 177, 199, 236–237
ROLLING THUNDER Armed Reconnaissance Coordinating Committee, 90–91, 97, 105, 107
Rommel, General, 40–41, 43–44, 275
RON, 193
Roosevelt, Franklin D., 25, 41, 45–46, 48
Rostow, Walter W., 10, 68

routes: See, highways.
route packages, 91, 95–98, 102, 104–106, 174, 270
Royal Air Force: See, RAF.
Royal Navy, 44, 62
rubber plantations, 294, 300, 330
rules of engagement, 20, 56, 133, 141, 147, 156, 158–159, 172, 174–176, 196, 207, 219, 223, 227, 237, 337–339
Rusk, Dean, 10, 14
Russian pilots, 114
Ryan, General John D., 27

SAC ADVON, 102, 104, 106
Sager, Captain, 300
Saigon, 31, 67, 83, 107, 118, 154, 218, 248, 295–296, 314, 330
Saigon Conference (1967), 25
SAM, 16, 20, 26–27, 30, 33, 118–119, 123, 126, 128, 131–133, 135, 150, 177, 179–181, 183, 199, 205, 214–215, 217–219, 222, 224–226, 228, 231, 233–234, 237–238, 240, 255, 283–284, 304, 317, 321, 325
"SAM busting", 131
Sam Neua, 178
SAM radar, 118, 131
sanctuary, 5, 23, 115, 141, 183, 302, 311
sapper attacks, 17, 313–314, 329
satellite operations, 228–231
Savannakhet, 305
Saville, Brig. Gen. Gordon P., 260
SA–2, 32, 118, 222, 326
SA–7 "Strella", 32, 317, 324, 326, 328–330, 332
Schweinfurt, 219
search and destroy operations, 277, 294
search and rescue, 97, 153, 227
SEATO, 7, 9, 11, 82
Secretary of Defense, 10, 14, 18–21, 25, 99, 248–250, 283
Secretary of State, 14, 18, 25, 99
Seine River, 165
seismic sensors, 308
Senate Preparedness Investigation Subcommittee, 25
Sharp, Admiral U.S. Grant, 15, 22–23, 25–27, 31, 76, 78, 83, 89, 98, 102, 237, 309
Shaw AFB, South Carolina, 250
shootdown: first U.S. aircraft over N. Vietnam, 20, 123
SHORAN, 180
"short rounds", 303, 310
SHRIKE missile, 131
Sicily, 1, 43–44, 51, 323
SIDEWINDER missile, 138, 155, 157
Sihanouk, Prince Noradom, 30
Sihanoukville, 295
single air commander, 50, 55, 71, 79, 83, 105–106, 285–287, 309, 317, 319, 322–323
single air manager: See, single air commander
SIOP, 145
Site 85, 178–179, 248
"smart bomb": See, laser weapons
Smart, General Jacob E., 70–71, 82
Sniper Ridge, 280
sortie rate, 217, 220, 278, 328–329
Soui Tre: See, Ap Soui Tre
South Vietnamese Air Force (VNAF), 11, 17, 30, 32, 66–67, 203, 214, 248, 250, 253–254, 258, 261, 263–264, 266, 272, 286, 312, 319, 325, 328, 332
South Vietnamese Army (ARVN), 11, 17, 20, 30, 32, 214, 250, 254, 258, 261, 268, 272, 274, 303, 305, 311–312, 314, 319, 321, 324–325, 328, 330
South Vietnamese Rangers, 67, 248, 250, 308, 310, 323

southwest monsoon: See, monsoon
Soviet aircraft, 250
Soviet Air Defense Force, 137
Soviet air defense system, 118, 126
Soviet Union, 2, 9, 13, 118, 137, 140, 183, 237, 325
Spaatz, General Carl A., 41, 43–45, 47–51, 112, 164
SPARROW missile, 155–158
Special Forces, U.S. Army, 9–10, 67, 248–250, 252, 309
specified command, 70, 76, 106
Speer, Albert, 113, 163, 190
spider hole, 280
Spring Offensive, Korea (1953), 171
Spring Offensive, Vietnam (1972): See, Easter
stalemate, 170
Stalingrad, 192
Standard Arm missile, 131
standdowns—Christmas & New Year, 107, 145, 203, 313
standoff jammers, 222
starlight scope, 204–205
State Department, U.S., 75, 183
STEEL TIGER, 85–87, 101, 196, 199–200
Stilwell, Maj. Gen. Joseph W., 74
Strategic Air Command (SAC), 55, 99, 101–103, 105–107, 126, 145, 310
Strategic Air Force Europe, U.S. (USSAFE), 41, 47, 54, 112
strategic air offensive, 16, 18–19, 21, 24, 47–49
strategic air power, 43
strategic bombing, 46, 117, 164
strategic reconnaissance, 101
strategy (enemy), 13, 302–303, 305, 311–312
strategy, (U.S.), 1, 3–5, 9, 13–15, 21–24, 29–31, 33–34, 39, 50, 80, 101, 173, 248, 256, 293, 307, 337
Stratemeyer, Lt. Gen. George E., 54–58, 170, 286, 307
Strike Command (STRICOM), 65–66
sub-limited war, 10–11
sub-unified command, 69, 73, 76, 79, 82, 271
Sui Ho power plant, 171
Sullivan, William A., 179
supply centers, enemy, 18, 21, 29, 30, 33, 163, 217–219, 236, 329
supply lines: See, lines of communication
Supreme Commander-In-Chief of Allied Expeditionary Forces, 45
Supreme Headquarters Allied Expeditionary Forces, 48, 50, 74
surface-to-air missile: See, SAM.
surrender (Japan), 34
suspension of bombing: See, bombing halt.
SWIFT STRIKE, Operation, 65
Symington, Stewart, 176–177
synthetic oil plants, 163

TACAN, 248
Tactical Air Command (TAC), 2, 28, 55, 71–72, 250
Tactical Air Control Center (TACC), 81, 154, 218, 258, 262, 278–280, 282
Tactical Air Control System (TACS), 67, 73, 81, 101, 203, 253–254, 262, 264, 278, 287, 310, 314, 338
Tactical Air Direction Center (TADC), 156
tactical airlift, 32
Tactical Air Support Element (TASE), 317, 319–320
Tactical Operations Center (TOC), 261, 272
Takhli AB, Thailand, 181, 220
TALLY HO, 199, 217
tankers: See, air refueling.
tanks, 218, 276, 309–310, 314, 321, 323–327, 330, 333

Viet Tri thermal power plant, 119
Vinh, 14, 31, 92, 119, 149, 188, 192, 195, 215, 231, 284
visual reconnaissance (VR), 267
Vogt, General John W., Jr., 106

"wars of liberation", 9, 248, 287
War Zone "C", 266, 294–295
War Zone "D", 266
"Washington", 23, 56, 75, 227, 233, 303
WATERBOY, 151, 155
weather operations, 27, 88, 175–177, 180–181, 189–190, 218–219, 225–228, 230–231, 236, 240, 280, 282, 302, 306, 310, 318–319
Wehrmacht, 166
Western Dester Air Force, 46, 275–276
Westmoreland, General William C., 14, 30, 75, 77, 80–81, 101–102, 118, 176, 288, 296, 309–310, 314, 316–317, 319
wet season, 324. See also, monsoon.
Weyland, Maj. Gen. Otto P., 2, 56–57, 59, 114, 170–171, 257, 286
Wheeler, General Earl G., 13, 22, 25–27, 31, 77, 237, 256, 309–310
Wild Weasel, 101, 107, 130–131, 145, 200, 205, 220, 240–241, 283
Wilson, General Sir Henry Maitland, 47–48
Wingate, Brig. Gen. Orde, 252
"Wingate's Force", 252
withdrawal, U.S., 12, 28–30, 206, 211, 215, 312, 320, 324, 333
Wonson, 168

Yalu River, 5, 56–57, 59, 62, 115, 119, 147, 156, 171, 180, 222, 234
YANKEE Station, 285, 326
Yen Bai, 123, 180, 241
Yen Vien, 227
Yugoslavia, 252

358

☆ U.S. GOVERNMENT PRINTING OFFICE : 1983 O - 403-892